Nitric Oxide in Plants: Metabolism and Role in Stress Physiology

M. Nasir Khan · Mohammad Mobin
Firoz Mohammad · Francisco J. Corpas
Editors

Nitric Oxide in Plants: Metabolism and Role in Stress Physiology

Springer

Editors
M. Nasir Khan
Mohammad Mobin
Faculty of Science,
Department of Biology
University of Tabuk
Tabuk
Saudi Arabia

Francisco J. Corpas
Department of Biochemistry,
Cell and Molecular Biology of Plants
Estación Experimental del Zaidín Consejo
Superior de Investigaciones Científicas
Granada
Spain

Firoz Mohammad
Plant Physiology Section,
Department of Botany
Aligarh Muslim University
Aligarh
India

ISBN 978-3-319-06709-4 ISBN 978-3-319-06710-0 (eBook)
DOI 10.1007/978-3-319-06710-0
Springer Cham Heidelberg New York Dordrecht London

Library of Congress Control Number: 2014940326

Printed on acid-free paper

Springer is part of Springer Science+Business Media (www.springer.com)

Professor M. M. R. K. Afridi (1931–2010)

Professor Muhammad Mahmudur Rahman Khan Afridi was born at Farrukhabad (U.P.), India on May 7, 1931. He did his graduation and post-graduation from Aligarh Muslim University (AMU), Aligarh, India. He completed his doctoral studies from University of Bristol, England. After completion of Ph.D., he joined AMU, as Lecturer of Botany in 1951. He became Reader in 1961 and Professor of Plant Physiology and Head of the Department of Botany in 1978. He assumed the charge of Dean, Faculty of Life Sciences, in 1989 and retired in 1991.

Professor Afridi started his research career in 1957 when he was awarded a Colombo Plan Fellowship for research in U.K. and joined the famous Long Ashton Research Station of the University of Bristol with renowned Professor T. Wallace F.R.S. and Dr. E. J. Hewitt F.R.S. His pioneering work on the inducible formation of nitrate reductase, the key enzyme known today for nitric oxide synthesis in plants, was soon recognized internationally.

On his return to India, Professor Afridi devoted his energies to the applied aspects of the mineral nutrition of crop plants and established the first school on macro-nutrition. He successfully guided the research work of many students and published several research papers in the journals of national and international repute. Professor Afridi was one of the founding members and life member of Indian Society for Plant Physiology (I.S.P.P.). He was elected as the President of I.S.P.P. in 1979, and was awarded the Distinguished Scientists Medal in 1983 by the Academy for the Advancement of Agricultural Sciences, India and the J. J. Chinoy Medal in 1985 by I.S.P.P. Professor Afridi left all of us on 6th January, 2010 for heavenly abode at the age of about 79 years.

We dedicate this book to Professor M. M. R. K. Afridi for his marvelous contribution in the area of physiology and mineral nutrition of higher plants.

Preface

Nitric oxide (NO), a versatile gaseous free radical that diffuses readily through biological membranes, plays important role in diverse physiological processes in plants. A plethora of NO-generated events encompasses through germination to flowering and fruit ripening in a plant's life cycle. It alters flowering, stimulates germination, induces pollen tube re-orientation, breaks seed dormancy, triggers mitogen-activated protein (MAP) kinase signaling pathways, modulates the activity of certain enzymes, regulates stomatal closure, photosynthesis, cellular trafficking, cell death, expression of cell cycle genes, and other key metabolic processes. NO plays a key role as signaling molecule in biotic and abiotic stress signal transduction pathways in plants. NO acts as an antioxidant and confers resistance against detrimental consequences of stresses.

Acknowledging NO as a significant modulator of biological processes, renewed attention has been given to the mechanism of NO synthesis in plants. The reaction pathway of NO synthesis in animals has been employed to investigate the likely parallel in plants. In animal systems, NO is synthesized predominantly by the enzyme NO synthase (NOS) that converts L-Arginine into L-citrulline in a NADPH-dependent reaction, which releases one molecule of NO for each molecule of L-Arginine. Assays for Arginine to citrulline conversion and compounds that inhibit mammalian NOS have been used on several occasions to draw an analogy that NO synthesis by a NOS-type enzyme also occurs in plants. But still no direct homologs of any of the animal enzymes have been found in any of the fully sequenced plant genomes. This leaves us with many questions than answers related to NO biosynthesis, detection and mode of action in plants.

The research field of NO biology has transcended rapidly over the last few years, and a huge wealth of information has been accumulated in NO research arena. As a result, it became tangible that NO affects far more fundamental biological processes in plants, than originally anticipated.

Therefore, in our opinion, an overview of detection, biosynthesis and metabolism of NO and its role in stress physiology of plants is well timed.

This book "Nitric Oxide in Plants: Metabolism and Role in Stress Physiology" comprises of 17 chapters that covers the key features of NO molecule in a sequential manner starting from its metabolism, identification and detection in plants (Part I) to current understanding of NO molecule and its derivatives in terms of chemical, physical, and biochemical properties, functional role, mode of action, signaling and interaction with phytohormones, mineral nutrients, biomolecules, ions and ion channels in plants under abiotic stresses (Part II).

Part I of the book comprises Chaps. 1–9. Chapter 1 presents an overview of NO metabolism with particular emphasis on the sources of NO in plants and their importance under abiotic stress conditions. Chapter 2 sheds light on the reductive and oxidative NO synthesis and their regulation. Chapter 3 discusses the peroxisomes as a source of NO and NO-derived species in response to abiotic stresses and detection of NO generation in peroxisomes. Chapter 4 is focussed on the role of mitochondrial NO homeostasis during hypoxic conditions. Chapter 5 deals with the detection methods and synthesis of NO in plants using marine unicellular red tide phytoplankton, *Chattonella marina,* as a model. Chapter 6 sheds light on the role of NO in nitrosylation of cystein thiol residues in proteins, and summarizes different methods developed to identify and quantify nitrosylated proteins. In this chapter authors also provided the first overview of plant nitrosylated proteome showing a wide range of functions and cellular compartments involved in NO signaling and/or targeting. Chapter 7 presents an overview of detection and measurement of NO and nitrosylated proteins, and various levels of regulation of NO on jasmonate signaling and biosynthesis pathway in response to abiotic stress. Chapter 8 sheds light on the function of *S*-nitrosoglutathione reductase (GSNO) as a natural reservoir of NO bioactivity and role of GSNO in plant development and stress response. Chapter 9 discusses nitro-fatty acids in the context of their biochemical activities and cell signaling actions.

Part II of the book includes Chaps. 10–17. Chapter 10 is focused on the properties of NO and its derivatives and their role as potent modulator of the redox regulation in various cell transduction pathways in response to abiotic stresses. Chapter 11 highlights the recent advances in NO signal transduction and its interactions with other signaling molecules in response to abiotic stress. Chapter 12 summarizes the role of exogenously applied NO on structural and functional parameters of plant cells under H_2O_2-induced oxidative stress. Chapter 13 focuses on the current knowledge of possible interactions between NO and phytohormones during plant abiotic stress responses. Whereas Chap. 14 presents an overview of the synergistic role of NO and calcium in the tolerance of plants to abiotic stress. Chapter 15 discusses functional links between the plant growth promoting action of humic substances and NO in response to abiotic stresses. Chapter 16 is focused on the role of chitosan-mediated induction of NO in plant defense responses against pathogen attack and crosstalk between abiotic and biotic stress responses is also discussed. Chapter 17 deals with the involvement of NO and other signaling molecules in signaling cascade and gene expression during biotic and abioitc stresses induced programmed cell death.

We collected contributions from various laboratories studying NO plant biology, and intended to present an overview of the contemporary challenges and possibilities in different areas of NO. We hope that this book will raise your interest in the field of NO research and will serve as a valuable reference.

We would like to express our gratitude to all the authors and reviewers who contributed to this book. Furthermore, we acknowledge Springer Science+Business Media, Heidelberg, with heartfelt gratitude to Dr. Christina Eckey, Editor, Plant Sciences and Dr. Andrea Schlitzberger, Project Coordinator for their professional support and cooperation during the preparation of the manuscript.

Tabuk, Saudi Arabia — M. Nasir Khan
Tabuk, Saudi Arabia — Mohammad Mobin
Aligarh, India — Firoz Mohammad
Granada, Spain — Francisco J. Corpas

Contents

Contributors

Dariusz Abramowski Department of Plant Physiology, Poznan University of Life Sciences, Poznan, Poland

Magdalena Arasimowicz-Jelonek Faculty of Biology, Department of Plant Ecophysiology, Adam Mickiewicz University, Poznan, Poland

E. Bacaicoa CIPAV-Roullier Group, Orcoyen, Spain

R. Baigorri CIPAV-Roullier Group, Orcoyen, Spain

Y. S. Bakakina Institute of Biophysics and Cell Engineering, National Academy of Sciences of Belarus, Minsk, Belarus

Juan B. Barroso Área de Bioquímica y Biología Molecular, Facultad de Ciencias Experimentales, Departamento de Biología Experimental, Universidad de Jaén, Jaén, Spain

Stefania Bonsegna CNR-ISPA, Institute of Sciences of Food Productions, Lecce, Italy

Mounira Chaki German Research Center for Environmental Health, Institute of Biochemical Plant Pathology, Helmholtz Zentrum München, Neuherberg, Germany

Francisco J. Corpas Department of Biochemistry, Cell and Molecular Biology of Plants, Estación Experimental del Zaidín, Consejo Superior de Investigaciones Científicas, Granada, Spain

Stefania De Domenico CNR-ISPA, Institute of Sciences of Food Productions, Lecce, Italy

L. V. Dubovskaya Institute of Biophysics and Cell Engineering, National Academy of Sciences of Belarus, Minsk, Belarus

Abasse Fares CEA Saclay DSV/iBiTec-S/SIMOPRO/LCV, Gif-sur-Yvette, France

Jolanta Floryszak-Wieczorek Department of Plant Physiology, Poznan University of Life Sciences, Poznan, Poland

Luciano Freschi Department of Botany, Institute of Biosciences, University of São Paulo, São Paulo, SP, Brazil

M. Fuentes CIPAV-Roullier Group, Orcoyen, Spain

Andrea Galatro School of Pharmacy and Biochemistry, Physical Chemistry-Institute of Biochemistry and Molecular Medicine (IBIMOL), University of Buenos Aires-CONICET, Buenos Aires, Argentina

J. M. Garcia-Mina CIPAV-Roullier Group, Orcoyen, Spain; Faculty of Sciences, Department of Chemistry and Soil Chemistry, University of Navarra, Pamplona, Spain

Ming-Hua Gu College of Agronomy, Guangxi University, Nanning, People's Republic of China

Hu-Yi He College of Agronomy, Guangxi University, Nanning, People's Republic of China

Long-Fei He College of Agronomy, Guangxi University, Nanning, People's Republic of China

Karolina Izbiańska Faculty of Biology, Department of Plant Ecophysiology, Adam Mickiewicz University, Poznan, Poland

M. Nasir Khan Faculty of Science, Department of Biology, University of Tabuk, Tabuk, Saudi Arabia

Daekyung Kim Jeju Center, Korea Basic Science Institute, Ara-dong, Jeju-Si, Jeju Special Self-Governing Province, Korea

E. V. Kolesneva Institute of Biophysics and Cell Engineering, National Academy of Sciences of Belarus, Minsk, Belarus

Yee-Min Kwan Faculty of Agriculture, Department of Plant Protection, Institute of Tropical Agriculture, Universiti Putra Malaysia, Serdang, Selangor, Malaysia

Wei-Biao Liao College of Agronomy, Gansu Agricultural University, Lanzhou, China

Christian Lindermayr German Research Center for Environmental Health, Institute of Biochemical Plant Pathology, Helmholtz Zentrum München, Neuherberg, Germany

Eduardo López-Huertas Department of Biochemistry, Cell and Molecular Biology of plants, Estación Experimental del Zaidín, Consejo Superior de Investigaciones Científicas, Granada, Spain

Helenice Mercier Department of Botany, Institute of Biosciences, University of São Paulo, São Paulo, SP, Brazil

Paulo T. Mioto Department of Botany, Institute of Biosciences, University of São Paulo, São Paulo, SP, Brazil

M. Mobin Faculty of Science, Department of Biology, University of Tabuk, Tabuk, Saudi Arabia

Firoz Mohammad Plant Physiology Section, Department of Botany, Aligarh Muslim University, Aligarh, India

V. Mora CIPAV-Roullier Group, Orcoyen, Spain

Tatsuya Oda Faculty of Fisheries, Division of Biochemistry, Nagasaki University, Nagasaki, Japan

M. Olaetxea Faculty of Sciences, Department of Chemistry and Soil Chemistry, University of Navarra, Pamplona, Spain

Halley Caixeta Oliveira Departamento de Biologia Animal e Vegetal, Centro de Ciências Biológicas, Universidade Estadual de Londrina, Londrina, PR, Brazil

José M. Palma Department of Biochemistry, Cell and Molecular Biology of plants, Estación Experimental del Zaidín, Consejo Superior de Investigaciones Científicas, Granada, Spain

Jean-Benoît Peltier Campus SupAgro-INRA, Montpellier, France

Palmiro Poltronieri CNR-ISPA, Institute of Sciences of Food Productions, Lecce, Italy

Susana Puntarulo Facultad de Farmacia y Bioquímica, Fisicoquímica-IBIMOL, Buenos Aires, Argentina

Luis A. del Río Department of Biochemistry, Cell and Molecular Biology of plants, Estación Experimental del Zaidín, Consejo Superior de Investigaciones Científicas, Granada, Spain

Michel Rossignol Campus SupAgro-INRA, Montpellier, France

Tamás Rőszer Centro Nacional de Investigaciones Cardiovasculares, Madrid, Spain; Zentrum für Biomedizinische Forschung, Institut für Allgemeine Zoologie und Endokrinologie, Universität Ulm, Ulm, Germany

Homero Rubbo Facultad de Medicina, Departamento de Bioquímica, Center for Free Radical and Biomedical Research, Universidad de la República, Montevideo, Uruguay

Mansour Salati Department of Plant Protection, Golestan Agricultural and Natural Resources Research Centre, Golestan, Iran

Ione Salgado Departamento de Biologia Vegetal, Instituto de Biologia, Universidade Estadual de Campinas, Campinas, SP, Brazil

Angelo Santino CNR-ISPA, Institute of Sciences of Food Productions, Lecce, Italy

M. Ali Saqib Faculty of Science, Department of Biology, University of Tabuk, Tabuk, Saudi Arabia

Marco Taurino CNR-ISPA, Institute of Sciences of Food Productions, Lecce, Italy

Andrés Trostchansky Facultad de Medicina, Departamento de Bioquímica, Center for Free Radical and Biomedical Research, Universidad de la República, Montevideo, Uruguay

I. D. Volotovski Institute of Biophysics and Cell Engineering, National Academy of Sciences of Belarus, Minsk, Belarus

Mui-Yun Wong Faculty of Agriculture, Department of Plant Protection, Institute of Tropical Agriculture, Universiti Putra Malaysia, Serdang, Selangor, Malaysia

Ji-Hua Yu College of Agronomy, Gansu Agricultural University, Lanzhou, China

A. M. Zamarreño CIPAV-Roullier Group, Orcoyen, Spain

Part I
Nitric Oxide: Metabolism, Identification and Detection

Chapter 1
An Update to the Understanding of Nitric Oxide Metabolism in Plants

Andrea Galatro and Susana Puntarulo

Abstract Nitric oxide (NO) is an inorganic free radical gaseous molecule which has been shown to play an unprecedented range of roles in biological systems. The potential reactions of NO are numerous and depend on many different factors. The site and source of production, as well as the concentration of NO collectively determine whether NO will elicit direct or indirect effects. In animals, NO is generated by the activity of nitric oxide synthase (NOS). In plants, neither the gene nor protein similar to known NOS has been found. However, different pathways producing NO in plants have been described, and can be classified as either oxidative or reductive steps. These sources of NO seem to cooperate to the growth and development, and to respond to several stress situations like abiotic stress. Chloroplasts are key organelles in plant metabolism and they seem to be involved in NO production, thus, proposed pathways for NO generation in chloroplasts are discussed.

Keywords Chloroplastic nitric oxide · Nitric oxide · Nitrogen active species · Nitric oxide sources

---

A. Galatro
School of Pharmacy and Biochemistry, Physical Chemistry-Institute of Biochemistry and Molecular Medicine (IBIMOL), University of Buenos Aires-CONICET, Buenos Aires, Argentina

S. Puntarulo (✉)
Facultad de Farmacia y Bioquímica, Fisicoquímica-IBIMOL,
Junín 956, C1113AAD Buenos Aires, Argentina
e-mail: susanap@ffyb.uba.ar

M. N. Khan et al. (eds.), *Nitric Oxide in Plants: Metabolism and Role in Stress Physiology*, DOI: 10.1007/978-3-319-06710-0_1,

1.1 Introduction

1.1.1 Brief Review of the Chemistry of Nitrogen-Active Species

The broader chemistry of nitric oxide (NO) involves a redox array of species such as nitrosonium (NO^+), NO radical (NO) and nitroxyl anion (NO^-) (Fig. 1.1) which exhibit distinctive properties and reactivities (Gisone et al. 2004).

Neutral NO has a single electron in its 2p-π antibonding orbital and the removal of this electron forms NO^+ while the addition of one more electron to NO forms NO^- (Stamler et al. 1992). The chemistry of NO^+ is characterized by addition and substitution reactions with nucleophiles such as electron-rich bases and aromatic compounds. Nitrosation in aqueous phase can occur at –S, –N, –O, and –C centers in organic molecules and appears to involve NO^+ or related NO^+ equivalents. The biological relevance of NO^+ under weakly acidic or physiological conditions had been disputed, however a variety of nitroso-compounds that form effectively under neutral physiological conditions (Stamler et al. 1992) can be interpreted as reactions with NO^+ carriers. Important examples of such compounds are metal-nytrosyl complexes, thionitrites (RS–NO), nitrosamines (RNH–NO), alkyl and aryl nitrites (RO–NO) and dinitrogen tri- and tetra-oxides (N_2O_3 and N_2O_4). In biological systems, there are numerous nucleophilic centers whose potential susceptibility to nitrosative attack has been shown in in vitro studies (Stamler et al. 1992). The chemistry of NO^- has received significantly less attention, particularly in aqueous solution. NO^- converts rapidly to N_2O through dimerization and dehydration (Basylinski and Hollocher 1985) and it is known to react with Fe (III) heme (Goretski and Hollocher 1988). NO^- also undergoes reversible addition to both low molecular weight and protein-associated thiols, leading to sulfhydryl oxidation. Electron transfer and collisional detachment reactions are common and generally yield NO radical (NO) as the major product. *S*-nitrosothiols are believed to be a (minor) product of the reaction of NO^- with disulfides (Stamler et al. 1992).

From a biological point of view, the important reactions of NO are those with O_2 and its various redox forms and with transition metal ions. The reaction of NO with O_2 in aqueous solution is a second-order reaction in [NO] ($v = k\,[NO]^2\,[O_2]$) (Stamler et al. 1992), thus the biological half life of NO, generally assumed to be in the order of seconds, strongly depends on its initial concentration. NO also reacts rapidly with O_2^- in aqueous solution, yielding peroxynitrite ($ONOO^-$) (Saran et al. 1990). When discussing the chemistry and physiological effects of NO, it should be considered that NO is a highly diffusible second messenger that can elicit effects relatively far from its site of production. The concentration and therefore the source of NO are the major factors determining its biological effects (Wink and Mitchell 1998). At low concentrations (<1 μM), the direct effects of NO predominate. At higher concentrations (>1 μM), the indirect effects mediated by reactive nitrogen species (RNS) prevail. The direct effects of NO most often involve the interaction of NO with metal complexes. NO forms complexes with

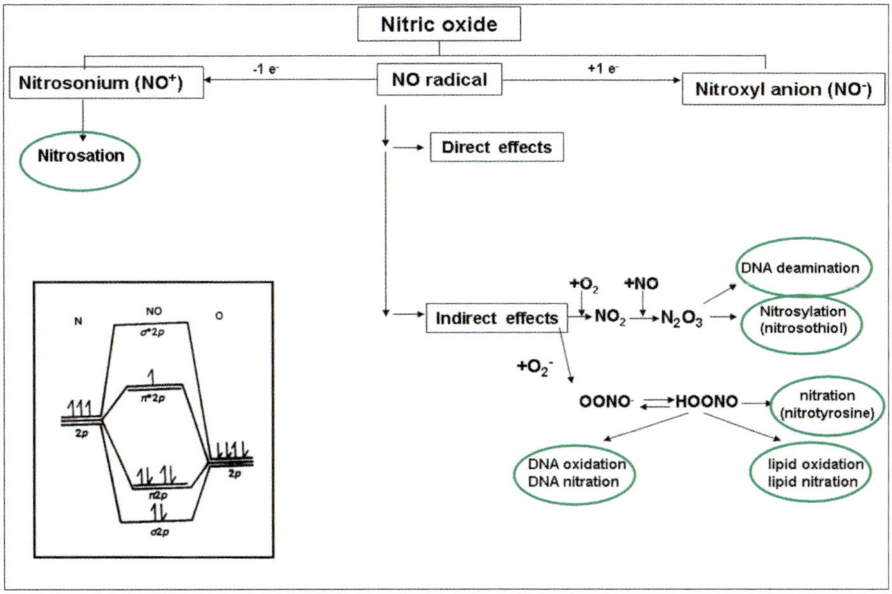

Fig. 1.1 Summary of chemistry of nitrogen-active species and some effects of the independent species. *Inset* Molecular orbital diagram for NO

the transition metal ions, including those regularly found in metalloproteins. The reactions with heme-containing proteins have been widely studied. NO also forms non-heme transition metal complexes and biochemical interest has been focused on its reactions toward Fe–sulfur centers in proteins, including several proteins involved in mitochondrial electron transport and enzymes (Henry et al. 1991). The reactions of NO with heme-containing proteins are the most physiologically relevant and include interactions with cytochrome P_{450} (Wink et al. 1993). Another established direct effect of NO on proteins is tyrosine nitration. Tyrosine nitration is selective and reversible and it has been shown that there are $ONOO^-$ dependent and independent pathways for the nitration in vivo (Davis et al. 2001). NO is also able to terminate lipid peroxidation (Rubbo et al. 1995). The indirect effects of NO, produced through the interaction of NO with either O_2 or O_2^-, include nitrosation (when NO^+ is added to an amine, thiol, or hydroxy aromatic group), oxidation (when one or two electrons are removed from the substrate), or nitration (when NO_2^+ is added to a molecule) (Wink et al. 1993). In aqueous solution NO can undergo autoxidation (i.e., reaction with O_2) to produce N_2O_3 and this compound can undergo hydrolysis to form nitrite (Ford et al. 1993). Since NO and O_2 are 6–20 times more soluble in lipid layers as compared to aqueous fractions, the rate of autoxidation is increased dramatically in the lipid phase (Ford et al. 1993) and the primary reactions of N_2O_3 are thought to occur primarily in the membrane fraction. In its reaction with O_2^-, NO generates $ONOO^-$ at a rate close to diffusion, and $ONOO^-$ acts as both nitrating agent and powerful oxidant to modify proteins (formation of nitrotyrosine), lipids (lipid oxidation, lipid nitration), and nucleic acids (DNA oxidation and DNA nitration).

In summary, the potential reactions of NO are numerous and depend on many different factors. The site and source of production, as well as the concentration of NO collectively determine whether NO will elicit direct or indirect effects. In addition, a relative balance between oxidative and nitrosative stress exists, and it is a main aspect that should be carefully evaluated for understanding the complexity of biological effects of NO.

1.2 Sources of NO in Plants: An Overview

In animals, NO is generated by the activity of nitric oxide synthase (NOS). NOSs catalyze the conversion of L-Arginine to L-Citrulline and NO. The reaction requires O_2 and NADPH (Wendehenne et al. 2001) (Eq. 1.1).

$$\text{L} - \text{Arginine} + \text{NADPH}_2 + O_2 \rightarrow \text{L} - \text{Citrulline} + \text{NO} + H_2O \qquad (1.1)$$

While these mammalian NOSs are long known and well characterized, the plant community has not been successful in identifying corresponding genes or enzymes in higher plants so far (Fröhlich and Durner 2011). In plants, neither the gene or cDNA, nor any protein with high sequence similarity to known NOS, have been found (Lamattina et al. 2003). Despite this, several efforts have been made to improve this knowledge. Chandok et al. (2003) described the purification and characterization of a pathogen inducible NOS-like activity from tobacco plants and its identification as a variant form of P subunit of the glycine decarboxylase complex. However, this work was retracted by Klessig et al. (2004) due to difficulties in reproducing some data related to NO-synthesizing activity of the recombinant variant P.

A second approach was developed by Guo et al. (2003), with the identification of a plant NOS gene involved in hormonal signaling (*Atnos1*). *Arabidopsis* mutant (*Atnos1*) had impaired NO production, organ growth, and abscisic acid-induced stomatal movements. According to Guo et al. (2003), purified AtNOS1 protein employed arginine and NADPH as substrates, and was activated by Ca^{2+} and calmodulin, like mammalian endothelial and neuronal NOS. Thus, AtNOS1 was proposed as a distinct enzyme, with no sequence similarities to any mammalian isoform, and with a role in growth and hormonal signaling in plants (Guo et al. 2003). Later, due to the failure in the detection of NOS activity in purified AtNOS1 protein (Crawford et al. 2006; Zemojtel et al. 2006), it was suggested renaming AtNOS1 to AtNOA1 (nitric oxide associated 1), because it seems to be important for NO generation in the cell, but it is not a real NOS as defined for animal system. Although different research groups have independently confirmed the presence of decreased NOS activity and NO levels in the *Arabidopsis* mutant (*Atnos1*), other reports found that NO accumulation in response to different hormones or oxidative stress was similar in wild-type and *nos1* plants (Gas et al. 2009). Besides, not all the phenotypes observed in the mutant can be rescued by NO supplementation (Gas et al. 2009). Thus, AtNOS1, renamed as AtNOA1, seems to have another function different from NO synthesis. Moreau et al. (2008) showed that AtNOA1 is a member of the circularly permuted GTPase family (cGTPase). AtNOA1 specifically binds GTP and hydrolyzes it. However, GTP hydrolysis is necessary but not sufficient for the physiological function of AtNOA1. Also, the C-terminal domain seems to play a crucial role *in planta*. cGTPases appear to be RNA-binding proteins, and the closest homolog of AtNOA1, the *Bacillus subtilis* YqeH, has been shown to participate in ribosome assembly and stability (Moreau et al. 2008).

Even though finally AtNOS1 is not a NOS, the discovery and development of the *Arabidopsis* mutant *Atnos1* was an important finding. The biological role of AtNOA1 or RIF1 (Flores-Pérez et al. 2008) is believed to be primarily associated with chloroplasts ribosome functions (Moreau et al. 2008; Gas et al. 2009; Liu et al. 2010). In *rif1* seedlings, not only chloroplast ultrastructure, but also the level of proteins encoded by the chloroplastic genome were affected (Flores-Pérez et al. 2008), suggesting that NOA1/RIF1 might bind plastidial ribosomes and is required for the normal function and proper protein synthesis in plastids (Gas et al. 2009). It has also been reported that NO accumulation in *Arabidopsis* is independent of NOA1 in the presence of sucrose (Van Ree et al. 2011). Thus, it is possible that the primary requirement for *noa1* activity is efficient chloroplast function to generate photosynthates. Provision of sucrose enables *noa1* to accumulate NO, raising the question why fixed carbon may be necessary for NO accumulation in *Arabidopsis* (Van Ree et al. 2011).

To add more complexity to this scenario, Foresi et al. (2010) have characterized the sequence, protein structure and biochemistry of NOS from the green alga *Ostreococcus tauri*. This NOS contains the main characteristics of animal NOS, and NO generation in this alga is dependent on light irradiance and growth phase. This single-cell alga is of particular interest because it shares a common ancestor with higher plants, providing compelling evidence that an active NOS functions in a photosynthetic organism belonging to the plant kingdom (Foresi et al. 2010).

NOS enzymes seem to be present in almost all organisms except plants. Despite the fact that NO plays a crucial role in plant physiology, higher plants seem to have lost the specific NOSs in the course of evolution (Fröhlich and Durner 2011). However, different pathways to produce NO in plants have been described, and they can be classified as either oxidative or reductive (Gupta et al. 2011a). Briefly, nitrate reductase (NR) as shown in Eq. 1.2, and mitochondrial or plasma membrane-associated NO production (NR:NiNOR system) are all reductive pathways and depend on nitrite as a primary substrate, whereas NO production from L-Arginine, polyamine or hydroxylamine are among the oxidative pathways (Gupta et al. 2011a).

$$2NO_2^- + NAD\,(P)H\ + 3H_3O^+ \overset{NR}{\rightarrow} 2NO\ + NAD(P)^+ + 5H_2O \tag{1.2}$$

Although no NOS enzyme has been identified in plants, a NOS-like activity has been extensively reported. We have described L-Arginine-dependent NO generation in soybean leaves (Galatro et al. 2004) and soybean chloroplasts (Jasid et al. 2006), which were evaluated employing electron paramagnetic resonance (EPR). In both cases, NO generation was NADPH dependent and inhibited by known NOS mammalian inhibitors. Corpas et al. (2006) also described NO production from L-Arginine (NOS activity) in leaves, stems, and roots of pea seedlings during plant development, using a chemiluminescence-based assay and confocal laser scanning microscopy. Peroxisomes, have also been proposed as cellular source of RNS. NOS activity in peroxisomes was described employing several approaches (for a review, see del Río 2011). Also EPR measurements, employing isolated peroxisomes from pea leaves, clearly indicated the generation of NO as a result of the L-Arginine-dependent NOS activity (del Río 2011). Another candidate for NO production is the peroxisomal enzyme xanthine oxidoreductase (XOR). XOR from animal origin can produce superoxide (O_2^-) and NO free radicals during its catalytic reaction (del Río 2011).

Regarding polyamine (PA)-mediated NO generation, Tun et al. (2006) observed that addition of PAs to *Arabidopsis thaliana* seedlings caused rapid release of NO. A speculation could be the conversion of PA by as yet unknown enzymes or by PA oxidases to generate NO. PA oxidases are not known to generate NO in animal systems, and PA oxidase could be inhibited by L-NAME (L-nitroarginine methyl ester) (Tun et al. 2006).

L-Arginine and NR-dependent pathways have been the most reported (Rasul et al. 2012). Rasul et al. (2012) have investigated NO production in *Arabidopsis* elicited by oligogalacturonides (OGs) and have suggested that L-Arginine and NR pathways are co-involved in NO production and do not work independently. Recently, we also observed that cotyledons from soybean plants growing in the presence of ammonia as the unique source of nitrogen were physiologically nondistinguishable from control (nitrate-fed) cotyledons, and showed a similar NO accumulation, indicating that cotyledons are able to produce similar amounts of NO independently of the source of nitrogen supplied. These results led us assumed that different sources of NO could operate for NO accumulation in soybean cotyledons, e.g., nitrite- and L-Arginine-dependent sources. Thus, it is likely that under different conditions, for example the lack of a substrate, one pathway could result more operative to maintain NO generation and support the required NO levels in the cell to allow a normal function and development (Galatro et al. 2013). In this sense, NO production in *Arabidopsis* plants following pathogen attack may result from the interplay of L-Arginine- and nitrite-dependent pathways (Modolo et al. 2005). Rasul et al. (2012), suggested that L-NAME-sensitive NO production also affect NR-dependent NO production. NO can stimulate NR activity at the pos-translational level through a direct interaction or, alternatively, by affecting the activity of proteins involved in NR regulation. Part of the NO produced by L-Arginine-dependent pathway could be oxidized to nitrite, thus providing substrate for NR-triggered NO synthesis. Polyamines seem to be involved in NR activity regulation. Rosales et al. (2012) studied the effect of PAs on NR activity in wheat leaves exposed to exogenously added PAs, and demonstrated that NO was involved in the inhibition or increase of NR activity. These findings point out the complexity of the study of NO generation in plants, as different pathways could be involved, and also work together for NO production in the plant cell under physiological or stress situations.

Evidence that plants oxidize hydroxylamines to NO has been described, open a new possibility for oxidative NO formation in plants. However, the existence and role of these reactions under physiological conditions are not clear (Rümer et al. 2009). Further experiments are required to find out whether any natural hydroxylamines can be formed under specific conditions by plants to serve as substrates for an endogenous oxidative NO generation (Rümer et al. 2009).

The mitochondrial electron transport chain is another proposed site for nitrite to NO reduction, operating significantly when the normal electron acceptor, O_2, is low or absent. Under these conditions, the mitochondrial NO production contributes to hypoxic survival by maintaining a minimal ATP formation (Gupta et al. 2011b).

1.2.1 Is Chloroplast a Source of NO?

The first reports describing chloroplasts as an NO source were based on studies developed with tobacco (Foissner et al. 2000; Gould et al. 2003). Foissner et al. (2000) described NO accumulation in epidermal tobacco leaf cells subjected to a proteinaceous elicitor from *Phytophthora cryptogea*. They evidenced an NO production in the cytosol, along plasma membrane, in chloroplasts, and organelles probably representing peroxisomes. NOS inhibitor N_G-mono-methyl-arginine monoacetate (L-NMMA) reduced NO levels but not as the NO scavenger cPTIO (2-(4-carboxyphenyl)-4,4,5,5-tetramethylimidazoline-1-oxyl-3-oxide). These results suggested that other sources of NO could be operative.

In our laboratory, we have identified two independent pathways for NO generation in soybean chloroplasts, one pathway was dependent of the activity of a NOS-like enzyme employing L-Arginine and NADPH, and another pathway was dependent of nitrite (Jasid et al. 2006). NO generation in isolated chloroplasts was evaluated employing EPR in the presence of the spin trap (sodium-*N*-methyl-D-glucamine dithiocarbamate [MGD])$_2$-Fe(II), and the required cofactors described for assaying the activity of plant NOS (Galatro et al. 2004). The EPR signal corresponding to NO–MGD–Fe adduct was inhibited if the chloroplasts were incubated with NOS inhibitors, such as N_{ϖ}-nitro-L-Arg methyl ester hydrochloride (L-NAME) or N_{ϖ}-nitro-L-Arg (L-NNA). It is interesting to point out that Arginine was shown to be an abundant amino acid in chloroplast stroma, and that the reported synthesis of NO was not affected either by omission or addition of Ca^{2+} or by supplementation with calmodulin (Jasid et al. 2006). On the other hand, intact chloroplasts incubated under light conditions in the presence of sodium nitrite also generated NO. However, this generation was detectable in the thylakoid fraction but not in the stroma, and was affected by the inhibition of photosynthetic electron flow by the herbicide 3-(3,4-dichlorophenyl)-1,1-dimethyl urea (DCMU), that binds plastoquinone and blocks electron flow at the quinone acceptors of photosystem II. These results suggested that thylakoids were the main component of chloroplast involved in nitrite reduction (Jasid et al. 2006). Thus, chloroplasts seem to be able to produce NO in vitro, with the supplementation of adequate substrates. However, other alternative sources could be relevant under certain physiological or pathological conditions. Further experiments are required to assess the relative contribution of different sources, such as NO release from endogenous GSNO (*S*-nitrosoglutathione) (Barroso et al. 2006).

Arnaud et al. (2006), described NO generation in chloroplasts from *Arabidopsis* cells. They reported that NO accumulated in the chloroplasts after Fe treatment, and acts downstream of Fe to promote an increase of *AtFer1* (*Arabidopsis* Ferritin 1) mRNA level. This increase was inhibited by L-NMMA indicating that a NOS activity is involved in the pathway. However, since inhibition was not complete other pathways may lead to NO production in response to Fe (Arnaud et al. 2006). Tewari et al. (2013) also described endogenous NO, and $ONOO^-$ generation in protoplasts chloroplasts from *Brassica napus* L. cv. Bronowski plants. The inhibition of DAF fluorescence in the presence of NOS inhibitors suggests the involvement of NOS-like activity in NO generation in these chloroplasts. Moreover, protoplasts from *Atnoa1* mutants exhibited weak signal of NO generation (Tewari et al. 2013). Thus, AtNOA1 seems to be important for NO generation also in chloroplasts.

Recently, we explore the hypothesis that the content of NO in soybean cotyledons is related to chloroplast functionality *in planta*. Employing confocal fluorescence microscopy and EPR techniques, Galatro et al. (2013) showed that chloroplasts contribute to NO synthesis in vivo. Moreover, the level of NO in the whole tissue was related to chloroplasts functionality. The detection of NO in coincidence with cotyledon maximum fresh weight, chlorophyll content, and quantum yield of PSII, supported the hypothesis of a strong link between NO levels and chloroplast functionality. In addition, seedlings exposed in vivo to herbicides showed deleterious effects on chloroplast function (loss of photosynthetic capacity), and an impaired NO accumulation. The employment of the herbicide DCMU supports a role for the integrity of the photosynthetic electron chain in chloroplasts NO production in vivo, as was previously observed by Jasid et al. (2006) in the in vitro experiments with isolated chloroplasts. These results are consistent with the requirement of chloroplasts for NO generation in soybean cotyledons, both as a result of the active synthesis of NO in the organelle and/or because of an indirect requirement of some chloroplast products for NO synthesis in other areas of the plant, as it was described by Van Ree et al. (2011). Overall, these findings strongly suggest that chloroplasts are the organelles that contribute to NO synthesis in vivo, and that their proper functionality is essential for maintaining NO levels in soybean cotyledons (Galatro et al. 2013).

Chloroplasts are key organelles in plant metabolism, and seem to be strongly involved in NO synthesis. NO may function in chloroplasts as a regulator of photosynthetic electron transport, and as an antioxidant preserving lipids, proteins (including D1) and nucleic acids from photooxidative damage (Jasid et al. 2006; Beligni et al. 2002), but also may be part of a complex network of regulation involved in processes that transcend chloroplasts, as its participation in Fe metabolism (Arnaud et al. 2006) through transcription of nuclear-encoded *AtFer1* gene.

1.2.2 NO Sources Under Abiotic Stress

Gould et al. (2003) have reported the impact of several abiotic stresses like, light, high temperatures, osmotic shock, salinity and mechanical injury on NO evolution from tobacco leaf cells. They tested the hypothesis that NO generation occurs as a general response to different environmental cues. However, they concluded that although different stressors can trigger NO synthesis (like high temperatures, osmotic stress, or salinity), it cannot be considered a universal plant stress response.

Several sources of NO would be involved in responses to abiotic stress. A NOS-like activity was detected in guard cells of *B. juncea*, which was enhanced by abiotic stress (Talwar et al. 2012). NOS-like activity has been involved in the induction of cadmium accumulation, cadmium-induced programmed cell-death, and protective responses against UV-B (Gupta et al. 2011a), salt stress, and phosphate deficiency (Fröhlich and Durner 2011). In addition, the NOS pathway is important for postharvest NO synthesis in tomato to avoid chilling injury (Zhao et al. 2011). NR as NO producer has been involved in cold, drought and osmotic stress (Gupta et al. 2011a; Fröhlich and Durner 2011). Ziogas et al. (2013) have studied nitrosative responses in citrus plants exposed to various abiotic stresses, including continuous light, continuous dark, heat, cold, drought, and salinity. They have shown that the expression of several genes potentially involved in NO production, was affected by the abiotic stress treatments, demonstrating that NO-derived nitrosative responses could be regulated by various pathways.

From these studies, it can be concluded that NO synthesis in response to abiotic stress could be achieved by different sources acting separately or jointly to deal with the stress for cell viability.

1.3 Concluding Remarks

It is clear that NO content in plants varies among tissues, and also depends on physiological status. The generated NO is widely accepted to cooperate for the growth and development of plants, and also to be a good candidate to participate in response to several stress conditions. Figure 1.2 briefly summarized proposed sources of NO in plants. Although the knowledge of NO functions in plants has been largely improved, the isolation and characterization of a single protein with

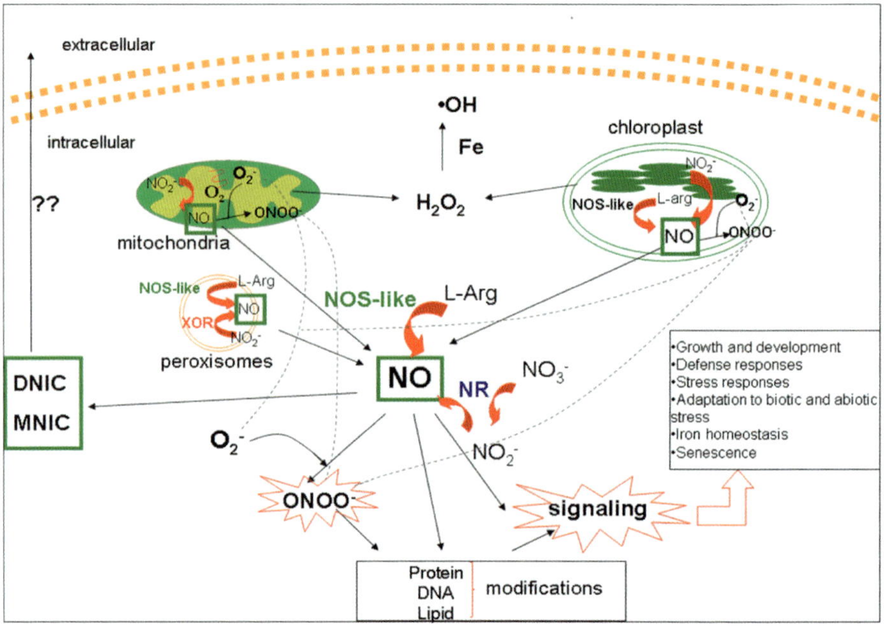

Fig. 1.2 Scheme of main proposed subcellular sources of NO in plants. Reaction with ROS and Fe, and biological effects are indicated. *NOS-like* nitric oxide synthase-like activity; *NR* nitrate reductase; *XOR* xanthine oxidoreductase; *MNIC* mononitrosyl Fe complexes, *DNIC* dinitrosyl Fe complexes; *L-Arg* L-Arginine; $ONOO^-$ peroxinitrite; H_2O_2 hydrogen peroxide. *Dotted lines* indicate the diffusion of the species. *Continuous lines* link species to their functions

NOS activity is still matter of active research and remains an issue to be fully elucidated. The complex scenario shown in the Fig. 1.2 reflects the participation of several organelles (chloroplasts, mitochondria, peroxisomes and cytosolic enzymatic activities) and reactive species that lead to the generation of not only NO but also $ONOO^-$. The dual effects of NO in the cellular biochemical steady state condition due to its capacity of both protect or damage bio-molecules require a careful analysis of each condition before designing any operative strategy. However, the possibility of affording laboratory protocols developed to change this versatile molecule functions in the inner of the cell could be considered as one of the intriguing issues and is nowadays the centre of an active debate and investigation.

Acknowledgments This study was supported by grants from the University of Buenos Aires (UBA) and National Council for Science and Technology (CONICET) and the National Agency of Scientific and Technological Promotion (ANPCyT). Susana Puntarulo and Andrea Galatro are career investigators from CONICET.

References

Arnaud N, Murgia I, Boucherez J et al (2006) An iron induced nitric oxide burst precedes ubiquitin-dependent protein degradation for Arabidopsis AtFer1 ferritin gene expression. J Biol Chem 281:23579–23588

Barroso JB, Corpas FJ, Carreras A et al (2006) Localization of *S*-nitrosoglutathione and expression of *S*-nitrosoglutathione reductase in pea plants under cadmium stress. J Exp Bot 57:1785–1793

Basylinski DA, Hollocher TC (1985) Evidence from the reaction between trioxodinitrate (II) and ^{15}NO that trioxidinitrate (II) decomposes into nitrosyl hydride and nitrite in neutral aqueous solution. Inorg Chem 24:4285–4288

Beligni MV, Fath A, Bethke PC et al (2002) Nitric oxide acts as an antioxidant and delays programmed cell death in barley aleurone layers. Plant Physiol 129:1642–1650

Chandok MR, Ytterberg AJ, van Wijk KJ, Klessig DF (2003) The pathogen-inducible nitric oxide synthase (iNOS) in plants is a variant of the P protein of the glycine decarboxylase complex. Cell 113:469–482

Corpas FJ, Barroso JB, Carreras A et al (2006) Constitutive arginine-dependent nitric oxide synthase activity in different organs of pea seedlings during plant development. Planta 224:246–254

Crawford NM, Galli M, Tischner R et al (2006) Response to Zemojtel et al.: Plant nitric oxide synthase: back to square one. Trends Plant Sci 11:526–527

Davis KL, Martin E, Turko IV, Murad F (2001) Novel effects of nitric oxide. Annu Rev Pharmacol Toxicol 41:203–236

del Río LA (2011) Peroxisomes as a cellular source of reactive nitrogen species signal molecules. Arch Biochem Biophys 506:1–11

Flores-Pérez Ú, Sauret-Güeto S, Gas E et al (2008) A mutant impaired in the production of plastome-encoded proteins uncovers a mechanism for the homeostasis of isoprenoid biosynthetic enzymes in *Arabidopsis* plastids. Plant Cell 20:1303–1315

Foissner I, Wendehenne D, Langebartels C, Durner J (2000) In vivo imaging of an elicitor-induced nitric oxide burst in tobacco. Plant J 23:817–824

Ford PC, Wink DA, Stanbury DM (1993) Autoxidation kinetics of aqueous nitric oxide. FEBS Lett 326:1–3

Foresi N, Correa-Aragunde N, Parisi G et al (2010) Characterization of a nitric oxide synthase from the plant kingdom: NO generation from the green alga *Ostreococcus tauri* is light irradiance and growth phase dependent. Plant Cell 22:3816–3830

Fröhlich A, Durner J (2011) The hunt for plant nitric oxide synthase (NOS): is one really needed? Plant Sci 181:401–404

Galatro A, Puntarulo S, Guiamet JJ, Simontacchi M (2013) Chloroplast functionality has a positive effect on nitric oxide level in soybean cotyledons. Plant Physiol Biochem 66:26–33

Galatro A, Simontacchi M, Puntarulo S (2004) Effect of nitric oxide exposure on antioxidant capacity of soybean leaves. Current Topics Plant Biol 5:69–79

Gas E, Flores-Pérez Ú, Sauret-Güeto S, Rodríguez-Concepción M (2009) Hunting for plant nitric oxide synthase provides new evidence of a central role for plastids in nitric oxide metabolism. Plant Cell 21:18–23

Gisone P, Dubner D, Pérez MR et al (2004) The role of nitric oxide in the radiation-induced effects in the developing brain. In vivo 18:281–292

Goretski J, Hollocher TC (1988) Trapping of nitric oxide produced during denitrification by extracellular hemoglobin. J Biol Chem 263:2316–2323

Gould KS, Lamotte O, Klinguer A et al (2003) Nitric oxide production in tobacco leaf cells: a generalized stress response? Plant, Cell Environ 26:1851–1862

Guo F-Q, Okamoto M, Crawford NM (2003) Identification of a plant nitric oxide synthase gene involved in hormonal signaling. Science 302:100–103

Gupta KJ, Fernie AR, Kaiser WM, van Dongen JT (2011a) On the origins of nitric oxide. Trends Plant Sci 16:160–168
Gupta KJ, Igamberdiev AU, Manjunatha G et al (2011b) The emerging roles of nitric oxide (NO) in plant mitochondria. Plant Sci 181:520–526
Henry Y, Ducrocq C, Drapier JC et al (1991) Nitric oxide, a biological effector. Electron paramagnetic resonance detection of nitrosyl-iron-protein complexes in whole cells. Eur Biophys J 20:1–15
Jasid S, Simontacchi M, Bartoli CG, Puntarulo S (2006) Chloroplasts as a nitric oxide cellular source. Effect of reactive nitrogen species on chloroplastic lipids and proteins. Plant Physiol 142:1246–1255
Klessig DF, Ytterberg AJ, van Wijk KJ (2004) The pathogen-inducible nitric oxide synthase (iNOS) in plants is a variant of the P protein of the glycine decarboxylase complex. Cell 119:445
Lamattina L, Garcia-Mata C, Graziano M, Pagnussat G (2003) Nitric oxide: the versatility of an extensive signal molecule. Annu Rev Plant Biol 54:109–136
Liu H, Lau E, Lam MP et al (2010) OsNOA1/RIF1 is a functional homolog of AtNOA1/RIF1: implication for a highly conserved plant cGTPase essential for chloroplast function. New Phytol 187:83–105
Modolo LV, Augusto O, Almeida IMG et al (2005) Nitrite as the major source of nitric oxide production by *Arabidopsis thaliana* in response to *Pseudomonas syringae*. FEBS Lett 579:3814–3820
Moreau M, Lee GI, Wang Y et al (2008) AtNOS/AtNOA1 is a functional *Arabidopsis thaliana* cGTPase and not a nitric-oxide synthase. J Biol Chem 283:32957–32967
Rasul S, Wendehenne D, Jeandroz S (2012) Study of oligogalacturonides-triggered nitric oxide (NO) production provokes new questioning about the origin of NO biosynthesis in plants. Plant Signal Behav 7:1031–1033
Rosales EP, Iannone MF, Groppa MD, Benavides MP (2012) Polyamines modulate nitrate reductase activity in wheat leaves: involvement of nitric oxide. Amino Acids 42:857–865
Rubbo H, Parthasarathy S, Barnes S et al (1995) Nitric oxide inhibition of lipoxygenase-dependent liposome and low density lipoprotein oxidation: termination of radical chain propagation reactions and formation of nitrogen containing oxidized lipid derivatives. Arch Biochem Biophys 324:15–25
Rümer S, Gupta KJ, Kaiser WM (2009) Plant cells oxidize hydroxylamines to NO. J Exp Bot 60:2065–2072
Saran M, Michel C, Bors W (1990) Reaction of NO with O_2^- implications for the action of endothelium-derived relaxing factor (EDRF). Free Radic Res Commun 10:221–226
Stamler JS, Singel DJ, Loscalzo J (1992) Biochemistry of nitric oxide and its redox-activated forms. Science 258:1898–1902
Talwar PS, Gupta R, Maurya AK, Deswal R (2012) *Brassica juncea* nitric oxide synthase like activity is stimulated by PKC activators and calcium suggesting modulation by PKC-like kinase. Plant Physiol Biochem 60:157–164
Tewari RK, Prommer J, Watanabe M (2013) Endogenous nitric oxide generation in protoplast chloroplasts. Plant Cell Rep 32:31–44
Tun NN, Santa-Catarina C, Begum T et al (2006) Polyamines induce rapid biosynthesis of nitric oxide (NO) in *Arabidopsis thaliana* seedlings. Plant Cell Physiol 47:346–354
Van Ree K, Gehl B, Wassim Chehab E et al (2011) Nitric oxide accumulation in *Arabidopsis* is independent of NOA1 in the presence of sucrose. Plant J 68:225–233
Wendehenne D, Pugin A, Klessig DF, Durner J (2001) Nitric oxide: comparative synthesis and signaling in animal and plant cells. Trends Plant Sci 6:177–183
Wink DA, Mitchell JB (1998) Chemical biology of nitric oxide: insights into regulatory, cytotoxic, and cytoprotective mechanisms of nitric oxide. Free Radic Biol Med 25:434–456
Wink DA, Osawa Y, Darbyshire JF et al (1993) Inhibition of cytochrome P_{450} by nitric oxide and a nitric oxide-releasing agent. Arch Biochem Biophys 300:115–123

Zemojtel T, Fröhlich A, Palmieri MC et al (2006) Plant nitric oxide synthase: a never-ending story? Trends Plant Sci 11:524–525
Zhao R, Sheng J, Lv S et al (2011) Nitric oxide participates in the regulation of LeCBF1 gene expression and improves cold tolerance in harvested tomato fruit. Postharvest Biol Technol 62:121–126
Ziogas V, Tanou G, Filippou P et al (2013) Nitrosative responses in citrus plants exposed to six abiotic stress conditions. Plant Physiol Biochem 68:118–126

Chapter 2
Biosynthesis of Nitric Oxide in Plants

Tamás Rőszer

Abstract Nitric oxide (NO) regulates important events in plant physiology, disease resistance and stress tolerance. In plants, distinct enzymatic and chemical processes can generate NO from nitrite (NO_2^-), L-Arginine and possibly other *N*-compounds. Reduction of NO_2^- to NO is catalyzed by nitrate reductase and the mitochondrial electron transport chain. Deoxygenated heme-proteins also facilitate NO production from NO_2^-. NO may also be released in nonenzymatic processes from nitrous acid and *S*-nitrosoglutathione. Whether plants have a specific enzyme with primary oxidative NO synthesizing activity is an open debate. Although, NO synthase-homolog genes are present in green algae, and a protein (AtNOS1/AtNOA1) with regulatory effects on oxidative NO synthesis is known in vascular plants, integration of the multiple NO producing processes requires a complex regulatory network in the plant cell. However, our insight into the underlying molecular mechanisms is still limited. Plant hormones, stress and injury signals, modulation of intracellular Ca^{2+} levels are the potential drivers of plant NO synthesis under physiological and stress conditions.

Keywords Cell signaling • Nitrate reductase • Nitric oxide synthase • Plant hormones

2.1 Introduction

Nitric oxide (NO) is a bioactive molecule with multifaceted physiological roles in plants (Rőszer 2012b). Endogenous NO synthesis has been identified in cyanobacteria (Sturms et al. 2011), green algae (Foresi et al. 2010), lichens (Catala et al.

T. Rőszer (✉)
Centro Nacional de Investigaciones Cardiovasculares, Madrid, Spain
e-mail: troszer@cnic.es; tamas.roeszer@uni-ulm.de

T. Rőszer
Zentrum für Biomedizinische Forschung, Institut für Allgemeine Zoologie und Endokrinologie, Universität Ulm, Helmholtz Strasse 8/1 89081 Ulm, Germany

M. N. Khan et al. (eds.), *Nitric Oxide in Plants: Metabolism and Role in Stress Physiology*, DOI: 10.1007/978-3-319-06710-0_2,

2010), species representing pteridophyta, gymnosperms, monocots, and eudicots (Salmi et al. 2007; Rőszer 2012b; Yu et al. 2012). As a signal molecule, NO is involved in germination, root morphogenesis, pollen tube growth, chloroplast biogenesis, transpiration, cell wall synthesis, and other biosynthetic pathways (Rőszer 2012b). NO is implicated in the control of oxidative phosphorylation and photosynthesis and can protect the cell organelles from oxidative damage and consequently delay senescence and cell death (Rőszer 2012b). In vascular plants, NO synthesis is an important element in acquiring disease resistance as well as adapting to distinct abiotic stressors such as salinity, cold, osmotic stress, hypoxia and excess absorption of minerals and heavy metals (Camejo et al. 2012; Chun et al. 2012; Lehotai et al. 2012; Sun and Li 2012; Tan et al. 2013). NO synthesis can also initiate programmed cell death in distinct species ranging from algae to vascular plants (Lombardi et al. 2010; Ma et al. 2010; Rosales et al. 2010; Yordanova et al. 2010; Pedroso et al. 2000).

To date, eight distinct enzymatic and nonenzymatic processes have been recognized which can elaborate NO in plants (Fig. 2.1). These include NO generation by the reduction of nitrite (NO_2^-), or by the oxidation of more reduced nitrogen compounds, such as the amino acid L-Arginine or hydroxylamine (Mur et al. 2013). Major sites of NO biosynthesis are the protoplasts and the chloroplasts, the mitochondria and the peroxisomes (Rőszer 2012a, b). The cytoplasm, the cell membrane, the endoplasmic reticulum, and the apoplast can also generate NO in vascular plants (Fröhlich and Durner 2011) (Fig. 2.2).

2.2 Mechanisms of Reductive NO Synthesis

The cytoplasm, the mitochondria, the chloroplasts, the peroxisomes and the apoplast are sites of reductive NO generation from NO_2^- (Rőszer 2012b). The NO_2^-/NO reduction can be catalyzed by assimilatory nitrate reductase (NR; EC 1.6.6.1, transferred to EC 1.7.1.1) or the mitochondrial electron transport chain (Fig. 2.1). Deoxygenated heme-containing proteins can also facilitate the reductive NO generation from NO_2^- (Rőszer 2012b). Nonenzymatic NO_2^-/NO reduction can also occur in acidic compartments of the plant tissues (Rőszer 2012b).

2.2.1 Reductive NO Synthesis by Nitrate Reductase

Nitrate reductase (NR), in addition to its primary nitrate (NO_3^-) oxidoreductase activity, is capable of reducing NO_2^- to NO with low efficacy (Rockel et al. 2002). The NR-catalyzed reduction of NO_2^- to NO is apparent in green algae and vascular plants (Rőszer 2012b). NR-mediated NO synthesis is involved in physiological processes, pathogen defense and stress response (Mur et al. 2013). The presence of NR-catalyzed NO synthesis in the cyanobacterium *Anabaena doliolum* (Mallick

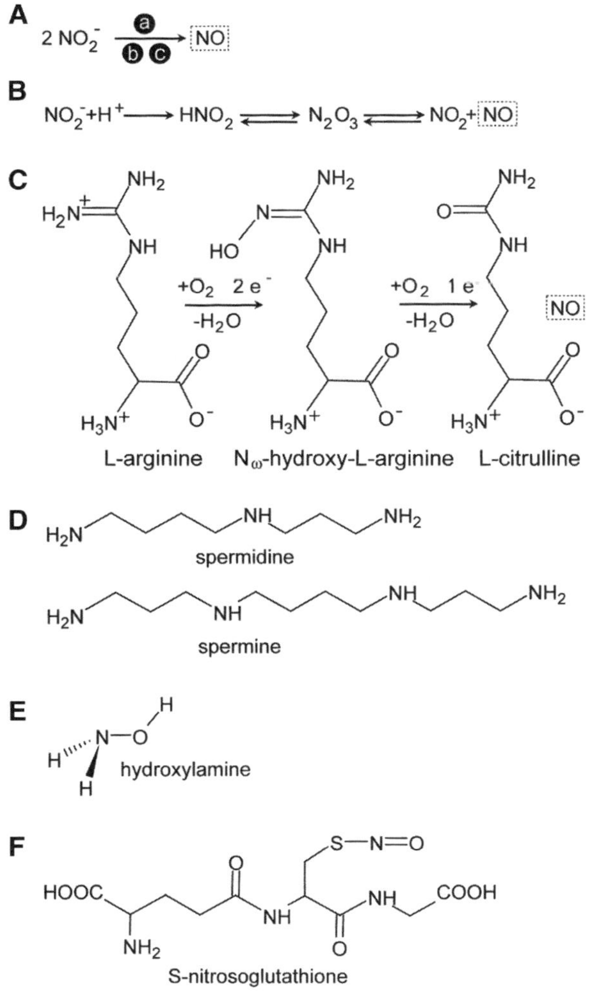

Fig. 2.1 Substrates and elicitors of NO synthesis in plants. **A** Reductive NO generation from nitrite can be catalyzed by nitrate reductase (a), the mitochondrial electron transport chain (b) or by deoxygenated heme-proteins (c). **B** Under acidotic conditions nitrous acid, the protonated form of nitrite can elaborate NO in a non-enzymatic process. **C** Oxidative NO synthesis from L-Arginine is catalyzed by a yet undefined NO-synthase in plants. To date the only plant-type NO-synthase encoding gene known is from green algae. **D, E** Polyamines and hydroxylamine can increase NO synthesis by unknown mechanisms. **F** NO can react with glutathione to form *S*-nitrosoglutathione, which can be a source of non-enzymatic NO liberation

et al. 1999) suggests that this mechanism may be one of the most ancient forms of NO generation in plants.

In green algae NR is associated with the pyrenoids and the thylakoid membranes of the chloroplasts and NR is responsible for the chloroplastic NO_2^-/NO reduction (Rőszer 2012b). In vascular plants, a NO_2^-/NO reduction has also been

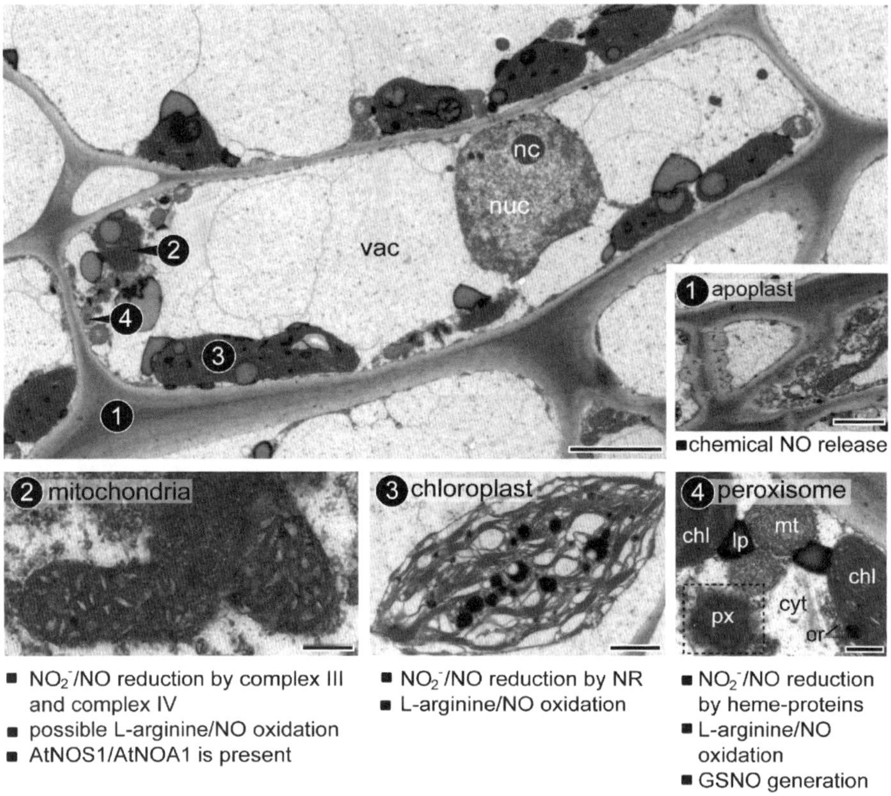

Fig. 2.2 Localization of distinct NO producing activities in the plant cell. The apoplast (*1*), the mitochondria (*2*), the chloroplasts (*3*) and the peroxisomes (*4*) are the most characterized NO-generating compartments in plants. The cytoplasm and the intracellular membranes can also be sites of enzymatic and chemical NO release. TEM images of *Scindapsus aureus* mesophyll cell. *chl* chloroplast, *cyt* cytoplasm, *er* endoplasmic reticule, *lp* lipid droplet, *mt* mitochondria, *nuc* nucleus, *nc* nucleolus, *px* peroxisome, *vac* vacuole. *Scale bar* 2 μm (*overview*), 0.5 μm (*inset* 1), 0.2 μm (*inset* 2), 2 μm (*inset* 3), 1 μm (*inset* 4). Author's images

assayed in the chloroplasts, and the responsible enzyme may be a thylakoid-associated NR (Jasid et al. 2006). However, the main pool of NR is the cytoplasm and chloroplast association of NR is debated in vascular plants (Rőszer 2012b).

Another possible NO_2^-/NO-reductase (NI-NOR) of vascular plants has been identified in the root plasma membrane of the tobacco, *Nicotiana tabacum* (Stohr et al. 2001). NI-NOR reduces NO_2^- to NO using reduced cytochrome c as an electron donor. NO generation of NI-NOR is comparable to the NO_3^--reducing activity of a root-specific NR, however, NO-NOR can be a distinct protein, and it still remains to be characterized as NO producing enzyme (Stohr et al. 2001).

2.2.2 Reductive NO Synthesis by the Mitochondrial Electron Transport Chain

Mitochondria in green algae and higher plants can use NO_2^- as an alternative electron acceptor to sustain ATP synthesis under O_2 deprivation (Tischner et al. 2004; Gupta and Igamberdiev 2011). The mitochondrial respiratory chain is able to reduce NO_2^- to NO at the complex III (cytochrome *bc*1) and the complex IV (cytochrome-c oxidase, CcO) (Igamberdiev et al. 2010; Gupta and Igamberdiev 2011; Castello et al. 2006). This mechanism results in mitochondrial NO generation in plant cells experiencing hypoxia.

Hypoxia increases the activity and the transcription of NR, which converts NO_3^- to NO_2^- and leads to NO_2^- accumulation in the cytoplasm. In cells suffering from hypoxia, the further reduction of NO_2^- is limited allowing a sustained NO_2^- supply for reductive NO synthesis (Rőszer 2012a, b). Due to the lack of a specific O_2 transporting system in plants, assimilating tissues in the leaf, stem or cells of rapidly growing tissues can be short of O_2 supply, making it possible that mitochondrial NO_2^-/NO reduction may be a common mechanism of NO synthesis in the plant tissues.

Mitochondrial NO_2^-/NO reduction may control the role in the initiation of seed germination. In many plants, seed dormancy is interrupted by imbibition, a process in which water penetrates the seed coat and generates a temporal hypoxic condition. Imbibition is associated with a rapid increase of NO levels (Liu and Zhang 2009), and possibly, favors the mitochondrial reductive NO synthesis (Gupta and Igamberdiev 2011). The NO generated within the mitochondria inhibits CcO, which eventually stimulates germination (Gniazdowska et al. 2010).

A recent model suggests that a recycling of NO_2^-/NO/NO_2^- exists between the mitochondria and the cytoplasm experiencing hypoxia. This mechanism improves the energy status of cells suffering from O_2 limitation. Since NO inhibits electron transport to O_2 at the CcO site, mitochondrial NO production reduces further O_2 consumption when O_2 availability is already limited (Igamberdiev et al. 2010; Gupta and Igamberdiev 2011). The reductive mitochondrial NO generation also inhibits the photorespiratory cycle (Gupta and Igamberdiev 2011) and the fermentative metabolism (Oliveira et al. 2013). NO released from the mitochondria to the cytosol undergoes oxidation to nitrate (NO_3^-) by plant hemoglobin (class 1 nonsymbiotic hemoglobin), which is expressed in response to hypoxia (Igamberdiev and Hill 2004). As mentioned above, cytoplasmic NR reduces NO_3^- to NO_2^-, which recycles to the mitochondria and is being reduced to NO (Gupta and Igamberdiev 2011). This NO/NO_2^- exchange between the mitochondria and the cytoplasm maintains the NO_2^- supply for the ATP synthesis under hypoxia (Gupta and Igamberdiev 2011). The cytoplasmic NO/NO_3^-/NO_2^- conversion keeps NADH/NAD^+ and NADPH/$NADP^+$ ratios low, ensuring a low redox level and helping the adaptation to O_2 limitation (Igamberdiev et al. 2010).

2.2.3 Reductive NO Generation by Heme Containing Proteins

Plant peroxisomes can also generate NO by NO_2^- reduction under hypoxic or anoxic conditions (Igamberdiev et al. 2010). The responsible mechanism may be NO_2^-/NO reducing ability of deoxygenated heme-containing proteins in the peroxisome matrix (Igamberdiev et al. 2010; Sturms et al. 2011). Similar reductive NO generation has been shown in the plant plasma membrane, cytosol and endoplasmic reticulum (Igamberdiev et al. 2010). Reduction of NO_2^- to NO by heme-proteins (e.g., hemoglobins) also occurs in cyanobacteria (Sturms et al. 2011) and mammalian tissues under O_2 limitation (Shiva et al. 2011; Tiso et al. 2011).

2.3 Mechanisms of Oxidative NO Synthesis

Oxidative NO synthesis from L-Arginine is also present in plant cells, although the responsible enzyme, the putative plant NO-synthase (NOS) has not yet been identified. In representatives of prokaryotes, unicellular eukaryotes, invertebrates, nonmammalian vertebrates and mammals, several NOS (EC 1.14.23.29) proteins and NOS-encoding genes have been identified (Rőszer 2012b). Higher plants however, are lacking homolog sequences to already known NOS-encoding genes (Mur et al. 2013).

2.3.1 Oxidative NO Synthesis from L-Arginine

Enzymatic oxidation of L-Arginine to NO and L-Citrulline has been identified in the chloroplasts and the leaf peroxisomes of the vascular plants and in green algae (Rőszer 2012b). The oxidation of L-Arginine to NO in the chloroplasts requires NADPH and is independent from Ca^{2+} supply (Jasid et al. 2006). In the leaf peroxisomes the L-Arginine/L-Citrulline conversion requires Ca^{2+}, calmodulin, FAD (flavin adenine dinucleotide), FMN (flavin mononucleotide), and NADPH (Barroso et al. 1999; del Río et al. 2003; del Río 2011). Peroxisomal oxidative NO synthesis has been measured in the presence of BH_4 (tetrahydrobiopterine), although, other studies have shown that it is not required for NO synthesis in vascular plants (Rőszer 2012b). It has been found recently, that oxidative NO synthesis from L-Arginine requires not only Ca^{2+} and NADPH but also BH_4 in the green algae *Ostreococcus* species (Foresi et al. 2010). Plant mitochondria may also oxidize L-Arginine to NO and the responsible enzyme may be present in the mitochondrial matrix or the intermembrane space (Guo and Crawford 2005). However, it is debatable whether plant mitochondria contain a specific enzyme which is responsible for the oxidative NO synthesis (Barroso et al. 1999).

2.3.2 The Enigmatic Plant-Type NOS

Homolog genes of mammalian NOS have been identified recently in the genome of the marine green algae *Ostreococcus tauri* and *Ostreococcus lucimarinus* (Foresi et al. 2010). The recombinant *O. tauri* NOS (OtNOS) protein shares 44 % sequence overlap with human NOS3, 45 % with human NOS1 and NOS2.

To date OtNOS is the only NOS found in plants. The enzyme responsible for the NOS-like activity in higher plants is still a subject of debate (Mur et al. 2013). A protein recognized by an antibody against mammalian NOS2 has been found in the leaf peroxisomes and the chloroplasts of *Pisum sativum* (Barroso et al. 1999; Corpas et al. 2001; del Río et al. 2003). NOS has already been characterized in many prokaryotes (bacteria and archaea). Since chloroplasts are descendants of ancient endosymbiotic cyanobacteria, one can assume that the NOS2 immunoreactive protein of the chloroplast stroma might be a cognate of a prokaryote NOS molecule (Rőszer 2012b). However, the NOS molecule responsible for L-Arginine dependent NO synthesis of the chloroplasts still remains unknown (Mur et al. 2013).

One possible plant-specific NOS has been described in the mitochondria of *Arabidopsis thaliana* (Guo et al. 2003). This putative NOS molecule has been identified based on its gene sequence homology (23 % identity, 39.5 % similarity) to a putative NOS of the snail *Helix pomatia* (Huang et al. 1997). This 561-amino acid *Arabidopsis* protein has been annotated as *A. thaliana* NOS-1 (AtNOS1), later renamed as *A. thaliana* NOS-associated protein-1 (AtNOA1). It has been shown that AtNOS1 can oxidize L-Arginine to NO in a NADPH and Ca^{2+} dependent mechanism and its activity is sensitive to mammalian NOS inhibitors (Guo et al. 2003). However, AtNOS1 does not show sequence similarities to mammalian NOS isoforms (Guo et al. 2003) and further studies have concluded that AtNOS1 is a GTPase protein (Moreau et al. 2008; Sudhamsu et al. 2008). Similarly, the putative NOS in *Helix pomatia* is more likely to be an NOS-associated protein rather than a NO producing enzyme (Rőszer et al. 2010).

Collectively, these data suggest that AtNOS1/AtNOA1 (*A. thaliana* NOS-associated protein 1) and its orthologs may be involved in NO synthesis only in an indirect way, by allowing proper NO synthesis of a yet undefined NO producing molecule. Since AtNOS1/AtNOA1-associated protein 1 (AtNOS1/AtNOA1) is a GTPase, it is possible that AtNOS1/AtNOA1 generates cGMP (cyclic guanosine monophosphate) and activates downstream NO signal pathways (Moreau et al. 2008). For instance, a mammalian AtNOS1-related protein is implicated in mitochondrial protein synthesis (Kolanczyk et al. 2011), thus it might have an indirect effect on the maintenance of NO production. Of note, AtNOS1/AtNOA1 is associated with the plant mitochondria, where reductive NO synthesis can overshadow a putative NOS-like activity.

2.3.3 Other Forms of Oxidative NO Synthesis

Recently it has been shown that polyamines and hydroxylamine can increase the oxidative NO synthesis in plant cells (Tun et al. 2006; Rumer et al. 2009; Wimalasekera et al. 2011). NO can mediate effects of polyamines in plants, however, the manner in which polyamines can increase NO synthesis is uncertain (Fröhlich and Durner 2011). Possible mechanisms include an interaction of polyamines with the NR-catalyzed NO production (Rosales et al. 2012) and the indirect effect of polyamine synthesis on L-Arginine metabolism (Zhang et al. 2011). Interestingly, polyamine synthesis is inhibited by NO, and *A. thaliana* plants lacking AtNOA1 accumulate polyamines, rendering a yet unexplored interplay between polyamines and NO biosynthesis in plants (Yamasaki and Cohen 2006; Majlath et al. 2011; Filippou et al. 2012). Hydroxylamine, an intermediate in the process of nitrification, can be oxidized to NO in tobacco cell cultures (Rumer et al. 2009). This mechanism may be an alternative of L-Arginine dependent oxidative NO synthesis. However, the underlying molecular mechanism is still unknown and the sufficient availability of hydroxylamine for NO synthesis is debated (Rumer et al. 2009). Other enzymes, such as xanthine oxidase, catalase) and horseradish peroxidase are able to elaborate NO under specific conditions (Huang et al. 2002; Igamberdiev et al. 2010; del Río 2011), however, their possible contribution to NO synthesis in plants is still yet to be ascertained.

2.4 Nonenzymatic NO Release

NO can be released from nitrous acid (HNO_2), a protonated form of NO_2^-. This type of chemical NO release is favored by acidic environments found, e.g., in the apoplast of germinating and thus hypoxic seeds (Yamasaki 2000; Bethke et al. 2004a). Accordingly, the NO liberation from NO_2^- has been shown in the apoplast of the aleuron layer of the barley, *Hordeum vulgare* (Bethke et al. 2004a). This nonenzymatic NO release is augmented by phenolics, compounds found in the aleuron apoplast and in the seed coat (Bethke et al. 2004a). In germinating seed, the NO release may provide an antimicrobial protection for the seeds in the soil (Bethke et al. 2004a). Moreover, seed dormancy is interrupted by NO, thus a NO generation from NO_2^- along with an enzymatic NO synthesis can contribute to the proper germination (Rőszer 2012b). NO-mediated programmed cell death also occurs during germination, when the aleuron cells are being eliminated (Lombardi et al. 2010). Collectively, NO release can act synergistically with the enzymatic NO_2^-/NO reduction to evoke a NO burst during germination.

Another possible but yet unexplored mechanism of nonenzymatic NO generation is the release of NO from *S*-nitrosoglutathione (GSNO) (del Río 2011). This compound is formed in the oxidative environment of the peroxisomes, where both NO and the NO-derived peroxynitrite can react with glutathione to generate GSNO

(Barroso et al. 2006). GSNO behaves as a NO-donor compound and can be a transportable NO reserve distributed in the plant tissues. Genesis of NO from GSNO is facilitated by ambient light and transition metals (Floryszak-Wieczorek et al. 2006). As was described above, hydroxylamine is a possible substrate of NO synthesis (Rumer et al. 2009), however, it is uncertain that GSNOR would support NO production with hydroxylamine.

2.5 Control of NO Synthesis in the Plant Cell

Deficiencies in genes implicated in NO homeostasis lead to severe alterations in plants, underlining the importance of balanced NO production and elimination (Fröhlich and Durner 2011). However, the mechanisms which control plant NO homeostasis are largely undefined. Chemical and enzymatic NO synthesis can occur simultaneously, for example in germinating seeds (Bethke et al. 2004a; Gupta and Igamberdiev 2011). The multiplicity of NO-producing mechanisms makes plant-type NO homeostasis a complex phenomenon. As a framework for understanding the control of NO levels, we provide an overview on the potential mechanisms which can control NO synthesis. These include the regulation of substrate and cofactor availability; the chemical environment which allows non-enzymatic NO release and certain upstream signaling events that can modulate transcription and activity of NO producing enzymes.

2.5.1 Control of Reductive and Oxidative NO Synthesis

Main sources of NO in plant cells are NO_2^- and L-Arginine, thus their levels are key determinants of NO synthesis. Stress conditions, including hypoxia, inhibition of the photosynthetic electron transport or increased NO_2^- absorption from the soil lead to excessive NO_2^- accumulation in the cytoplasm, which favors NO_2^- reduction to NO (Gupta et al. 2010; Mur et al. 2013). Cytoplasmic NO_2^- can be removed through increased influx into the vacuole or efflux from the cell, however, it is yet uncertain how these mechanisms can be integrated to control NO synthesis (Mur et al. 2013). Light exposure promotes the chloroplastic reduction of NO_2^- to NH_4^+, which impedes reductive NO generation (Sakihama et al. 2002; Rőszer 2012b).

When L-Arginine is abundant and NO_2^- availability is limited, oxidative NO synthesis can be the dominant form of NO generation. Accordingly, *A. thaliana* mutants which accumulate L-Arginine in the chloroplast display increased NO synthesis (Streatfield et al. 1999; He et al. 2004). Increasing L-Arginine availability in *Arabidopsis* plants by inhibiting arginase activity also leads to an increased NO production (Flores et al. 2008; Shi et al. 2013). Importantly, the carbohydrate and ATP supply of L-Arginine synthesis is provided by the photosynthetic light reactions, therefore L-Arginine production positively correlates with photosynthetic activity (Krueger and Kliewer 1995). The light reactions of the photosynthesis also sustain the appropriate NADPH and O_2 supply for the oxidative NO synthesis (Jasid et al. 2006). Active photosynthesis also favors the consumption of NO_2^- in amino acid synthesis through reduction to NH_4^+. Interestingly, L-Arginine inhibits chloroplastic NO_2^- uptake (Ferrario-Mery et al. 2008). These findings suggest that light exposure and photosynthesis increases the L-Arginine pool and reduces NO_2^- levels within the chloroplast, thus favoring oxidative and inhibiting reductive NO synthesis. Although a recent study proposes that a NOS-like activity may be the only source of NO in the chloroplast (Tewari et al. 2013), several others provide evidence that both reductive and oxidative NO generation takes place in the chloroplast (Rőszer 2012b). Oxidative and reductive NO synthesis may be temporally separated, i.e., due to a photoperiodic change in L-Arginine and NO_2^- availability (Rőszer 2012b).

2.5.2 Hormonal Control of NO Synthesis

To date, some chemical signals have already been identified as elicitors of NO synthesis (Table 2.1). The plant hormone auxin can increase both reductive and oxidative NO synthesis (Kolbert et al. 2008; Jin et al. 2011). It implies that the same signal can impact distinct forms of NO generation. NO is a downstream mediator of other plant hormones, such as cytokinins, abscisic acid and brassinosteroids (Beligni and Lamattina 2000; Tun et al. 2001; Ötvös et al. 2005; Kolbert et al. 2008; Zhang et al. 2010; Romera et al. 2011; Liu et al. 2013) but it is unknown how these hormones regulate NO synthesis. Salicylic acid, which has prominent roles in host defense against fungal and oomycete pathogens enhances NO synthesis in *A. thaliana* (Zottini et al. 2007) and tomato *Solanum lycopersicum* (Poór and Tari 2012). Expression of NR is increased in response to salicylic acid (Caamal-Chan et al. 2011). Increased NR expression can explain the elevated NO synthesis under stress conditions; however, salicylic acid induced NO synthesis can be associated with oxidative NO production (Zottini et al. 2007). In the green alga *Chlamydomonas reinhardtii* ethylene has been described as elicitor of NO synthesis under stress conditions (Yordanova et al. 2010). In vascular plants recent findings also point to the possible involvement of ethylene, a mediator produced under stress conditions and injury (Poór et al. 2013), however, the molecular link between abiotic stressors and increased NO synthesis is yet undefined.

Some mechanisms which lower NO levels have also been described in plants (Table 2.1). Oxygenated nonsymbiotic hemoglobins and glutathione are important sinks for NO (Igamberdiev and Hill 2004). A recent study shows that zeatin, a prevalent cytokinin in *Arabidopsis* can interact with NO and reduce intracellular NO levels (Liu et al. 2013). Similarly, in cadmium toxicity, gibberellic acid reduces NO accumulation (Zhu et al. 2012). GSNOR activity is also important in

Table 2.1 Signaling mechanisms affecting NO levels in the plant cell

Signal or stimulus	Effect on NO production	References
Auxin	Increases oxidative NO synthesis and NR-mediated reductive NO generation	Ötvös et al. (2005), Kolbert et al. (2008), Jin et al. (2011), Beard et al. (2012)
Abscisic acid	Increases NR-dependent NO synthesis	Bethke et al. (2004b), Bright et al. (2006)
Brassinosteroids	Increase NO levels, possibly stimulate oxidative NO synthesis	Zhang et al. (2010)
Cytokinins	Increase NO levels, possibly through oxidative NO synthesis	Tun et al. (2001)
Salicylic acid	Upregulates NR expression and increases NO synthesis, possibly stimulating oxidative NO synthesis	Zottini et al. (2007), Sun et al. (2010), Caamal-Chan et al. (2011)
Ethylene	Stimulates NO synthesis in green algae, possibly plays a similar role in vascular plants	Yordanova et al. (2010), Poór et al. (2013)
Carbon dioxide	Increases oxidative NO synthesis	Wang et al. (2013)
Ca^{2+}, PKC	Increase oxidative NO synthesis	Zottini et al. (2007), Talwar et al. (2012)
Lipopolysaccharides	Evoke NO burst	Zeidler et al. (2004), Sun and Li (2012)
Oligogalacturonides	Induce NR-dependent NO synthesis	Rasul et al. (2012)
Hypoxia	Increases NR expression, sustains NO_2^- supply for NO production, favors mitochondrial reductive NO production, promotes non-enzymatic NO release	Liu and Zhang (2009), Igamberdiev et al. (2010), Gupta and Igamberdiev (2011), Sturms et al. (2011)
Cold, osmotic stress	Increase NO synthesis	Wang et al. (2012), Tan et al. (2013)
Gibberellin	Can reduce NO levels under stress conditions	Fernandez-Marcos et al. (2012)
Zeatin	Scavenges NO	Liu et al. (2013)
Oleic acid	Increases AtNOA1 degradation	Mandal et al. (2012)

eliminating NO in plant cells under stress conditions (Barroso et al. 2006; Lee et al. 2008). Turnover of NO generating proteins can also affect cellular NO homeostasis, however, this possibility is not analyzed in details in plants. A recent study shows that oleic acid can bind to AtNOA1 and increase its degradation in a protease-dependent manner (Mandal et al. 2012). Oleic acid is involved in pathogen defense signaling (Kachroo et al. 2008), thus increased oleic acid levels can moderate NO synthesis in infected plants (Mandal et al. 2012). It is also possible, that NO can diminish its own production, as suggested by the inhibition of NR activity by NO (Rosales et al. 2010).

2.6 Summary and Open Debates

NO plays important roles in plant physiology, disease resistance, and stress tolerance. Various enzymatic and chemical processes elaborate NO in plants; however, there are significant gaps in our understanding of plant-type NO homeostasis. A well characterized NO producing plant enzyme is NR, which generates NO as a secondary activity. Similarly, the mitochondrial electron transport chain and deoxygenated heme-proteins also facilitate NO generation from NO_2^-, although, they are not dedicated NO synthesizing enzymes. Future research should identify the enzymes responsible for oxidative NO synthesis from L-Arginine and answer the open debate whether plants have a specific enzyme with primary NO synthesizing activity. The mechanism of NO synthesis from various *N*-compounds should also be defined, as they can provide alternatives of L-Arginine dependent NO synthesis in plants. Nonenzymatic processes contribute to NO generation, however, their physiological relevance still remains elusive. Plant hormones, stress and injury signals, modulation of intracellular Ca^{2+} levels have the potential to drive NO synthesis in the plant cell. Integration of the distinct NO producing processes requires a complex regulatory network; however, our insight into the underlying molecular mechanisms is still limited.

Acknowledgment Livia I. Lelkes provided editorial assistance; her timely and accurate work is highly appreciated.

References

Barroso JB, Corpas FJ, Carreras A et al (2006) Localization of *S*-nitrosoglutathione and expression of *S*-nitrosoglutathione reductase in pea plants under cadmium stress. J Exp Bot 57:1785–1793

Barroso JB, Corpas FJ, Carreras A et al (1999) Localization of nitric-oxide synthase in plant peroxisomes. J Biol Chem 274:36729–36733

Beard RA, Anderson DJ, Bufford JL, Tallman G (2012) Heat reduces nitric oxide production required for auxin-mediated gene expression and fate determination in tree tobacco guard cell protoplasts. Plant Physiol 159:1608–1623

Beligni MV, Lamattina L (2000) Nitric oxide stimulates seed germination and de-etiolation, and inhibits hypocotyl elongation, three light-inducible responses in plants. Planta 210:215–221
Bethke PC, Badger MR, Jones RL (2004a) Apoplastic synthesis of nitric oxide by plant tissues. Plant Cell 16:332–341
Bethke PC, Gubler F, Jacobsen JV, Jones RL (2004b) Dormancy of *Arabidopsis* seeds and barley grains can be broken by nitric oxide. Planta 219:847–855
Bright J, Desikan R, Hancock JT et al (2006) ABA-induced NO generation and stomatal closure in *Arabidopsis* are dependent on H_2O_2 synthesis. Plant J 45:113–122
Caamal-Chan MG, Souza-Perera R, Zuniga-Aguilar JJ (2011) Systemic induction of a *Capsicum chinense* nitrate reductase by the infection with *Phytophthora capsici* and defence phytohormones. Plant Physiol Biochem 49:1238–1243
Camejo D, Romero-Puertas MD, Rodriguez-Serrano M et al (2012) Salinity-induced changes in S-nitrosylation of pea mitochondrial proteins. J Proteomics 79C:87–99
Castello PR, David PS, McClure T et al (2006) Mitochondrial cytochrome oxidase produces nitric oxide under hypoxic conditions: implications for oxygen sensing and hypoxic signaling in eukaryotes. Cell Metab 3:277–287
Catala M, Gasulla F, Pradas del Real AE et al (2010) Fungal-associated NO is involved in the regulation of oxidative stress during rehydration in lichen symbiosis. BMC Microbiol 10:297
Corpas FJ, Barroso JB, del Río LA (2001) Peroxisomes as a source of reactive oxygen species and nitric oxide signal molecules in plant cells. Trends Plant Sci 6:145–150
Chun HJ, Park HC, Koo SC et al (2012) Constitutive expression of mammalian nitric oxide synthase in tobacco plants triggers disease resistance to pathogens. Mol Cells 34:463–471
Fernandez-Marcos M, Sanz L, Lorenzo O (2012) Nitric oxide: an emerging regulator of cell elongation during primary root growth. Plant Signal Behav 7:196–200
Ferrario-Mery S, Meyer C, Hodges M (2008) Chloroplast nitrite uptake is enhanced in *Arabidopsis* PII mutants. FEBS Lett 582:1061–1066
Filippou P, Antoniou C, Fotopoulos V (2012) The nitric oxide donor sodium nitroprusside regulates polyamine and proline metabolism in leaves of *Medicago truncatula* plants. Free Radic Biol Med 56:172–183
Flores T, Todd CD, Tovar-Mendez A et al (2008) Arginase-negative mutants of *Arabidopsis* exhibit increased nitric oxide signaling in root development. Plant Physiol 147:1936–1946
Floryszak-Wieczorek J, Milczarek G, Arasimowicz M, Ciszewski A (2006) Do nitric oxide donors mimic endogenous NO-related response in plants? Planta 224:1363–1372
Foresi N, Correa-Aragunde N, Parisi G et al (2010) Characterization of a nitric oxide synthase from the plant kingdom: NO generation from the green alga *Ostreococcus tauri* is light irradiance and growth phase dependent. Plant Cell 22:3816–3830
Fröhlich A, Durner J (2011) The hunt for plant nitric oxide synthase (NOS): is one really needed? Plant Sci 181:401–404
Gniazdowska A, Krasuska U, Debska K et al (2010) The beneficial effect of small toxic molecules on dormancy alleviation and germination of apple embryos is due to NO formation. Planta 232:999–1005
Guo FQ, Crawford NM (2005) *Arabidopsis* nitric oxide synthase1 is targeted to mitochondria and protects against oxidative damage and dark-induced senescence. Plant Cell 17:3436–3450
Guo FQ, Okamoto M, Crawford NM (2003) Identification of a plant nitric oxide synthase gene involved in hormonal signaling. Science 302:100–103
Gupta KJ, Fernie AR, Kaiser WM, van Dongen JT (2010) On the origins of nitric oxide. Trends Plant Sci 16:160–168
Gupta KJ, Igamberdiev AU (2011) The anoxic plant mitochondrion as a nitrite: NO reductase. Mitochondrion 11:537–543
He Y, Tang RH, Hao Y et al (2004) Nitric oxide represses the *Arabidopsis* floral transition. Science 305:1968–1971
Huang J, Sommers EM, Kim-Shapiro DB, King SB (2002) Horseradish peroxidase catalyzed nitric oxide formation from hydroxyurea. J Am Chem Soc 124:3473–3480

Huang S, Kerschbaum HH, Engel E, Hermann A (1997) Biochemical characterization and histochemical localization of nitric oxide synthase in the nervous system of the snail, Helix pomatia. J Neurochem 69:2516–2528
Igamberdiev AU, Bykova NV, Shah JK, Hill RD (2010) Anoxic nitric oxide cycling in plants: participating reactions and possible mechanisms. Physiol Plant 138:393–404
Igamberdiev AU, Hill RD (2004) Nitrate, NO and haemoglobin in plant adaptation to hypoxia: an alternative to classic fermentation pathways. J Exp Bot 55:2473–2482
Jasid S, Simontacchi M, Bartoli CG, Puntarulo S (2006) Chloroplasts as a nitric oxide cellular source. Effect of reactive nitrogen species on chloroplastic lipids and proteins. Plant Physiol 142:1246–1255
Jin CW, Du ST, Shamsi IH et al (2011) NO synthase-generated NO acts downstream of auxin in regulating Fe-deficiency-induced root branching that enhances Fe-deficiency tolerance in tomato plants. J Exp Bot 62:3875–3884
Kachroo A, Fu DQ, Havens W, Navarre D, Kachroo P, Ghabrial SA (2008) An oleic acid-mediated pathway induces constitutive defense signaling and enhanced resistance to multiple pathogens in soybean. Mol Plant Microbe Interact 21:564–575
Kolanczyk M, Pech M, Zemojtel T et al (2011) NOA1 is an essential GTPase required for mitochondrial protein synthesis. Mol Biol Cell 22:1–11
Kolbert Z, Bartha B, Erdei L (2008) Exogenous auxin-induced NO synthesis is nitrate reductase-associated in *Arabidopsis thaliana* root primordia. J Plant Physiol 165:967–975
Krueger R, Kliewer M (1995) Arginine synthesis in grapevine leaves and berries: diurnal and seasonal patterns, environmental and physiological influences. Am J Enol Vitic 46:6
Lee U, Wie C, Fernandez BO et al (2008) Modulation of nitrosative stress by *S*-nitrosoglutathione reductase is critical for thermotolerance and plant growth in *Arabidopsis*. Plant Cell 20:786–802
Lehotai N, Kolbert Z, Peto A et al (2012) Selenite-induced hormonal and signalling mechanisms during root growth of *Arabidopsis thaliana* L. J Exp Bot 63:5677–5687
Liu WZ, Kong DD, Gu XX et al (2013) Cytokinins can act as suppressors of nitric oxide in *Arabidopsis*. Proc Natl Acad Sci USA 110:1548–1553
Liu Y, Zhang J (2009) Rapid accumulation of NO regulates ABA catabolism and seed dormancy during imbibition in *Arabidopsis*. Plant Signal Behav 4:905–907
Lombardi L, Ceccarelli N, Picciarelli P et al (2010) Nitric oxide and hydrogen peroxide involvement during programmed cell death of *Sechium edule* nucellus. Physiol Plant 140:89–102
Ma W, Xu W, Xu H et al (2010) Nitric oxide modulates cadmium influx during cadmium-induced programmed cell death in tobacco BY-2 cells. Planta 232:325–335
Majlath I, Szalai G, Papp I et al (2011) Atnoa1 mutant *Arabidopsis* plants induce compensation mechanisms to reduce the negative effects of the mutation. J Plant Physiol 168:1184–1190
Mallick N, Rai LC, Mohn FH, Soeder CJ (1999) Studies on nitric oxide (NO) formation by the green alga *Scenedesmus obliquus* and the diazotrophic cyanobacterium *Anabaena doliolum*. Chemosphere 39:1601–1610
Mandal MK, Chandra-Shekara AC, Jeong RD et al (2012) Oleic acid-dependent modulation of nitric oxide associated1 protein levels regulates nitric oxide-mediated defense signaling in *Arabidopsis*. Plant Cell 24:1654–1674
Moreau M, Lee GI, Wang Y et al (2008) AtNOS/AtNOA1 is a functional *Arabidopsis thaliana* cGTPase and not a nitric-oxide synthase. J Biol Chem 283:32957–32967
Mur LA, Mandon J, Persijn S et al (2013) Nitric oxide in plants: an assessment of the current state of knowledge. AoB Plants 5:pls052. doi:10.1093/aobpla/pls052
Oliveira H, Salgado I, Sodek L (2013) Nitrite decreases ethanol production by intact soybean roots submitted to oxygen deficiency: a role for mitochondrial nitric oxide synthesis? Plant Signal Behav 8:e23578
Ötvös K, Pasternak TP, Miskolczi P et al (2005) Nitric oxide is required for, and promotes auxin-mediated activation of, cell division and embryogenic cell formation but does not influence cell cycle progression in alfalfa cell cultures. Plant J 43:849–860

Pedroso MC, Magalhaes JR, Durzan D (2000) A nitric oxide burst precedes apoptosis in angiosperm and gymnosperm callus cells and foliar tissues. J Exp Bot 51:1027–1036
Poór P, Kovacs J, Szopko D, Tari I (2013) Ethylene signaling in salt stress—and salicylic acid-induced programmed cell death in tomato suspension cells. Protoplasma 250:273–284
Poór P, Tari I (2012) Regulation of stomatal movement and photosynthetic activity in guard cells of tomato abaxial epidermal peels by salicylic acid. Funct Plant Biol 39:1028–1037
Rasul S, Wendehenne D, Jeandroz S (2012) Study of oligogalacturonides-triggered nitric oxide (NO) production provokes new questioning about the origin of NO biosynthesis in plants. Plant Signal Behav 7:1031–1033
del Río LA (2011) Peroxisomes as a cellular source of reactive nitrogen species signal molecules. Arch Biochem Biophys 506:1–11
del Río LA, Corpas FJ, Sandalio LM et al (2003) Plant peroxisomes, reactive oxygen metabolism and nitric oxide. IUBMB Life 55:71–81
Rockel P, Strube F, Rockel A et al (2002) Regulation of nitric oxide (NO) production by plant nitrate reductase in vivo and in vitro. J Exp Bot 53:103–110
Romera FJ, Garcia MJ, Alcantara E, Perez-Vicente R (2011) Latest findings about the interplay of auxin, ethylene and nitric oxide in the regulation of Fe deficiency responses by Strategy I plants. Plant Signal Behav 6:167–170
Rosales EP, Iannone MF, Groppa MD, Benavides MP (2010) Nitric oxide inhibits nitrate reductase activity in wheat leaves. Plant Physiol Biochem 49:124–130
Rosales EP, Iannone MF, Groppa MD, Benavides MP (2012) Polyamines modulate nitrate reductase activity in wheat leaves: involvement of nitric oxide. Amino Acids 42:857–865
Rőszer T (2012a) Nitric oxide synthesis in leaf peroxisomes and in plant-type mitochondria. In: Rőszer T (ed) The biology of subcellular nitric oxide. Springer, New York, pp 67–80
Rőszer T (2012b) Nitric oxide synthesis in the chloroplast. In: Rőszer T (ed) The biology of subcellular nitric oxide. Springer, New York, pp 49–66
Rőszer T, Kiss-Tóth E, Rózsa D et al (2010) Hypothermia translocates nitric oxide synthase from cytosol to membrane in snail neurons. Cell Tissue Res 342:191–203
Rumer S, Kapuganti JG, Kaiser WM (2009) Oxidation of hydroxylamines to NO by plant cells. Plant Signal Behav 4:853–855
Sakihama Y, Nakamura S, Yamasaki H (2002) Nitric oxide production mediated by nitrate reductase in the green alga *Chlamydomonas reinhardtii*: an alternative NO production pathway in photosynthetic organisms. Plant Cell Physiol 43:290–297
Salmi ML, Morris KE, Roux SJ, Porterfield DM (2007) Nitric oxide and cGMP signaling in calcium-dependent development of cell polarity in *Ceratopteris richardii*. Plant Physiol 144:94–104
Shi H, Ye T, Chen F et al (2013) Manipulation of arginase expression modulates abiotic stress tolerance in *Arabidopsis*: effect on arginine metabolism and ROS accumulation. J Exp Bot 64:1367–1379
Shiva S, Rassaf T, Patel RP, Gladwin MT (2011) The detection of the nitrite reductase and NO-generating properties of haemoglobin by mitochondrial inhibition. Cardiovasc Res 89:566–573
Stohr C, Strube F, Marx G et al (2001) A plasma membrane-bound enzyme of tobacco roots catalyses the formation of nitric oxide from nitrite. Planta 212:835–841
Streatfield SJ, Weber A, Kinsman EA et al (1999) The phosphoenolpyruvate/phosphate translocator is required for phenolic metabolism, palisade cell development, and plastid-dependent nuclear gene expression. Plant Cell 11:1609–1622
Sturms R, Dispirito AA, Hargrove MS (2011) Plant and cyanobacterial hemoglobins reduce nitrite to nitric oxide under anoxic conditions. Biochem 50:3873–3878
Sudhamsu J, Lee GI, Klessig DF, Crane BR (2008) The structure of YqeH. An AtNOS1/AtNOA1 ortholog that couples GTP hydrolysis to molecular recognition. J Biol Chem 283:32968–32976
Sun A, Li Z (2012) Regulatory role of nitric oxide in lipopolysaccharides-triggered plant innate immunity. Plant Signal Behav 8:e22554

Sun LR, Hao FS, Lu BS, Ma LY (2010) AtNOA1 modulates nitric oxide accumulation and stomatal closure induced by salicylic acid in *Arabidopsis*. Plant Signal Behav 5:1022–1024

Talwar PS, Gupta R, Maurya AK, Deswal R (2012) *Brassica juncea* nitric oxide synthase like activity is stimulated by PKC activators and calcium suggesting modulation by PKC-like kinase. Plant Physiol Biochem 60:157–164

Tan J, Zhuo C, Guo Z (2013) Nitric oxide mediates cold- and dehydration-induced expression of a novel MfHyPRP that confers tolerance to abiotic stress. Physiol Plant 149:310–320

Tewari RK, Prommer J, Watanabe M (2013) Endogenous nitric oxide generation in protoplast chloroplasts. Plant Cell Rep 32:31–44

Tischner R, Planchet E, Kaiser WM (2004) Mitochondrial electron transport as a source for nitric oxide in the unicellular green alga *Chlorella sorokiniana*. FEBS Lett 576:151–155

Tiso M, Tejero J, Basu S et al (2011) Human neuroglobin functions as a redox-regulated nitrite reductase. J Biol Chem 286:18277–18289

Tun NN, Holk A, Scherer GF (2001) Rapid increase of NO release in plant cell cultures induced by cytokinin. FEBS Lett 509:174–176

Tun NN, Santa-Catarina C, Begum T et al (2006) Polyamines induce rapid biosynthesis of nitric oxide (NO) in *Arabidopsis thaliana* seedlings. Plant Cell Physiol 47:346–354

Wang H, Xiao W, Niu Y et al (2013) Nitric oxide enhances development of lateral roots in tomato (*Solanum lycopersicum L.*) under elevated carbon dioxide. Planta 237:137–144

Wang YH, Li XC, Zhu-Ge Q et al (2012) Nitric oxide participates in cold-inhibited *Camellia sinensis* pollen germination and tube growth partly via cGMP in vitro. PLoS ONE 7:e52436

Wimalasekera R, Tebartz F, Scherer GF (2011) Polyamines, polyamine oxidases and nitric oxide in development, abiotic and biotic stresses. Plant Sci 181:593–603

Yamasaki H (2000) Nitrite-dependent nitric oxide production pathway: implications for involvement of active nitrogen species in photoinhibition in vivo. Philos Trans R Soc Lond B Biol Sci 355:1477–1488

Yamasaki H, Cohen MF (2006) NO signal at the crossroads: polyamine-induced nitric oxide synthesis in plants? Trends Plant Sci 11:522–524

Yordanova ZP, Iakimova ET, Cristescu SM et al (2010) Involvement of ethylene and nitric oxide in cell death in mastoparan-treated unicellular alga *Chlamydomonas reinhardtii*. Cell Biol Int 34:301–308

Yu LZ, Wu XQ, Ye JR et al (2012) NOS-like-mediated nitric oxide is involved in *Pinus thunbergii* response to the invasion of *Bursaphelenchus xylophilus*. Plant Cell Rep 31:1813–1821

Zeidler D, Zahringer U, Gerber I, Dubery I, Hartung T, Bors W, Hutzler P, Durner J (2004) Innate immunity in *Arabidopsis thaliana*: lipopolysaccharides activate nitric oxide synthase (NOS) and induce defense genes. Proc Natl Acad Sci USA 101:15811–15816

Zhang A, Zhang J, Ye N et al (2010) Nitric oxide mediates brassinosteroid-induced ABA biosynthesis involved in oxidative stress tolerance in maize leaves. Plant Cell Physiol 52:181–192

Zhang X, Shen L, Li F et al (2011) Methyl salicylate-induced arginine catabolism is associated with up-regulation of polyamine and nitric oxide levels and improves chilling tolerance in cherry tomato fruit. J Agric Food Chem 59:9351–9357

Zhu XF, Jiang T, Wang ZW et al (2012) Gibberellic acid alleviates cadmium toxicity by reducing nitric oxide accumulation and expression of IRT1 in *Arabidopsis thaliana*. J Hazard Mater 15:302–307

Zottini M, Costa A, De Michele R et al (2007) Salicylic acid activates nitric oxide synthesis in *Arabidopsis*. J Exp Bot 58:1397–1405

Chapter 3
Function of Peroxisomes as a Cellular Source of Nitric Oxide and Other Reactive Nitrogen Species

Luis A. del Río, Francisco J. Corpas, Juan B. Barroso, Eduardo López-Huertas and José M. Palma

Abstract Peroxisomes are subcellular organelles bounded by a single membrane and devoid of DNA, with an essentially oxidative type of metabolism and are probably the major sites of intracellular H_2O_2 production. These organelles also generate superoxide radicals ($O_2^{\cdot-}$) and besides catalase they have a complex battery of antioxidative enzymes. The existence of L-Arginine-dependent Nitric oxide synthase (NOS) activity and the generation of the reactive nitrogen species (RNS) nitric oxide (NO) have been demonstrated in plant peroxisomes. Besides NO, the presence in peroxisomes of the RNS *S*-nitrosoglutathione (GSNO) and the generation of peroxynitrite ($ONOO^-$) have also been reported. This implies that peroxisomes can function in plant cells as a source of the signaling molecules NO and GSNO, besides $O_2^{\cdot-}$ and H_2O_2. As a result of the presence of NO and GSNO, and the production of the powerful oxidant and nitrating chemical $ONOO^-$, important post-translational modifications can take place in peroxisomes, such as *S*-nitrosylation and nitration of proteins which could have an impact on the peroxisomal and cellular metabolism of plants. The important physiological functions carried out by NO and other RNS in intra- and inter-cellular communication in different organisms evidence the key role displayed by peroxisomes in plant cellular metabolism as a source of these signaling molecules.

Keywords Nitric oxide (NO) · Peroxisomes · Peroxynitrite ($ONOO^-$) · Protein nitration · Reactive nitrogen species (RNS) · RNS signaling · *S*-nitrosylation

L. A. del Río (✉) · F. J. Corpas · E. López-Huertas · J. M. Palma
Department of Biochemistry, Cell and Molecular Biology of plants,
Estación Experimental del Zaidín, Consejo Superior de Investigaciones Científicas (CSIC),
Profesor Albareda 1, 18080 Granada, Spain
e-mail: luisalfonso.delrio@eez.csic.es

J. B. Barroso
Área de Bioquímica y Biología Molecular, Facultad de Ciencias Experimentales,
Departamento de Biología Experimental, Universidad de Jaén,
Campus Universitario "Las Lagunillas" s/n, 23071 Jaén, Spain

M. N. Khan et al. (eds.), *Nitric Oxide in Plants: Metabolism and Role in Stress Physiology*, DOI: 10.1007/978-3-319-06710-0_3,

3.1 Introduction

Peroxisomes are subcellular organelles bounded by a single membrane that contain as basic enzymatic constituents catalase and H_2O_2-producing flavin oxidases, and occur in almost all eukaryotic cells (Fahimi and Sies 1987; Baker and Graham 2002; del Río 2013). These organelles are devoid of DNA and have an essentially oxidative type of metabolism. When these organelles where first isolated and characterized from mammalian tissues by Christian de Duve, it was thought that their main function was the removal by catalase of toxic hydrogen peroxide generated in the peroxisomal respiratory pathway by different oxidases (De Duve and Baudhuin 1966). However, in recent years it has become increasingly clear that peroxisomes are involved in a range of important cellular functions in most eukaryotic cells (Waterham and Wanders 2012; Islinger et al. 2012; Baker and Graham 2002; Hu et al. 2012; del Río 2013).

The peroxisome of plant cells is a highly dynamic compartment that is dependent upon the actin cytoskeleton, not microtubules, for its subcellular distribution and movements (Mathur et al. 2002; Mano et al. 2002; Rodríguez-Serrano et al. 2009). Today, it is known that fatty acid β-oxidation is a general feature of virtually all types of peroxisomes, but in plants, peroxisomes carry out different functions, apart from fatty acid β-oxidation (Baker et al. 2006), mainly including: photorespiration; metabolism of reactive oxygen, nitrogen, and sulfur species (ROS, RNS, and RSS, respectively); photomorphogenesis; biosynthesis of phytohormones (auxin, jasmonic acid, salicylic acid); senescence; and defense against pathogens and hervibores (for a review see del Río 2011, 2013; Hu et al. 2012). Three important characteristic properties of peroxisomes are their oxidative type of metabolism, their capacity of sharing metabolic pathways with other cell compartments, and their metabolic plasticity, because their enzymatic content can vary depending on the organism, cell/tissue-type and environmental conditions (Fahimi and Sies 1987; Baker and Graham 2002; del Río 2013).

3.2 Functions of NO in Plants

The gaseous free radical nitric oxide (NO) is a widespread intra- and inter-cellular messenger with a broad spectrum of regulatory functions in many physiological processes of animal and plant systems (Martínez-Ruiz and Lamas 2009; del Río et al. 2004; Neill et al. 2008). The use of NO by higher plants was first reported in 1960 (Fewson and Nicholas 1960) much earlier than in animals, and the NO emission from plants was first observed in 1975, in soybean plants treated with herbicides (Klepper 1979). In recent years, NO was demonstrated to have an important function in plant growth and development, including seed germination, primary and lateral root growth, development of functional nodules, flowering, pollen tube growth regulation, fruit ripening and senescence, pathogen response and abiotic stress (for a review see Corpas et al. 2013a). In most of these physiological processes NO participates as a key signaling molecule at intra-cellular or inter-cellular level (Shapiro 2005; Wilson et al. 2008; Baudouin 2011; Corpas et al. 2013a).

3.3 Generation of NO in Plants and Subcellular Sites of Production

In animal systems most of the NO produced is due to the enzyme Nitric oxide synthase (NOS; EC 1.14.13.39) (Alderton et al. 2001). This enzyme catalyzes the oxygen- and NADPH-dependent oxidation of L-Arginine to NO and citrulline in a complex reaction requiring FAD, FMN, tetrahydrobiopterin (BH_4), calcium and calmodulin (Knowles and Moncada 1994; Alderton et al. 2001). There are three distinct isoforms of NOS designated as neuronal NOS (nNOS or NOS-1), inducible NOS (iNOS or NOS-2), and endothelial NOS (eNOS or NOS-3). These isoforms are also classified on the basis of either their constitutive (eNOS and nNOS) or inducible (iNOS) expression, and their dependence (eNOS and nNOS) or independence (iNOS) on calcium (Alderton et al. 2001).

However, in plants a gene or a protein with homology to mammalian NOS enzymes has not been found in *Arabidopsis thaliana* (The *Arabidopsis* genome initiative, 2000). The different molecular approaches developed so far to clone a higher plant NOS, based on the sequence of animal NOS, have always given negative results (Zemojtel et al. 2006; Neill et al. 2008; Wilson et al. 2008; Gas et al. 2009; del Río 2011). The only case reported so far in the plant kingdom of a NOS characterized is that of a unicellular species of marine green alga, *Ostreoccus tauri* (Foresi et al. 2010). The-length sequence of *O. tauri* NOS showed a similarity of 42, 43, and 34 % with respect to eNOS, iNOS, and nNOS, respectively. The authors suggested that the active form of *O. tauri* NOS is a dimer with a subunit of 119 kDa, which is close to the molecular mass of the animal NOS subunits (Foresi et al. 2010).

In plants there are several potential sources of NO including enzymatic and nonenzymatic systems (del Río et al. 2004; Wilson et al. 2008; Gupta et al. 2011; Mur et al. 2012; Hancock 2012). A list of some established and potential enzymatic sources of NO in plant cells, with indication of the different substrates used is presented in Table 3.1. Nitrate reductase is a well-established enzymatic generator of NO in plants (Dean and Harper 1988; Yamasaki and Sakihama 2000; Rockel et al. 2002; Gupta and Kaiser 2010). Other enzyme that has been shown to produce NO is a plasma membrane-bound enzyme of tobacco roots, nitrite-NO oxidoreductase (Stöhr et al. 2001; Stöhr and Stremlau 2006).

In addition there are numerous reports of L-Arginine-dependent NOS activity in different plant extracts (del Río et al. 2004; Corpas et al. 2006, 2009a). A summary of the different plant species where L-Arginine-dependent NOS activity has been

Table 3.1 Some established and potential enzymatic sources of NO in plant cells

Source	Substrates	References
Different crude extracts	L-Arg and NOS cofactors	Reviewed by Corpas et al. (2009a)
Plant peroxisomes	L-Arg and NOS cofactors	Barroso et al. (1999) Corpas et al. (2004a)
Nitrate reductase	NO_2^- and NADH	Dean and Harper (1988) Yamasaki et al. (1999) Gupta and Kaiser (2010)
Plasma membrane-bound enzyme	NO_2^- + reduced Cyt *c*	Stöhr et al. (2001, 2006)
Xanthine oxidoreductase	NO_2^- and NADH	Reviewed by Harrison (2002)
Catalase	NaN_3	Nicholls (1964)
Horseradish peroxidase	Hydroxyurea + H_2O_2 NOHA + H_2O_2	Huang et al. (2002) Boucher et al. (1992a)
Hemeproteins	NOHA + H_2O_2/ROOH	Boucher et al. (1992a)
Cytochrome P450	NOHA + NADPH + O_2	Boucher et al. (1992b)

NOHA *N*-hydroxyarginine; ROOH alkylhydroperoxides

detected is shown in Table 3.2. Three different approaches have been used to demonstrate the existence of L-Arginine-dependent NOS activity in plants, based on biochemical and physicochemical methods, immunological and molecular methods. Cueto et al. (1996) and Ninnemann and Maier (1996) were the first to show the existence of NOS activity in higher plants by using the conversion of radiolabelled arginine, the substrate for NOS, into radiolabelled citrulline. Another method which has been widely used is the measurement by fluorometry or chemiluminescence of the L-Arginine dependent NO production sensitive to NOS inhibitors. In crude extracts from sorghum the NOS activity-derived production of NO has been determined by spin trapping electron paramagnetic resonance (EPR) spectroscopy (Simontacchi et al. 2004). In addition, there are different evidences obtained by using physiological and/or pharmacological approaches with inhibitors analogous to L-Arginine, such as N^G-nitro-L-Arginine methyl ester (L-NAME), N^G-nitro-L-Arginine (L-NNA) or L-N^G-monomethyl-arginine monoacetate (L-NMMA), that have shown a decrease in NO production, thus supporting the involvement of an L-Arginine-dependent NOS activity in the generation of NO (for a review see Corpas et al. 2009a). In conclusion, in plants there is a body of evidence supporting the existence L-Arginine-dependent NOS activity in, at least, 11 different plant species (Corpas et al. 2009a; del Río 2011).

In plant systems, there is little information on the subcellular sites where NO is produced. Besides peroxisomes the only cell compartments where the generation of NO has been clearly demonstrated are mitochondria, and chloroplasts. In root mitochondria the production of NO appears to be due to the reduction of nitrite by the electron transport chain (Gupta et al. 2005; Gupta and Kaiser 2010), whereas in chloroplasts nitrite and L-Arginine were both substrates for NO generation (Jasid et al. 2006). The presence of NOS activity in peroxisomes was first demonstrated in plant tissues (Barroso et al. 1999), and in this review the different evidences

Table 3.2 Different plant species where L-Arginine-dependent NOS activity has been detected and its subcellular localization studied

Species/Tissue or cell type	NOS activity (pmol min^{-1} mg^{-1} protein)	Cellular localization
Arabidopsis thaliana/Leaves	4.5[a]	nd
Cowitch or velvetbean (*Mucuna hassjoo*)/Roots	2.7[a]	nd
Hibiscus moscheutos/Roots	0.7[a]	nd
Maize (*Zea mays*)/Root tips and young leaves	0.18[a]	nd
Seedlings	410[d]	nd
Olive (*Olea europaea*)/Leaves	294[b]	nd
Pea (*Pisum sativum*)/Leaves	5.0×10^3 [a,b,c]	Peroxisomes
Roots, Stems, Leaves	240, 630, 120[b]	nd
Sorghum bicolor		
Seed embryonic axes	2.2[c]	nd
Soybean (*Glycine max*)		
Cotyledons	7.7[a]	nd
Leaves	760[c]	Chloroplasts
Sunflower (*Helianthus annuus*)/Hypocotyls	280[b]	nd
TMV-infected tobacco (*Nicotiana tabacum*)/Leaves	6[a]	nd
White lupine (*Lupinus albus*)/Roots and nodules	296[a]	nd

[a] Arginine-citrulline assay
[b] Ozone chemiluminiscence assay
[c] Spin trapping electron paramagnetic resonance (EPR) spectroscopy
[d] BIOXYTECHÒ Nitric Oxide Synthase Assay Kit. *nd* not determined
Reproduced from Corpas et al. (2009a) New Phytol 184: 9–14

available on the properties of the NOS activity and the generation of NO and other RNS in peroxisomes are presented. The demonstration of NO production in vivo together with findings of the presence of *S*-nitrosylated and nitrated proteins in peroxisomes are analyzed in the context of a new subcellular relationship between these oxidative organelles and the signaling molecule NO.

3.4 Presence of NOS Activity in Peroxisomes

The first biochemical characterization of a NOS activity in higher plants was accomplished in isolated peroxisomes (Barroso et al. 1999). In peroxisomes purified from pea leaves the NOS activity was determined using L-Arginine as substrate plus all the NOS cofactors. Four different assays were employed: (a) monitoring the conversion of L-[^{3}H]Arginine to L-[^{3}H]Citrulline; (b) fluorometric detection with 4,5-diaminofluorescein diacetate (DAF-2 DA) of NO produced in the enzymatic reaction; (c) ozone chemiluminescence detection of NO produced with a nitric oxide analyzer (NOA); and (d) spin trapping EPR spectroscopy of NO generated during the enzymatic reaction, using the spin trap $Fe(MGD)_2$ (Barroso et al. 1999; Corpas et al. 2004a, 2009a; del Río 2011).

By using the arginine-citrulline method, it was found that the NOS activity was strictly dependent on L-Arginine and NADPH, and required Ca^{2+}, calmodulin, FAD, FMN, and BH_4, the same cofactors necessary for the animal NOS (Alderton et al. 2001). Likewise, the peroxisomal NOS activity was sensitive to archetype inhibitors of the three NOS isoforms (Barroso et al. 1999; del Río 2011). As the validity of the arginine-citrulline method has been questioned in some plant extracts due to interferences by the enzymes arginase and arginine decarboxylase -two enzymes which use L-Arginine as substrate and mimick NOS activity (Tischner et al. 2007)- two alternative methods of NOS activity determination were set up. The NOS activity was assayed by a spectrofluorometric method using the fluorescence probe DAF-2 DA, and also by an ozone chemiluminescence assay (Corpas et al. 2004a, 2008). The biochemical characterization of NOS activity in peroxisomes purified from pea leaves using the ozone chemiluminescence method is shown in Fig. 3.1. Results obtained showed that the peroxisomal NOS activity required the same cofactors as those found by the arginine-citrulline assay, where the conversion of L-[^{3}H]Arginine into L-[^{3}H]Citrulline was monitored.

The localization of NOS in peroxisomes was also studied using immunological methods. Using a polyclonal antibody to murine iNOS, by Western blotting the presence in peroxisomes from pea leaves of an immunoreactive polypeptide of about 130 kDa was demonstrated (Barroso et al. 1999). The electron microscopy (EM) immunolocalization of NOS activity showed the presence of the enzyme in the matrix of peroxisomes and also in chloroplasts, whereas no immunogold labeling was detected in mitochondria as can be seen in Fig. 3.2. Using the same immunogold EM method, NOS was also found in peroxisomes from olive leaves and sunflower hypocotyls (Corpas et al. 2004b). The peroxisomal localization of NOS was confirmed by confocal laser scanning microscopy (CLSM) using antibodies against catalase, a characteristic marker enzyme of peroxisomes, and murine iNOS. The punctuate patterns of both immunofluorescent markers colocalized indicating that NOS was present in peroxisomes (del Río et al. 2003; Corpas et al. 2004a).

Therefore, on the basis of results obtained with the different experimental approaches mentioned, it can be concluded that there are clear and unequivocal evidence of the presence of L-Arginine-dependent NOS activity in plant peroxisomes (Corpas et al. 2001, 2009a; del Río 2011). Some years later the results obtained on the presence of NOS in plant peroxisomes were extended to animal peroxisomes. In rat hepatocytes the occurrence of inducible nitric oxide synthase (iNOS) was reported in these organelles (Stolz et al. 2002) and this was only the monomeric form of the enzyme whereas in the cytosol both the iNOS active dimer and monomer exist (Loughran et al. 2005).

As to the possible identity of the L-Arginine-dependent NOS activity detected in plant peroxisomes, it is clear that this activity is not a canonical NOS enzyme and perhaps in higher plant cells NOS activity is carried out by several proteins that

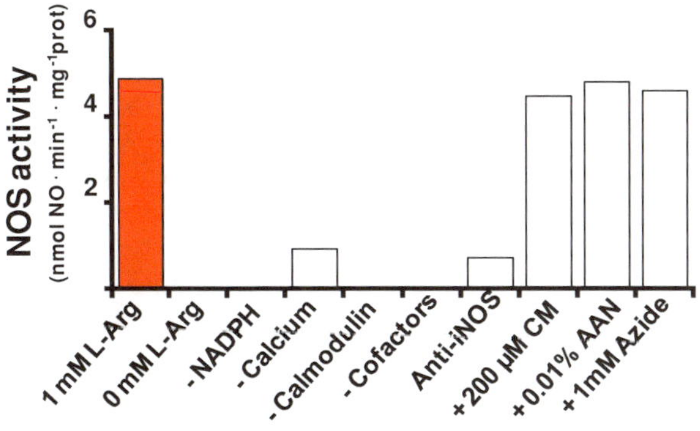

Fig. 3.1 Biochemical characterization of L-Arginine-dependent NOS activity in peroxisomes purified from pea leaves using the ozone chemiluminescence assay. Reaction mixtures containing peroxisomal fractions were incubated in the absence and presence of L-Arginine (1 mM), NADPH (1 mM), EGTA (0.5 mM), calmodulin (10 μg/ml), cofactors (10 μM FAD, 10 μM FMN and 10 μM BH_4), antibody against iNOS, 200 μM CM, 0.01 % AAN, and 1 mM azide. Then the NO production was assayed using a 1 mM L-Arginine concentration and an incubation time of 30 min. The NO generated was quantified by ozone chemiluminescence using a nitric oxide analyzer (Corpas et al. 2008). *CM* carboxymethoxylamine. *ANN* aminoacetonitrile. Reproduced from Corpas et al. (2009a) New Phytol 184: 9–14

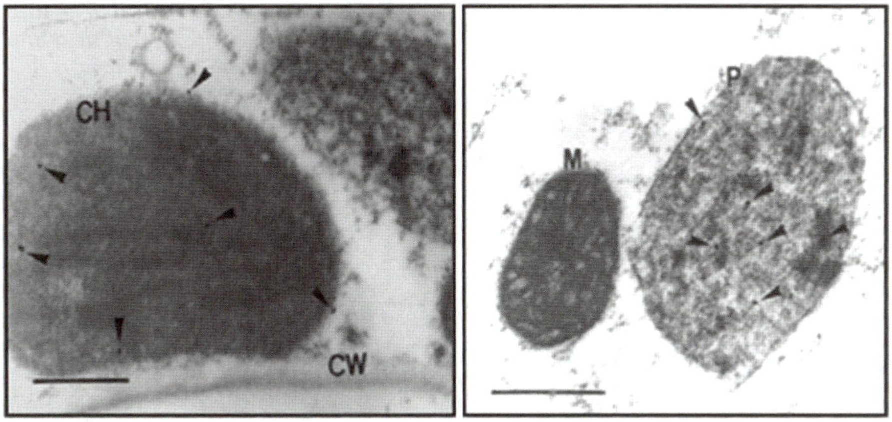

Fig. 3.2 Immunogold electron microscopy localization of NOS in pea leaves. The electron micrographs are representative of thin sections of pea leaves showing immunolocalization of NOS. Cell sections were probed with an antibody against iNOS. Arrows indicate 15 nm gold particles. *P* peroxisome. *M* mitochondrion. *CH* chloroplast. *CW* cell wall. Bar = 1.0 μ. Reproduced from Barroso et al. (1999). Copyright American Society of Biochemistry and Molecular Biology

could function together from L-Arginine, using the same substrate and cofactors as the animal NOS. However, the possibility of a peroxisomal enzyme that generates NO from L-Arginine as a by-product of a still unknown secondary reaction cannot be ruled out. Two candidate peroxisomal enzymes for NO production are XOR and catalase. XOR from animal origin can produce the O_2^{-} and NO free radicals during its catalytic reaction, depending on whether the oxygen tensions are high and low, respectively (Harrison 2002). On the other hand, catalase is associated with NADPH and can bind calmodulin in a Ca^{2+}-dependent way (Kirkman et al. 1999; Yang et al. 2002; Costa et al. 2010). And it is known that the heme-enzyme catalase can generate NO from sodium azide (NaN_3) (Nicholls 1964), and it has been also reported that the oxidation of N^{w}-hydroxy-L-Arginine by heme-proteins generates nitrogen oxides and citrulline (Boucher et al. 1992a). This suggests that catalase in the presence of H_2O_2 perhaps could transform peroxidatically L-Arginine into NO, using NADPH and Ca^{2+} as cofactors (del Río 2011).

To try to identify the protein responsible for the L-Arginine-dependent NOS activity detected in peroxisomes, the activity of three enzymes known to be present in plant peroxisomes was assayed: commercial animal XOR; catalase purified from pea leaf peroxisomes; and recombinant monodehydroascorbate reductase from cucumber (MDAR), an enzyme of the ascorbate-glutathione cycle which is known to produce superoxide radicals (Miyake et al. 1998; del Río et al. 2003). The NOS activity was determined by either the ozone chemiluminiscence method (Corpas et al. 2008) or spin trapping EPR (see Sect. 3.5; Corpas et al. 2004a), and the results obtained are shown in Table 3.3. Under the experimental conditions used, xanthine oxidase, catalase, and MDAR did not show any NO production in the L-Arginine-dependent NOS reaction and, therefore, they do not appear to be responsible for the NO generation in peroxisomes, at least in the experimental conditions used in this assay.

The purification of L-Arginine-dependent NOS activity from isolated peroxisomes of pea leaves was initiated in our laboratory. After a first purification step using anion-exchange Fast Protein Liquid Chromatography (FPLC), a single peak of L-Arginine-dependent NOS activity was obtained, although the protein yields were extremely low and proteomic analysis to identify the protein responsible could not be carried out yet.

3.5 Detection of NO Generation in Peroxisomes

Although the presence of NOS activity in peroxisomes implied the generation of NO in these organelles, the direct demonstration of L-Arginine-dependent production of NO in peroxisomes was carried out. Two different but complementary approaches were used, including: (1) spin trapping electron paramagnetic resonance (EPR) spectroscopy, in isolated peroxisomes, using as spin trap the Fe(II) complex of a dithiocarbamate $[Fe(MGD)_2]$ which reacts with the free radical NO forming a stable complex, NO–$Fe(MGD)_2$, with a characteristic three-line EPR spectrum (Caro and Puntarulo 1999); and (2) colocalization assays by confocal laser scanning microscopy (CLSM) in *Arabidopsis* plants expressing green

Table 3.3 Assay of NO generation by several pure peroxisomal enzymes in the presence of L-Arginine and the cofactors of the NOS reaction using the ozone chemiluminiscence method or spin trapping electron paramagnetic resonance (EPR) spectroscopy

Enzyme	Source	Substrate(s) used	NOS activity (nmol NO min^{-1} mg^{-1} protein)
Peroxisomal matrices (270 μg)	Pea leaves	L-Arginine (1 mM)	5.0[a]
Catalase (1.4–3.5 μg)	Purified enzyme from pea leaf peroxisomes	L-Arginine (1 mM) + H_2O_2 (50 and 100 μM)	0.0[a]
	Aspergillus niger (Sigma)	L-Arginine (1 mM) + H_2O_2 (10 and 20 μM)	0.0[a]
MDAR (2–10 μg)	Purified recombinant enzyme from cucumber	L-Arginine (1 mM)	nd[b]
Xanthine oxidase (XOD) (20 μg)	Cow milk (Calbiochem)	L-Arginine (1 mM)	nd[b]

[a] Ozone chemiluminescence method
[b] spin trapping EPR spectroscopy
In the ozone chemiluminiscence method, peroxisomal matrices were added to the NOS reaction mixture, containing 1 mM L-Arginine, 1 mM NADPH, 1.25 mM $CaCl_2$, 10 μg/ml calmodulin, 10 μM FMN, 10 μM FAD, and 10 μM BH_4. For the assay of NO-producing activity of catalase, the assay mixture was the same and the reaction was started by adding H_2O_2. The NO was measured by ozone chemiluminescence (Corpas et al. 2008). For the NO detection by spin trapping EPR, MDAR and XOD were added to a reaction mixture containing 1 mM L-Arginine and all the cofactors of the NOS reaction, plus the NO-spin trap $Fe(MGD)_2$, and were incubated for 1 h at 37 °C. Then samples were analyzed by EPR (Corpas et al. 2004a). Catalase was purified from isolated pea leaf peroxisomes, according to Corpas et al. (1999). Recombinant MDAR from cucumber was supplied by Sano et al. (1995). nd, no EPR signal detected

fluorescent protein (GFP) through the addition of a peroxisomal targeting signal type 1 (PTS1) (GFP-PTS1) (Mano et al. 2002).

A representative EPR spectrum of the NO–$Fe(MGD)_2$ spin adduct produced by pure neuronal nitric oxide synthase (nNOS), used as positive control of NO production, is shown in Fig. 3.3a with the triplet signal of this spin adduct ($g = 2.05$ and $a_N = 12.8$ G) (Corpas et al. 2004a). When crude extracts and isolated peroxisomes from pea leaves were assayed for NO production, in the presence of L-Arginine and all the NOS cofactors, similar three-line signals, with the same values for g and a_N were found (Fig. 3.3b, c, respectively). The preincubation of peroxisomal samples with the NOS inhibitor L-NAME (Fig. 3.3d) or without NADPH (Fig. 3.3e) produced spectra with lower signal intensities than control peroxisomes (Fig. 3.3c). The EPR results clearly indicated the generation of NO in peroxisomes as a result of the L-Arginine-dependent NOS activity present in these cell organelles (Corpas et al. 2004a).

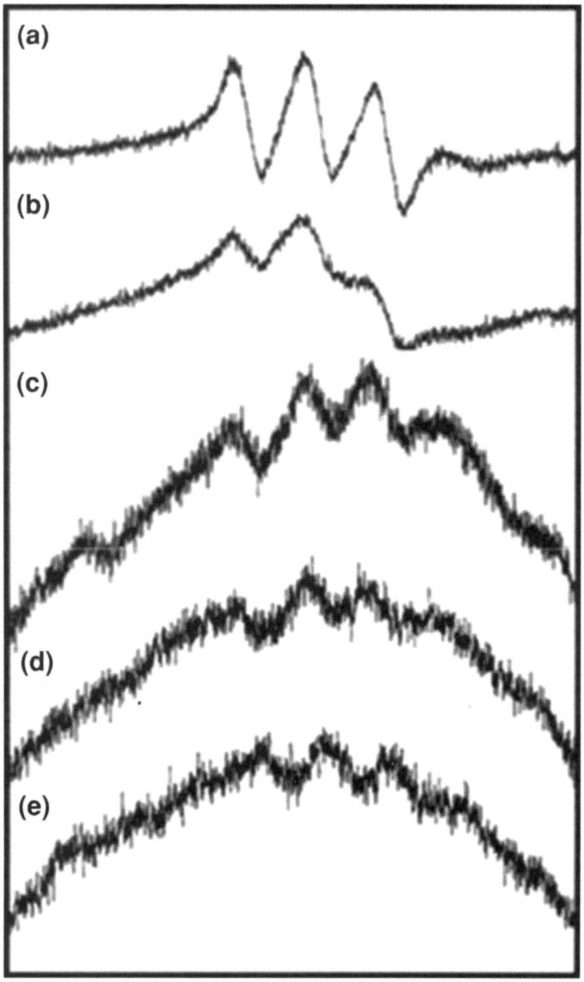

Fig. 3.3 Electron paramagnetic resonance (EPR) spectra of the NO-spin adduct of the $Fe(MGD)_2$ complex. For the detection of NO, samples were added to a reaction mixture containing the substrate and all the cofactors of the NOS reaction, plus the NO-spin trap $Fe(MGD)_2$, and were incubated for 1 h at 37°C. Then, samples were analyzed by EPR. **a** Pure nNOS (15 μg) from Merck Biosciences. **b** Crude extracts of pea leaves (2.5 mg protein). **c** Peroxisomes purified from pea leaves (270 μg protein). **d** Peroxisomes purified from pea leaves (270 μg protein) preincubated with 1 mM L-NAME. **e** Peroxisomes purified from pea leaves (270 μg protein) without NADPH in the reaction mixture. The EPR parameters of the spectra were $g = 2.05$ and $a_N = 12.8$ G. Reproduced from Corpas et al. (2004a) (www.plantphysiol.org). Copyright American Society of Plant Biologists

With regard to the detection of endogenous NO at subcellular level, Fig. 3.4 shows the in vivo CLSM visualization of NO in root peroxisomes from transgenic *Arabidopsis* plants expressing the green fluorescent protein (GFP) by the

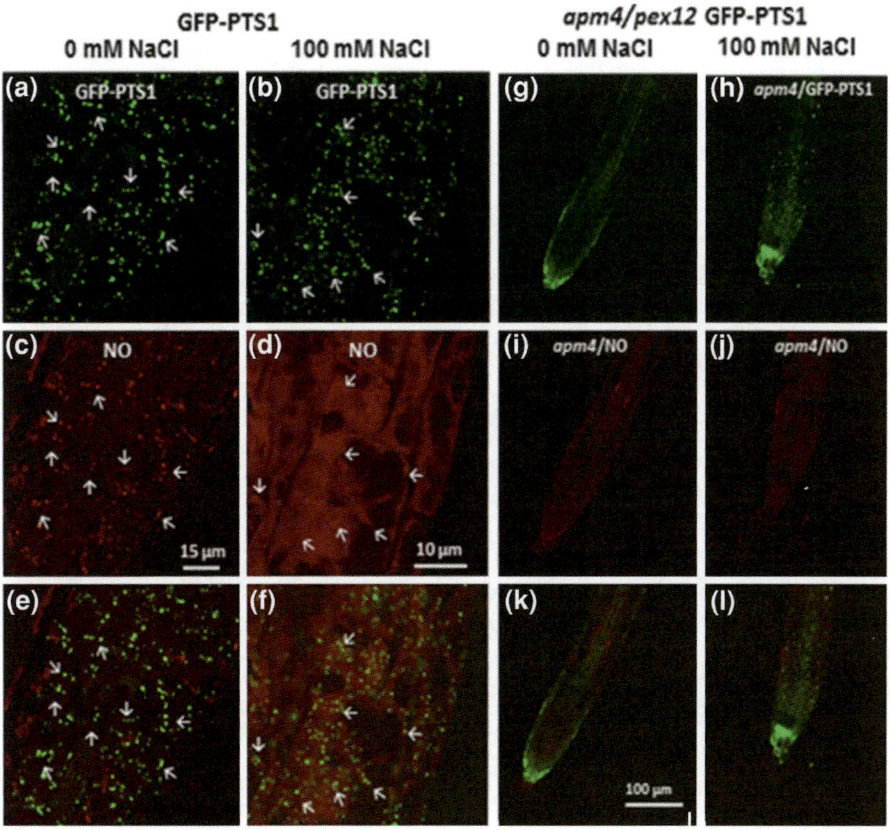

Fig. 3.4 In vivo CLSM localization of NO in root peroxisomes from transgenic *Arabidopsis* plants expressing the fused protein GFP-PTS1, which allows peroxisomes to be visualized, and subjected to salt stress. **a** and **b** Fluorescence punctuates attributable to GFP-PTS1, indicating the localization of peroxisomes in control (0 mM NaCl) and salt-treated plants, respectively. **c** and **d** Fluorescence punctuates attributable to NO detection in control (0 mM NaCl) and salt-treated plants, respectively. **e** Merged image of **a** and **c** showing colocalized fluorescence punctates (*yellow*). **f** Merged image of **b** and **d** showing colocalized fluorescence punctates (*yellow*). **g** and **h** Fluorescence punctuates attributable to GFP-PTS1, indicating the location of peroxisomes in control (0 mM NaCl) and salt-treated *apm4/pex12* mutants, respectively. **i** and **j** Fluorescence punctuates attributable to NO detection in the same root area of **g** and **h**, respectively. **k** Merged image of g and i showing colocalized fluorescence punctuates (*yellow*). **l**, Merged image of **h** and **j** showing colocalized fluorescence punctuates (*yellow*). NO was detected with DAR-4 M AM (excitation 543 nm; emission 575 nm) and peroxisomes with GFP (excitation 495 nm; emission 515 nm). *Arrows* indicate representative punctuate spots corresponding to NO and peroxisome localization. Reproduced with permission from Corpas et al. (2009b) (www.plantphysiol.org). Copyright American Society of Plant Biologists

incorporation of a peroxisomal targeting signal (PTS1) which allows peroxisomes to be visualized (Corpas et al. 2009b). Peroxisomes appeared as green spherical spots in all root tip cells (Fig. 3.4a). Using the fluorescent probe diaminorhodamine-4 M acetoxymethyl ester (DAR-4 M AM) it was possible to detect NO as an intense red fluorescence in spots with a similar punctuate pattern to that of GFP-PTS1 (Fig. 3.4c). The merged image of Fig. 3.4a, c showed a complete overlap of the two punctuate patterns (Fig. 3.4e), what corroborated the presence of NO in *Arabidopsis* root peroxisomes (Corpas et al. 2009b). The presence of NO in peroxisomes was also identified by CLSM in root tips of transgenic *Arabidopsis* seedlings expressing cyan fluorescent protein (CFP) through the addition of a peroxisomal targeting signal (CFP-PTS1) which allows peroxisomes to be visualized in vivo (Corpas and Barroso 2014).

Therefore, the use of these two complementary experimental approaches, EPR spectroscopy and CLSM colocalization in transgenic *Arabidopsis* plants expressing GFP and CFP in peroxisomes, demonstrated unequivocally that NO is generated in plant peroxisomes.

3.5.1 Effect of Senescence

Pioneer works on NO showed that exogenous NO can retard the flower and leaf senescence, as well as fruit ripening (Leshem 2000; Hung and Kao 2003). And there are different reports suggesting that NO and other reactive nitrogen species (RNS) are involved in plant senescence (Procházková and Wilhelmová 2011). In plants it has been proposed that peroxisomes have a ROS-mediated role in the oxidative reactions characteristic of senescence (del Río et al. 1998, 2006) where NO could also be involved. In peroxisomes isolated from senescent pea leaves it was demonstrated that the L-Arginine-dependent NOS activity was down-regulated by 72 % and this was accompanied by a reduction of the NO content (Corpas et al. 2004a). This led to the proposal that peroxisomal NO could be involved in the process of senescence of pea leaves (Corpas et al. 2004a).

3.5.2 Effect of Metal Stress

Leaf peroxisomes are also involved in the toxicity produced by heavy-metals, like Cd and Cu. Different evidence obtained in our lab in plants treated with cadmium have suggested that peroxisomes can have a role in the response to the heavy-metal toxicity del (Río et al. 2003, 2006). In leaf peroxisomes from plants grown with cadmium, an enhancement of the H_2O_2 concentration as well as the oxidative modification of some endogenous proteins was found (Romero-Puertas et al. 1999, 2002), and a slight increase in the peroxisomal population of pea plants by cadmium was also observed (Romero-Puertas et al. 1999). Cadmium induces senescence symptoms and, probably, a metabolic transition of leaf peroxisomes into glyoxysomes, with a participation of the peroxisomal proteases in all these metabolic changes (Palma et al. 2002). Peroxisomes responded to cadmium toxicity by increasing the activity of antioxidative enzymes involved in the ascorbate-glutathione cycle and the NADP-dehydrogenases located in these organelles (del Río et al. 2006).

In a study carried out with an *Arabidopsis thaliana* mutant expressing the GFP-SKL peptide targeted to peroxisomes, it was found that Cd increased the cellular speed of movement (dynamics or motility) of peroxisomes (Rodríguez-Serrano et al. 2009). More recently, the effect of cadmium stress (150 μM $CdCl_2$) on the cellular NO level was investigated by CLSM in *Arabidopsis thaliana* transgenic plants expressing CFP-PTS1 (Corpas and Barroso 2014). Results showed the presence of NO in peroxisomes and the cytosol, and the growth of plants with Cd produced an increase in the production of NO in peroxisomes, as shown in Fig. 3.5. Under the same conditions of Cd stress in pea plants, overproduction of superoxide radicals and hydrogen peroxide in leaf peroxisomes, as well as oxidative modification of endogenous proteins, have also been reported (Romero-Puertas et al. 1999, 2002).

3.6 Demonstration of in vivo NO Production in Peroxisomes

Although in peroxisomes the existence of L-Arginine-dependent NOS activity and the generation of NO has been demonstrated, as indicated in the previous sections, an important question is to know whether NO can be released from peroxisomes into the cytoplasm to get involved in different physiological functions of plant cells. This information is very important to substantiate the hypothesis postulating that peroxisomes are a cellular source of NO signal molecules (Corpas et al. 2001; del Río 2011).

Evidence recently obtained in *Arabidopsis* plants subjected to abiotic stress by salinity have drawn light on the NO production in peroxisomes and its release into the cytosol. In previous studies, in olive plants it has been demonstrated that salinity induces both oxidative and nitrosative stress, characterized by an increase in the production of reactive oxygen species and reactive nitrogen species, respectively (Valderrama et al. 2007; Corpas et al. 2011). For the experiments of in vivo NO production in peroxisomes, an *Arabidopsis* mutant defective in PTS1-dependent protein transport to peroxisomes and with a lower NO content was used (Fig. 3.4) (Mano et al. 2006; Corpas et al. 2009b). In roots of control (without NaCl) *Arabidopsis* plants expressing GFP-PTS1, analysis by CLSM showed that NO was clearly produced in peroxisomes and the cytosol did not show NO fluorescence (Fig. 3.4a, c, e). However, when plants were subjected to salt stress (100 mM NaCl) the NO generation was significantly increased in peroxisomes and now also appeared and distributed in the cytosol (Fig. 3.4b, d, f). This suggested that salinity induced the release of NO from the peroxisomes into the cytosol. To confirm this hypothesis, the effect of salt stress was also studied in an *Arabidopsis*

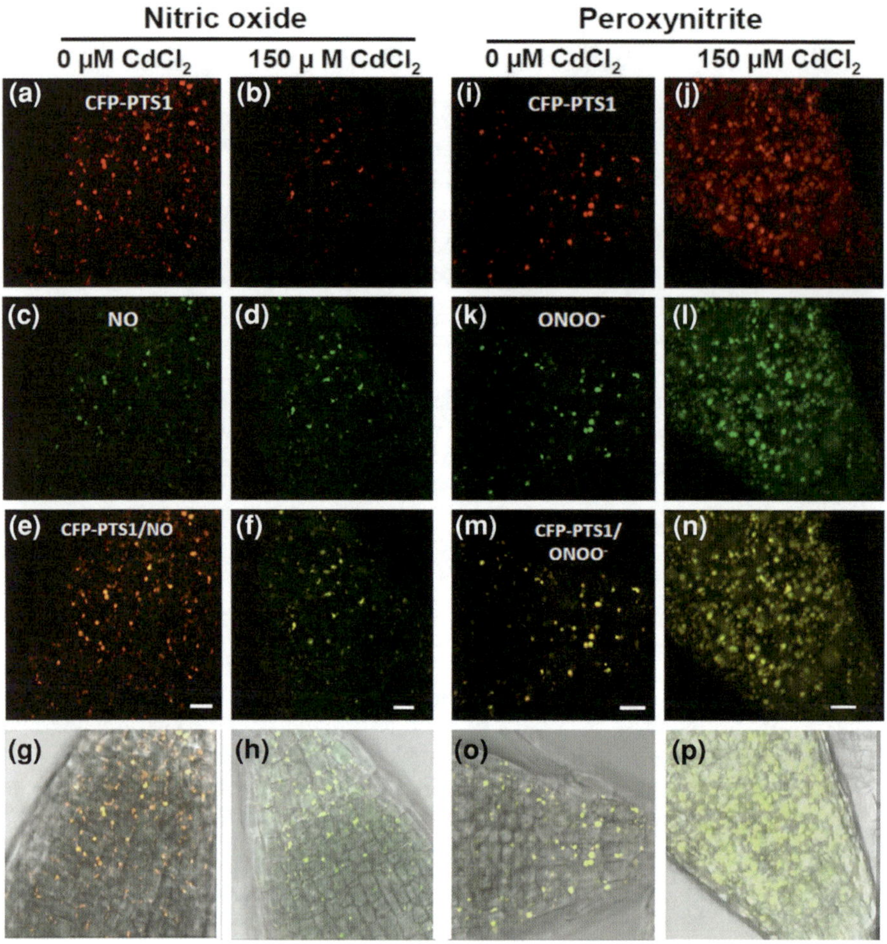

Fig. 3.5 Effect of Cd stress on the NO production (*green color*) in peroxisomes of root tips of transgenic *Arabidopsis* plants expressing the fused protein CFP-PTS1 (*red color*), and localization of peroxynitrite (*green color*) in peroxisomes using the fluorescent probe APF. **a–h** show peroxisomes and NO detection in roots of control and Cd-treated seedlings. **i–p** show peroxisomes and $ONOO^-$ detection in roots of control and Cd-treated seedlings. **e**, **f**, **m** and **n** show merged images showing co-localized fluorescence punctuates (*yellow*). **g**, **h**, **o** and **p** show the bright-field plus the merged image observed in the corresponding samples. *Red* fluorescence punctuates is attributable to CFP-PTS1, indicating the localization of peroxisomes in the corresponding panels. *Green* fluorescence is attributable to NO or peroxynitrite detection, with its specific fluorescent probes. Bar = 10 µm. Reproduced with permission from Corpas and Barroso (2014)

mutant (*apm4/pex12*) defective in the PTS1-dependent protein transport machinery into peroxisomes. As shown in Fig. 3.4g, h peroxisomes appear as green spots in the roots of control and salt-stressed plants. But in the same root area now NO was nearly totally absent (Fig. 3.4i). Salt stress increased slightly the NO content (Fig. 3.4j), and when the merged images of control and stressed roots were compared it was observed that NO was mainly present in the cytosol (Fig. 3.4k, l). This indicated that the peroxisomal protein responsible for NO production was not imported into the peroxisomes due to the defect in the protein targeting mechanism. In conclusion, the data obtained with *Arabidopsis* mutants suggest that in plant roots peroxisomes are the main source of NO, and the NO overproduced under salt stress was originated in peroxisomes and from these organelles released into the cytosol (Corpas et al. 2009b). This emphasizes the key role of peroxisomes as a source of NO signal molecules to be used in cellular metabolism for carrying out different plant physiological functions.

3.7 *S*-Nitrosylation and Nitration of Proteins in Peroxisomes

The presence of glutathione and ascorbate together with the whole antioxidative ascorbate-glutathione cycle enzymes has been demonstrated in leaf peroxisomes (Jiménez et al. 1997). It is known that NO in the presence of O_2 can react with reduced glutathione (GSH) to form *S*-nitrosoglutathione (GSNO), a NO-derived compound (Wink et al. 1996). This interaction with sulfhydryl-containing molecules is designated as *S*-nitrosylation and plays an important role in NO-mediated signaling (Stamler et al. 2001) and is also a key redox-based post-translational modification (Martínez-Ruiz and Lamas 2009). The presence of GSNO in peroxisomes has been demonstrated by EM immunocytochemistry, using a commercial antibody to GSNO, in sunflower hypocotyls (Chaki 2007) and pea plants (Rodríguez-Serrano 2007; Ortega-Galisteo et al. 2012; Barroso et al. 2013). GSNO is considered as an important mobile reservoir of NO bioactivity (Durner and Klessig 1999; Noble et al. 1999; Stamler et al. 2001) and this molecule can mediate the signaling pathway through specific post-translational modifications of redox-sensitive proteins by a reaction of transnitrosylation from GSNO to Cys-proteins, leading to *S*-nitrosylated proteins which are considered as high molecular mass *S*-nitrosothiols.

In recent years, different studies in plants under physiological or stress conditions have shown that a significant number of proteins are *S*-nitrosylated and have paved the way to research on the regulation of protein function by *S*-nitrosylation (Lindermayr et al. 2005; Lyndermayr and Durner 2009; Romero-Puertas et al. 2008; Abat and Deswal 2009; Palmieri et al. 2010). In rat liver, analysis of mitochondrial extracts treated with *S*-nitrosoglutathione (GSNO) allowed to identify *S*-nitrosylated catalase and malate dehydrogenase, two peroxisomal enzymes which copurified with the mitochondrial fractions (Foster and Stamler 2004). A similar result was obtained in *Arabidopsis* mitochondrial extracts where catalase was also identified as a target of *S*-nitrosylation (Palmieri et al. 2010). The presence of

GSNO-induced *S*-nitrosylated proteins has been studied in leaf peroxisomes purified from pea plants subjected to abiotic stress by Cd and herbicide 2,4-D (Ortega-Galisteo et al. 2012). Six peroxisomal proteins were identified as putative targets of *S*-nitrosylation involved in photorespiration, β-oxidation, and ROS detoxification, including: hydroxypyruvate reductase, glycolate oxidase, serine-glyoxylate aminotransferase, amino transferase 1, catalase, and malate dehydrogenase. The activity of catalase, glycolate oxidase and malate dehydrogenase was inhibited by NO donors. The *S*-nitrosylation levels of catalase and glycolate oxidase in leaf peroxisomes changed in pea plants treated with cadmium and 2,4-D, suggesting that this post-translational modification could be involved in the regulation of the H_2O_2 level under abiotic stress (Ortega-Galisteo et al. 2012). The inhibition of catalase activity by NO donors agrees with results of in vitro assays of purified tobacco catalase incubated with different NO donors (SNAP, GSNO, NOC-9) which produced decreases in enzyme activity of 70–90 %, and the activity was partly restored when the NO donors were removed (Clark et al. 2000).

Peroxynitrite ($ONOO^-$) is a class of reactive nitrogen species which is formed by a rapid chemical reaction between superoxide radicals ($O_2^{\cdot-}$) and NO ($k = 1.9 \times 10^{10}$ M^{-1} s^{-1}) (Kissner et al. 1997). Peroxynitrite is a powerful oxidant/nitrating species that produces the oxidation and nitration of proteins and other biomolecules (Radi 2013) and is considered to be responsible of the protein nitration process, an irreversible post-translational modification that can affect protein function (Radi 2004). Peroxynitrite has a very short half-life and its action must take place at the site of generation of both superoxide and NO (Szabó et al. 2007). Tyrosine nitration is at present the most studied protein modification by peroxynitrite, although other amino acids, such as cysteine, methionine and tryptophan can also be nitrated. In higher plants, tyrosine nitration is usually associated with environmental stress processes, and the identification of nitrated proteins is increasingly growing (Corpas et al. 2007, 2011; Chaki et al. 2009, 2011; Lozano-Juste et al. 2011; Begara-Morales et al. 2014). The peroxisomal enzymes catalase, glycolate oxidase and malate dehydrogenase are among the potential targets of protein nitration although the specific effect of nitration on their enzymatic activities remains to be determined (Lozano-Juste et al. 2011).

The presence of peroxynitrite in peroxisomes has been recently studied by CLSM in *Arabidopsis thaliana* transgenic plants expressing CFP-PTS1 which allows the in vivo visualization of peroxisomes (Corpas and Barroso 2014). Using the specific fluorescent probe APF it was demonstrated that peroxynitrite was generated endogenously in peroxisomes of root and guard cells. The localization of peroxynitrite in peroxisomes of root tips of *Arabidopsis* mutants expressing CFP-PTS1 is shown in Fig. 3.5. When plants were grown under cadmium stress (150 μM $CdCl_2$), an increase in the production of peroxynitrite was found in peroxisomes and cytosol of root tips (Fig. 3.5l). Under the same conditions of Cd stress, an overproduction of NO in peroxisomes and cytosol of *Arabidopsis* root tips was also observed (Fig. 3.5d) in comparison with plants under control conditions (Fig. 3.5c) (Corpas and Barroso 2014).

Results on the occurrence of peroxynitrite in peroxisomes suggest that this oxidizing RNS could play a regulatory role, by protein tyrosine nitration, on some peroxisomal enzymes. Recent results obtained in pea plants by EM immunogold-labeling have shown the presence of nitrated proteins in different subcellular compartments of leaf cells, including peroxisomes, chloroplasts, mitochondria, and cytosol (Barroso et al. 2013). Moreover, proteomic analysis of isolated pea leaf peroxisomes has shown that peroxisomal NADH-dependent hydroxypyruvate reductase is a target of nitration, and this reaction by peroxynitrite produced a loss of function in the enzyme (Corpas et al. 2013c).

3.8 Conclusions

In plants there are multiple sources of NO, both enzymatic and nonenzymatic, and there is a body of evidence supporting the existence of L-Arginine-dependent Nitric oxide synthase activity (NOS). However, the protein or gene responsible for this activity in higher plants has not been identified and characterized yet. Intensive efforts are necessary to try to identify the protein(s) responsible for the L-Arginine-dependent NOS activity detected in peroxisomes.

The existence of a RNS and ROS metabolism in plant peroxisomes and the presence in these organelles of a complex battery of antioxidative enzymes, remarks the importance of these organelles in cellular oxidative metabolism. The main ROS and RNS whose presence has been demonstrated in plant peroxisomes include: H_2O_2, $O_2^{\cdot-}$ radicals, NO, *S*-nitrosoglutathione (GSNO) and peroxynitrite ($ONOO^-$). The main components involved in the metabolism of NO in plant peroxisomes are summarized in Fig. 3.6. The demonstration of the presence of L-Arginine-dependent NOS activity in peroxisomes, and the generation of NO and GSNO in these oxidative organelles, implies that single membrane-bound peroxisomes can function in plant cells as a source of RNS signaling molecules, besides $O_2^{\cdot-}$ and H_2O_2. Plant peroxisomes are known to have a RNS- and ROS-mediated metabolic function in leaf senescence and certain types of abiotic stress. These organelles could act as subcellular indicators or sensors of plant stress by releasing the signaling molecules NO and GSNO, as well as $O_2^{\cdot-}$ and H_2O_2, to the cytoplasm and triggering specific changes in defense gene expression (stress signaling).

As a result of the presence of NO and GSNO, and the generation of $ONOO^-$, important post-translational modifications can take place in peroxisomes, such as *S*-nitrosylation and nitration of proteins. The demonstrated *S*-nitrosylation of catalase and glycolate oxidase in peroxisomes could regulate the level of key signaling molecules like H_2O_2 (Ortega-Galisteo et al. 2012). On the other hand, the generation of $ONOO^-$ in peroxisomes can produce tyrosine nitration of peroxisomal proteins and originate nitrosative damages in plant cells although a basal endogenous nitration also seems to have a regulatory function. The study of the *S*-nitrosylated and nitrated proteome of peroxisomes can provide important information not only

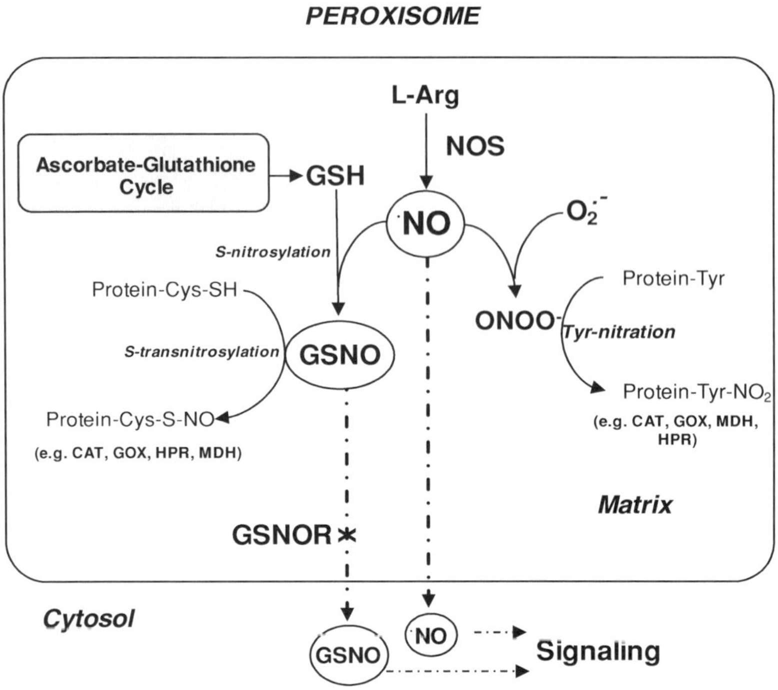

Fig. 3.6 Hypothetical model proposed for the metabolism and signalling function of nitric oxide (NO) and *S*-nitrosoglutathione (GSNO) in plant peroxisomes. L-Arginine-dependent nitric oxide syntase (NOS) generates NO which can react with reduced glutathione (GSH) in the presence of O_2 to form *S*-nitrosoglutathione (GSNO), a process named as *S*-nitrosylation. This metabolite can interact with SH-containing proteins by a reaction of *S*-transnitrosylation, affecting their function, or can be metabolized by the enzyme GSNO reductase (GSNOR). Nitric oxide can also react with superoxide radicals ($O_2^{\cdot-}$) to generate the powerful oxidant and nitrating species peroxynitrite ($ONOO^-$) which can produce the tyrosine nitration of proteins. And, interestingly, NO and GSNO can be released to the cytosol to participate in signaling cascades. *CAT* catalase. *GOX* glycolate oxidase. *MDH* malate dehydrogenase. *HPR* hydroxypyruvate reductase. Modified from Acta Physiol Plant (2013) 35: 2635–2640

on nitro-oxidative damages to peroxisomal proteins by abiotic and biotic stress situations but also on the possible regulation of peroxisomal metabolism by these two protein post-translational modifications (Corpas et al. 2013b, c; Begara-Morales et al. 2014; Romero-Puertas et al. 2013).

Acknowledgments This work was supported by ERDF-cofinanced grants from the Ministries of Science and Innovation, and Economy, and the *Junta de Andalucía* (Groups BIO-192 and BIO-286), Spain. Authors are indebted to Prof. Kozi Asada, Kyoto University, Japan, for supplying the pure recombinant enzyme monodehydroascorbate reductase (MDAR) for the assays of NO production by spin trapping EPR.

References

Abat JK, Deswal R (2009) Differential modulation of *S*-nitrosoproteome of *Brassica juncea* by low temperature: change in *S*-nitrosylation of rubisco is responsible for the inactivation of its carboxylase activity. Proteomics 9:4368–4380

Alderton WK, Cooper CE, Knowles RG (2001) Nitric oxide synthases: structure, function and inhibition. Biochem J 357:593–615

Baker A, Graham I (eds) (2002) Plant peroxisomes. Biochemistry, cell biology and biotechnological applications. Kluwer, Dordrecht

Baker A, Graham IA, Holdsworth M et al (2006) Chewing the fat: β-oxidation in signalling and development. Trends Plant Sci 11:124–132

Barroso JB, Corpas FJ, Carreras A et al (1999) Localization of nitric oxide synthase in plant peroxisomes. J Biol Chem 274:36729–36733

Barroso JB, Valderrama R, Corpas FJ (2013) Immunolocalization of *S*-nitrosoglutathione, *S*-nitrosoglutathione reductase and tyrosine nitration in pea leaf organelles. Acta Physiol Plant 35:2635–2640

Baudouin E (2011) The language of nitric oxide signaling. Plant Biol (Stuttg) 13:233–242

Begara-Morales JC, Sánchez-Calvo B, Chaki M et al (2014) Dual regulation of cytosolic ascorbate peroxidase (APX) by tyrosine nitration and *S*-nitrosylation. J Exp Bot 65:527–538

Boucher JL, Genet A, Vadon S et al (1992a) Formation of nitrogen oxides and citrulline upon oxidation of Nw-hydroxy-L-arginine by hemeproteins. Biochem Biophys Res Commun 184: 1158–1164

Boucher JL, Genet A, Valdon S et al (1992b) Cytochrome P450 catalyzes the oxidation of Nw-hydroxy-L-arginine by NADPH and O_2 to nitric oxide and citrulline. Biochem Biophys Res Commun 187: 880–886

Caro A, Puntarulo S (1999) Nitric oxide generation by soybean embryonic axes, possible effect on mitochondrial function. Free Radic Res 31:S205–S212

Chaki M (2007) Function of reactive nitrogen species in sunflower (*Helianthus annuus*) in response to abiotic and biotic stresses. PhD Thesis, University of Jaén, Spain

Chaki M, Fernández-Ocaña AM, Valderrama R et al (2009) Involvement of reactive nitrogen and oxygen species (RNS and ROS) in sunflower-mildew interaction. Plant Cell Physiol 50:265–279

Chaki M, Valderrama R, Fernández-Ocaña AM et al (2011) High temperatura triggers the metabolism of *S*-nitrosothiols in sunflower mediating a process of nitrosative stress which provokes the inhibition of ferredoxin-NADP reductase by tyrosine nitration. Plant, Cell Environ 34:1803–1818

Clark D, Durner J, Navarre DA, Klessig DF (2000) Nitric oxide inhibition of tobacco catalase and ascorbate peroxidase. Mol Plant Microbe Interact 13:1380–1384

Corpas FJ, Palma JM, Sandalio LM et al (1999) Purification of catalase from pea leaf peroxisomes: identification of five different isoforms. Free Radic Res 31:235–242

Corpas FJ, Barroso JB, del Río LA (2001) Peroxisomes as a source of reactive oxygen species and nitric oxide signal molecules in plant cells. Trends Plant Sci 6:145–150

Corpas FJ, Barroso JB, Carreras A et al (2004a) Cellular and subcellular localization of endogenous nitric oxide in young and senescent pea plants. Plant Physiol 136: 2722–2733

Corpas FJ, Barroso JB, León AM et al (2004b) Peroxisomes as a source of nitric oxide. In: Magalhaes JR, Singh RP, Passos LP (eds) Nitric oxide signaling in higher plants, Studium Press, LLC, Houston, pp 111–129

Corpas FJ, Barroso JB, Carreras A et al (2006) Constitutive arginine-dependent nitric oxide synthase activity in different organs of pea seedlings during plant development. Planta 224:225–246

Corpas FJ, del Río LA, Barroso JB (2007) Need of biomarkers of nitrosative stress in plants. Trends Plant Sci 12:436–438

Corpas FJ, Carreras A, Esteban FJ et al (2008) Localization of *S*-nitrosothiols and assay of nitric oxide synthase and *S*-nitrosoglutathione reductase activity in plants. Methods Enzymol 437:559–572

Corpas FJ, Palma JM, del Río LA, Barroso JB (2009a) Evidence supporting the existence of L-arginine-dependent nitric oxide synthase activity in plants. New Phytol 184: 9–14

Corpas FJ, Hayashi M, Mano S et al (2009b) Peroxisomes are required for in vivo nitric oxide accumulation in the cytosol following salinity stress of *Arabidopsis* plants. Plant Physiol 151: 2083–2094

Corpas FJ, Leterrier M, Valderrama R et al (2011) Nitric oxide imbalance provokes a nitrosative response in plants under abiotic stress. Plant Sci 181:604–611

Corpas FJ, Barroso JB (2014) Peroxynitrite ($ONOO^-$) is endogenously produced in *Arabidopsis* peroxisomes and is overproduced under cadmium stress. Ann Bot 113:87–96

Corpas FJ, Barroso JB, Palma JM, del Río LA (2013a) Peroxisomes as cell generators of reactive nitrogen species (RNS) signal molecules. In: del Río LA (ed) Peroxisomes and their key role in cellular signaling and metabolism. Subcellular Biochemistry 69. Springer, Dordrecht, pp 283–298

Corpas FJ, Palma JM, del Río LA, Barroso JB (2013b) Protein tyrosine nitration in higher plants plants under natural and stress conditions. Front Plant Sci 4: 1–4

Corpas FJ, Leterrier M, Begara-Morales JC et al (2013c) Inhibition of peroxisomal hydroxypyruvate reductase (HPR1) by tyrosine nitration. Biochim Biophys Acta 11:4981–4989

Costa A, Drago IO, Behera S et al (2010) H_2O_2 in plant peroxisomes: an in vivo analysis uncovers a Ca^{2+}-dependent scavenging system. Plant J 62:760–772

Cueto M, Hernández-Perera O, Martín R et al (1996) Presence of nitric oxide synthase activity in roots and nodules of *Lupinus albus*. FEBS Lett 398:159–164

Dean JV, Harper JE (1988) The conversion of nitrite to nitrogen oxide(s) by the constitutive NAD(P)H-nitrate reductase enzyme from soybean. Plant Physiol 88:389–395

De Duve C, Baudhuin P (1966) Peroxisomes (microbodies and related particles). Physiol Rev 46:323–357

Durner J, Klessig DF (1999) Nitric oxide as a signal in plants. Curr Opin Plant Biol 2:369–374

Fahimi HD, Sies H (eds) (1987) Peroxisomes in biology and medicine. Springer-Verlag, Berlin

Fewson CA, Nicholas DJD (1960) Utilization of nitric oxide by microorganisms and higher plants. Nature 188:794–796

Foresi N, Correa-Aragunde N, Parisi G et al (2010) Characterization of a nitric oxide synthase from the plant kingdom: NO generation from the green alga *Ostreococcus tauri* is light irradiance and growth phase dependent. Plant Cell 22:3816–3830

Foster MW, Stamler JS (2004) New insights into protein *S*-nitrosylation. Mitochondria as a model system. J Biol Chem 279:25891–25897

Gas E, Flores-Pérez U, Sauret-Güeto S, Rodríguez-Concepción M (2009) Hunting for plant nitric oxide synthase provides new evidence of a central role for plastids in nitric oxide metabolism. Plant Cell 21:18–23

Gupta KJ, Stoimenova M, Kaiser WM (2005) In higher plants, only root mitochondria, but not leaf mitochondria reduce nitrite to NO, in vitro and in situ. J Exp Bot 56:2601–2609

Gupta KJ, Kaiser WM (2010) Production and scavenging of nitric oxide by barley root mitochondria. Plant Cell Physiol 51:576–584

Gupta KJ, Fernie AR, Kaiser WM, van Dongen JT (2011) On the origins of nitric oxide. Trends Plant Sci 16:160–168

Hancock JT (2012) NO synthase? Generation of nitric oxide in plants. Periodicum Biologorum 114:19–24

Harrison R (2002) Structure and function of xanthine oxidoreductase: where are we now? Free Radical Biol Med 33:774–797

Hu J, Baker A, Bartel B et al (2012) Plant peroxisomes: biogenesis and function. Plant Cell 24:2279–2303

Huang J, Sommers EM, Kim-Shapiro DB, King SB (2002) Horseradish peroxidase catalyzed nitric oxide formation from hydroxyurea. J Am Chem Soc 124:3473–3480

Hung KT, Kao CH (2003) Nitric oxide counteracts the senescence of rice leaves induced by abscisic acid. J Plant Physiol 160:871–879

Islinger M, Grille S, Fahimi HD, Schrader M (2012) The peroxisome: an update on mysteries. Histochem Cell Biol 137:547–574

Jasid S, Simontacchi M, Bartoli CG, Puntarulo S (2006) Chloroplasts as a nitric oxide cellular source. Effect of reactive nitrogen species on chloroplastic lipids and proteins. Plant Physiol 142:1246–1255

Jiménez A, Hernández JA, del Río LA, Sevilla F (1997) Evidence for the presence of the ascorbate-glutathione cycle in mitochondria and peroxisomes of pea leaves. Plant Physiol 114:275–284

Kirkman HN, Rolfo M, Ferraris AM, Gaetani GF (1999) Mechanisms of protection of catalase by NADPH. Kinet Stoichiometry J Biol Chem 274:30451–30458

Kissner R, Nauser T, Bugnon P et al (1997) Formation and properties of peroxynitrite as studied by laser flash photolysis, high pressure stopped flow technique, and pulse radiolysis. Chem Res Toxicol 10:1285–1292

Klepper LA (1979) Nitric oxide (NO) and nitrogen dioxide (NO_2) emissions from herbicide-treated soybean plants. Atmos Environ 13:537–542

Knowles RG, Moncada S (1994) Nitric oxide synthases in mammals. Biochem J 298:249–258

Leshem YY (2000) Nitric oxide in plants: occurrence, function and use. Kluwer Academic Publishers, Dordrecht, The Netherlands

Lindermayr C, Saalbach G, Durner J (2005) Proteomic identification of *S*-nitrosylated proteins in *Arabidopsis*. Plant Physiol 137:921–930

Lindermayr C, Durner J (2009) *S*-Nitrosylation in plants: pattern and function. J Proteomics 73:1–9

Loughran PA, Stolz DB, Vodovotz Y et al (2005) Monomeric inducible nitric oxide synthase localizes to peroxisomes in hepatocytes. Proc Natl Acad Sci USA 102:13837–13842

Lozano-Juste J, Colom-Moreno R, León J (2011) In vivo protein tyrosine nitration in *Arabidopsis thaliana*. J Exp Bot 62:3501–3517

Mano S, Nakamori C, Hayashi M et al (2002) Distribution and characterization of peroxisomes in *Arabidopsis* by visualization with GFP: dynamic morphology and actin-dependent movement. Plant Cell Physiol 43:331–341

Mano S, Nakamori C, Nito K et al (2006) The *Arabidopsis* pex12 and pex13 mutants are defective in both PTS1- and PTS2-dependent protein transport to peroxisomes. Plant J 47:604–618

Martínez-Ruiz A, Lamas S (2009) Two decades of new concepts in nitric oxide signaling: from the discovery of a gas messenger to the mediation of nonenzymatic posttranslational modifications. IUBMB Life 61:91–98

Mathur J, Mathur N, Hülskamp M (2002) Simultaneous visualization of peroxisomes and cytoskeletal elements reveals actin and not microtubule-based peroxisome motility in plants. Plant Physiol 128:1031–1045

Miyake C, Schreiber U, Hormann H, Sano S, Asada K (1998) The FAD-enzyme monodehydroascorbate radical reductase mediates photoproduction of superoxide radicals in spinach thylakoid membranes. Plant Cell Physiol 39:821–829

Mur LAJ, Mandon J, Persijn S et al (2012) Nitric oxide in plants: an assessment of the current state of knowledge. AoB Plants 5: pls052

Neill S, Bright J, Desikan R et al (2008) Nitric oxide evolution and perception. J Exp Bot 59:25–35

Nicholls P (1964) The reactions of azide with catalase and their significance. Biochem J 90:331–343

Ninnemann H, Maier J (1996) Indications for the occurrence of nitric oxide synthases in fungi and plants and the involvement in photoconidiation of *Neurospora crassa*. Photochem Photobiol 64:393–398

Noble DR, Swift HR, Williams DLH (1999) Nitric oxide release from *S*-nitrosoglutathione (GSNO). Chem Commun 18:2317–2318

Ortega-Galisteo AP, Rodríguez-Serrano M, Pazmiño DM et al (2012) *S*-Nitrosylated proteins in pea (*Pisum sativum* L.) leaf peroxisomes: changes under abiotic stress. J Exp Bot 63:2089–2103

Palma JM, Sandalio LM, Corpas FJ et al (2002) Plant proteases, protein degradation, and oxidative stress: role of peroxisomes. Plant Physiol Biochem 40:521–530

Palmieri MC, Lindermayr C, Bauwe H et al (2010) Regulation of plant glycine decarboxylase by *S*-nitrosylation and glutathionylation. Plant Physiol 152:1514–1528

Procházková D, Wilhelmová N (2011) Nitric oxide, reactive nitrogen species and associated enzymes during plant senescence. Nitric Oxide: Biol Chem 24:61–65

Radi R (2004) Nitric oxide, oxidants, and protein tyrosine nitration. Proc Natl Acad Sci USA 101:4003–4008

Radi R (2013) Protein tyrosine nitration: biochemical mechanisms and structural basis of functional effects. Acc Chem Res 46:550–559

del Río LA, Pastori GM, Palma JM et al (1998) The activated oxygen role of peroxisomes in senescence. Plant Physiol 116:1195–1200

del Río LA, Corpas FJ, Sandalio LM et al (2003) Plant peroxisomes, reactive oxygen metabolism and nitric oxide. IUBMB Life 55:71–81

del Río LA, Corpas FJ, Barroso JB (2004) Nitric oxide and nitric oxide synthase activity in plants. Phytochemistry 65:783–792

del Río LA, Sandalio LM, Corpas FJ et al (2006) Reactive oxygen species and reactive nitrogen species in peroxisomes. Production, scavenging, and role in cell signaling. Plant Physiol 141:330–335

del Río LA (2011) Peroxisomes as a source of reactive nitrogen species signal molecules. Arch Biochem Biophys 506:1–11

del Río LA (ed) (2013) Peroxisomes and their key role in cellular signaling and metabolism. Subcellular biochemistry 69. Springer, Dordrecht, The Netherlands

Rockel P, Strube F, Rockel A et al (2002) Regulation of nitric oxide (NO) production by plant nitrate reductase in vivo and in vitro. J Exp Bot 53:103–110

Rodríguez-Serrano M (2007) Molecular mechanisms of response to cadmium in *Pisum sativum* L. plants: function of reactive oxygen and nitrogen species. PhD Thesis, University of Granada, Spain

Rodríguez-Serrano M, Romero-Puertas MC, Sparkes I et al (2009) Peroxisome dynamics in *Arabidopsis* plants under oxidative stress induced by cadmium. Free Radic Biol Med 47:1632–1639

Romero-Puertas MC, McCarthy I, Sandalio LM et al (1999) Cadmium toxicity and oxidative metabolism of pea leaf peroxisomes. Free Radic Res 31(Suppl):S25–S32

Romero-Puertas MC, Palma JM, Gómez M et al (2002) Cadmium causes the oxidative modification of proteins in pea plants. Plant, Cell Environ 25:677–686

Romero-Puertas MC, Campostrini N, Matte A et al (2008) Proteomic analysis of *S*-nitrosylated proteins in *Arabidopsis thaliana* undergoing hypersensitive response. Proteomics 8:1459–1460

Romero-Puertas MC, Rodríguez-Serrano M, Sandalio LM (2013) Protein *S*-nitrosylation in plants under abiotic stress: an overview. Front Plant Sci 4:1–6

Sano S, Miyake C, Mikami B, Asada K (1995) Molecular characterization of monodehydroascorbate radical reductase from cucumber highly expressed in *Escherichia coli*. J Biol Chem 270:21354–21361

Shapiro AD (2005) Nitric oxide signaling in plants. Vitam Horm 72:339–398

Simontacchi M, Jasid S, Puntarulo S (2004) Nitric oxide generation during early germination of sorghum seeds. Plant Sci 167:839–847

Stamler JS, Lamas S, Fang FC (2001) Nitrosylation the prototypic redox based signalling mechanism. Cell 106:675–683

Stöhr C, Strube F, Marx G et al (2001) A plasma membrane-bound enzyme of tobacco roots catalyses the formation of nitric oxide from nitrite. Planta 212:835–841

Stöhr C, Stremlau S (2006) Formation and possible roles of nitric oxide in plant roots. J Exp Bot 57:463–470

Stolz DB, Zamora R, Vodovotz Y et al (2002) Peroxisomal localization of inducible nitric oxide synthase in hepatocytes. Hepatology 36:81–93

Szabó C, Ischiropoulos H, Radi R (2007) Peroxynitrite: biochemistry, pathophysiology and development of therapeutics. Nat Rev Drug Discov 6:662–680

Tischner R, Galli M, Heimer YM et al (2007) Interference with the citrulline-based nitric oxide synthase assay by argininosuccinate lyase activity in *Arabidopsis* extracts. FEBS J 274:4238–4245

Valderrama R, Corpas FJ, Carreras A et al (2007) Nitrosative stress in plants. FEBS Lett 581:453–461

Waterham HR, Wanders RJA (2012) Metabolic functions and biogenesis of peroxisomes in health and disease. BBA Mol Basis Dis 1822:1325–1508

Wilson ID, Neill SJ, Hancock JT (2008) Nitric oxide synthesis and signalling in plants. Plant, Cell Environ 31:622–631

Wink DA, Hanbauer I, Grisham MB et al (1996) Chemical biology of nitric oxide: regulation and protective and toxic mechanisms. Curr Top Cell Regul 34:159–187

Yamasaki H, Sakihama Y, Takahashi S (1999) An alternative pathway for nitric oxide production in plants: new features of an old enzyme. Trends Plant Sci 4:128–129

Yang T, Poovaiah BW (2002) Hydrogen peroxide homeostasis: activation of plant catalase by calcium/calmodulin. Proc Nat Acad Sci USA 99:4097–4102

Zemojtel T, Fröhlich A, Palmieri MC et al (2006) Plant nitric oxide synthase: a never-ending story? Trends Plant Sci 11:524–525

Chapter 4
Role of Plant Mitochondria in Nitric Oxide Homeostasis During Oxygen Deficiency

Halley Caixeta Oliveira and Ione Salgado

Abstract During their life cycle, plants may be exposed to situations of reduced oxygen availability, such as those imposed by soil flooding, in which their tissues have to cope with restrictions of aerobic metabolism. The limited availability of oxygen for reduction by the mitochondrial respiratory chain has many effects on plant metabolism and physiology, negatively affecting the growth and productivity of economically important species. Nitrite has been considered a major alternative terminal acceptor of the respiratory chain under oxygen deprivation. The gaseous radical nitric oxide (NO) produced from mitochondrial nitrite reduction has emerged as an important mediator of plant tolerance to low oxygen tensions, regulating mitochondrial bioenergetics, gene expression and the pathways of plant hormones. In particular, a recent study has indicated the involvement of mitochondrial NO synthesis from nitrite in the nitrate-mediated response of soybean roots to hypoxia. The importance of processes for NO degradation in maintaining mitochondrial functionality and controlling root metabolism during an oxygen shortage has also been highlighted. In this regard, the involvement of respiratory proteins and non-symbiotic hemoglobins in NO degradation has been demonstrated. In the present chapter, advances in this area will be discussed with a special focus on the role of nitrogen nutrition and mitochondrial NO homeostasis for plant tolerance to oxygen deficiency.

Keywords Mitochondrial respiration · Nitric oxide · Nitrite reduction · Oxygen deficiency · Root hypoxia

H. C. Oliveira
Departamento de Biologia Animal e Vegetal, Centro de Ciências Biológicas, Universidade Estadual de Londrina (UEL), CP 10011, Londrina, PR 86057-970, Brazil
e-mail: halley@uel.br

I. Salgado (✉)
Departamento de Biologia Vegetal, Instituto de Biologia, Universidade Estadual de Campinas (UNICAMP), CP 6109, Campinas, SP 13083-970, Brazil
e-mail: ionesm@unicamp.br

M. N. Khan et al. (eds.), *Nitric Oxide in Plants: Metabolism and Role in Stress Physiology*, DOI: 10.1007/978-3-319-06710-0_4,

4.1 Introduction

Molecular oxygen (O_2), which acts as the primary final electron acceptor of the mitochondrial respiratory chain, and is also essential for the activity of diverse plant enzymes, has a primordial role in the growth and metabolism of higher plants (Geigenberger 2003). However, during their life cycle, plants may be subjected at a certain frequency to conditions of O_2 deficiency, which may occur as a normal feature of plant development and ontogeny (Geigenberger 2003). For example, tissues with high-metabolic activity or restricted capacity for O_2 diffusion may be frequently exposed to limited internal O_2 concentrations, as reported in developing seeds (Borisjuk et al. 2007), bulky storage organs (Geigenberger et al. 2000), root meristems (Ober and Sharp 1996) and phloem cells (van Dongen et al. 2003). O_2 deficiency may also be a consequence of changes in the external O_2 supply, such as those prevalent after excessive rainfall or irrigation in which the soil may become waterlogged, depending on its drainage capacity. In such flooding situations, the O_2 supply to submerged tissues is limited because gas diffusion is approximately 10,000 times slower in water than in air (Armstrong 1979). Roots are particularly prone to flooding, although aerial parts may also be submerged (Drew 1997; Bailey-Serres and Voesenek 2008).

A decrease in O_2 availability resulting from endogenous or exogenous constraints induces a wide range of biochemical changes in plant tissues. The primary effect of this stressful condition is the inhibition of mitochondrial electron transport with a consequent decrease of ATP synthesis by oxidative phosphorylation and an increase of ATP production by cytosolic glycolysis (Bailey-Serres and Voesenek 2008). A restriction in O_2 availability can lead to hypoxia, when oxygen levels are sufficiently low to limit mitochondrial respiration, or to anoxia, when the ATP produced by oxidative phosphorylation is negligible relative to that produced by glycolysis due to the complete absence of O_2 (Drew 1997). Under both conditions, due to the lower efficiency of ATP production by anaerobic glycolysis compared with oxidative phosphorylation, there is a reduction in the energetic charge that affects diverse aspects of cellular metabolism (Geigenberger 2003). To adapt to these conditions, a global inhibition of ATP-consuming biosynthetic processes has been observed (Geigenberger 2003). Accordingly, the large-scale analysis of gene expression has demonstrated that low O_2 tensions repress transcripts related to highly energy-demanding processes, such as transport, lipid biosynthesis and secondary metabolism, and induce the expression of genes encoding proteins related to anaerobic metabolism (Lasanthi-Kudahettige et al. 2007; Kreuzwieser et al. 2009; van Dongen et al. 2009; Narsai et al. 2011). Indeed, an early proteomic analysis in maize roots demonstrated the induction of a specific group of anaerobic proteins (ANPs) involved in carbohydrate mobilisation, glycolysis and fermentative pathways (Sachs et al. 1980). More recent proteomic approaches applied to the roots of different plant species have revealed novel proteins belonging to many functional classes as being differentially expressed during O_2 deficiency (Ahsan et al. 2007; Alam et al. 2010).

Consistent with such changes at the transcriptomic and proteomic levels, an extensive reprogramming of the metabolic network in plant tissues submitted to low O_2 tensions has been reported (Kreuzwieser et al. 2009; van Dongen et al. 2009; Narsai et al. 2011; Shingaki-Wells et al. 2011). Anaerobic glycolysis is stimulated by the fermentation of pyruvate to end products such as lactate and ethanol, allowing NAD^+ regeneration under O_2 deficiency (Bailey-Serres and Voesenek 2008). Ethanol, formed by the sequential action of pyruvate decarboxylase and alcohol dehydrogenase enzymes, is the main fermentative end product in plant roots subjected to O_2 deficiency (Gibbs and Greenway 2003). Ethanol production is often preceded by lactate dehydrogenase-catalysed lactic acid fermentation, which results in a transient lactate accumulation (Gibbs and Greenway 2003). In many species, O_2 depletion stimulates the accumulation of other metabolites, especially those of amino acid metabolism, such as alanine and γ-aminobutyric acid (GABA), and some organic acids that are intermediates of the Krebs' cycle, particularly succinate (Rocha et al. 2010; Narsai et al. 2011; Shingaki-Wells et al. 2011; Oliveira and Sodek 2013). Additionally, O_2 deficiency has been associated with cytosolic acidosis, which would result from ATP hydrolysis, the inhibition of H^+-ATPases pumps and lactate production (Gout et al. 2001).

Thus, by diverting aerobic to anaerobic metabolism, O_2 deficiency would trigger many effects on diverse aspects of plant growth and physiology, thereby affecting crop yield and the distribution of plant species in natural ecosystems (Bailey-Serres and Voesenek 2008). Diverse studies have focused on the mechanisms developed by plants to minimise the deleterious effects of O_2 deficiency and on the development of techniques that could allow a higher tolerance of plants to this stress. The relevance of these issues has increased due to the current scenario of global climatic change, which predicts an increase of heavy rainfall and flooding situations in various regions of the planet (Bailey-Serres and Voesenek 2008).

Nitric oxide (NO) is a gaseous free radical, and its importance as a signaling molecule in diverse processes of plant physiology has been increasingly recognised (see Wendehenne and Hancock 2011 and references therein). Moreover, the important role of NO in the response of plants to various types of biotic and abiotic stresses has been demonstrated by many studies (reviewed by Siddiqui et al. 2011). The present chapter will focus on recent evidence regarding the involvement of NO in the adaptive response of plant roots to hypoxic and anoxic conditions. The relevance of the mechanisms of NO synthesis and degradation for the metabolic response of plants to low O_2 tensions will be discussed. In particular, the involvement of mitochondria (the primary target of O_2 deficiency) in these mechanisms that control NO homeostasis will be highlighted. Figure 4.1 illustrates the roles of plant mitochondria on NO metabolism under O_2 deprivation, which are outlined in more detail below. Finally, the important link between nitrogen nutrition and NO homeostasis during the plant hypoxic response (also depicted in Fig. 4.1) will be discussed.

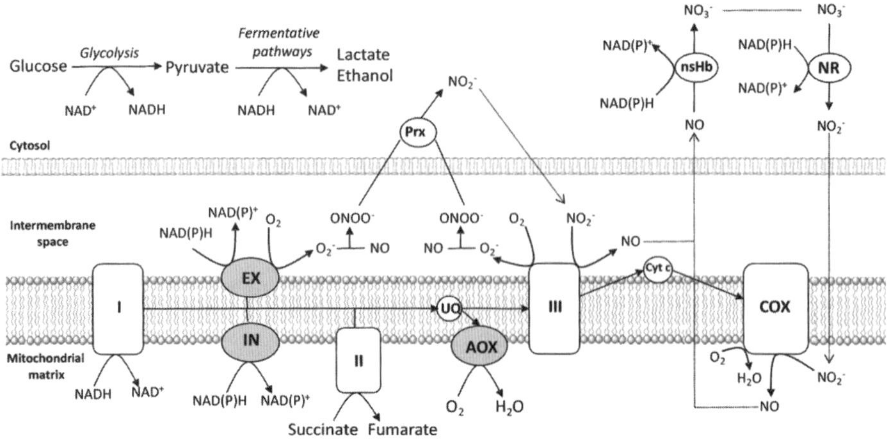

Fig. 4.1 Schematic model of NO homeostasis and its relevance for NAD(P)$^+$ regeneration during hypoxia. Under low O_2 conditions, nitrite (NO_2^-) is reduced to NO by Complex III or cytochrome *c* oxidase (COX), which allows the continuous operation of the mitochondrial electron transport chain and cytosolic NAD(P)H oxidation by external NAD(P)H dehydrogenases (EX). These enzymes are also related to the superoxide (O_2^-)-dependent degradation of NO to peroxynitrite ($ONOO^-$), which is further metabolised back to NO_2^- by peroxiredoxins (Prx). NO can also be degraded to nitrate (NO_3^-) by non-symbiotic hemoglobins (nsHb); the NO_3^- is then reduced to NO_2^- by nitrate reductase (NR), closing the cycle. Both reactions are associated with NAD(P)H oxidation. Thus, the operation of NO homeostasis cycle contributes to NAD(P)$^+$ regeneration to sustain the glycolytic pathway during hypoxia, constituting an alternative to fermentative pathways

4.2 Signaling Functions of NO During O_2 Deficiency: Plant Mitochondria As Important NO Targets

The wide range of actions of NO in biological systems results from its physico-chemical properties, which make it one of the most versatile signaling molecules. NO is an uncharged molecule with a relatively long half-life (approximately 5 s) compared with other free radicals (Stamler et al. 1992). As one of the smallest diatomic molecules in nature, it is highly diffusible, able to easily migrate through hydrophobic and hydrophilic cellular compartments, such as membranes and the cytosol (Stamler et al. 1992). Most of the effects of NO derive from the post-translational modification of proteins caused by the direct attachment of this radical or its derived compounds to protein residues (reviewed by Leitner et al. 2009). NO may interact with iron in heme or Fe–S groups, forming nitrosyl complexes in target molecules (Ramirez et al. 2011), and it may also react with –SH groups, resulting in the formation of *S*-nitrosothiols (SNOs) (Astier et al. 2011). Additionally, NO may indirectly modulate protein function by the nitration of tyrosine residues through the action of its derivate compound peroxynitrite (Corpas et al. 2013).

Mitochondrial proteins are important targets of NO during O_2 depletion in diverse organisms. In mammals, NO binds at nanomolar concentrations to the O_2-binding site of cytochrome *c* oxidase (COX), the terminal electron acceptor of the mitochondrial respiratory chain (Cleeter et al. 1994). The O_2-binding site of the enzyme is an iron/copper (heme a_3/Cu_B) binuclear centre, and NO binds reversibly to the ferrous heme a_3 to form a ferrous heme nitrosyl complex, resulting in a reversible inhibition of mitochondrial respiration (Cleeter et al. 1994). This competitive inhibitory effect of NO on COX is dependent on O_2 tension and becomes more intense as the O_2 levels decrease (Brown and Cooper 1994). The modulation of COX activity by NO has been considered to be a physiologically relevant mechanism of the regulation of mitochondrial respiration and anoxia avoidance (Cooper and Giulivi 2007).

Plant mitochondria constitute important targets for the action of NO during O_2 deficiency (Blokhina and Fagerstedt 2010). Many studies have shown that NO inhibits plant COX by a mechanism similar to that described for mammals (Millar and Day 1996; Yamasaki et al. 2001; Zottini et al. 2002; Martí et al. 2013). There is strong evidence regarding the role of the reversible and competitive inhibition of COX by NO in low O_2 sensing in plants (Borisjuk and Rolletschek 2009). In a study with soybean and pea seeds, hypoxia was shown to increase the endogenous NO content, with a consequent inhibition in the O_2 consumption and a local decrease in ATP production and biosynthetic activity (Borisjuk et al. 2007). The consequent increase of the O_2 levels would alleviate both mitochondrial and metabolic inhibition and reduce the NO concentration (Borisjuk et al. 2007). Therefore, this auto-regulatory mechanism would allow an adjustment of global metabolism in response to changes in O_2 availability, thus avoiding the establishment of the more deleterious anoxic conditions (Borisjuk and Rolletschek 2009). Consistent with this hypothesis, Borisjuk et al. (2007) detected increased internal O_2 concentrations in pea seeds after the exogenous addition of NO, similar to that reported in assays with isolated mitochondria under hypoxia (Benamar et al. 2008). In a recent study of maize roots (Mugnai et al. 2012), both an increase in the NO emission and a decrease in the O_2 consumption in the root apex transition zone were observed in response to O_2 deprivation. These local responses were considered to be essential for low O_2 sensing and for the hypoxic acclimation of the entire root, which was corroborated by the observation that the treatment with NO donors during hypoxic pre-treatment increased root survival in subsequent anoxic conditions (Mugnai et al. 2012).

Despite this important role of NO in promoting adaptive respiratory responses of plant tissues to O_2 deprivation, NO may also exert deleterious effects in plant mitochondria under certain circumstances. The prolonged exposure to NO donors was documented to induce the death of *Citrus sinensis* and carrot cells in culture by affecting the normal function of the mitochondria (Saviani et al. 2002; Zottini et al. 2002). The functional and morphological changes caused by the NO exposition of *Citrus* cells were characteristic of mitochondrial dependent-programmed cell death, including the loss of the mitochondrial membrane electrical potential, alteration in mitochondrial membrane permeability, and chromatin condensation and fragmentation (Saviani et al. 2002). In this scenario, a role for the alternative oxidase (AOX) of the plant respiratory chain has emerged in response to various stresses in which COX is inhibited (Wulff et al. 2009; Vanlerberghe et al. 2009). AOX catalyses electron transfer from ubiquinol directly to O_2, bypasses complex III and COX without translocating protons across the inner membrane and thus does not contribute to ATP production (Millar et al. 2011). In contrast to COX, AOX is practically insensitive to NO, which allows respiration to continue even when an excess of this radical is produced (Millar and Day 1996; Yamasaki et al. 2001; Martí et al. 2013). In addition to its resistance to NO inhibition, AOX decreases the production of reactive oxygen species by electron leakage from the respiratory chain, which is largely stimulated during reoxygenation (Maxwell et al. 1999). NO was known to increase AOX transcripts (Huang et al. 2002), and *Arabidopsis thaliana* plants with deficient NO production failed to increase AOX levels during hypoxia, suggesting that NO acts as a mediator of the hypoxic induction of AOX genes (Gupta et al. 2012). Interestingly, O_2 deficiency has also been shown to induce AOX expression and activity (Amora et al. 2000; Skutnik and Rychter 2009; Gupta et al. 2012), and a contribution of AOX to root respiration at low O_2 levels has been suggested (Zabalza et al. 2009). As demonstrated by large-scale transcriptomic analysis, in addition to AOX transcripts, NO modulates the expression of various genes related to the plant stress response (Parani et al. 2004; Palmieri et al. 2008; Besson-Bard et al. 2009). Thus, the involvement of this radical in the hypoxia-induced transcriptional reprogramming in plant tissues is promising.

The modulation of the tricarboxylic acid cycle by NO has also been suggested. NO inhibits aconitase by forming a metal-nitrosyl complex with the Fe-S cluster of this enzyme (Navarre et al. 2000). Recently, the involvement of NO in the modulation of aconitase activity during hypoxia was demonstrated (Gupta et al. 2012). Citrate accumulation in *Arabidopsis thaliana* roots, resulting from aconitase inhibition by NO, was shown to cause AOX induction and a shift of hypoxic metabolism towards amino acid biosynthesis. Consistent with this hypothesis, an increase in the free amino acid levels is a common response of roots subjected to hypoxia (see Oliveira and Sodek 2013 and references therein). In addition to aconitase, respiratory chain-linked succinate dehydrogenase activity was recently shown to be negatively affected by NO in assays with isolated potato tuber mitochondria (Simonin and Galina 2013). This effect was suggested to result from an interaction of NO with ubiquinone or Fe-S centres of the enzyme, but it was observed only in the presence of excess ADP (Simonin and Galina 2013). In contrast, succinate dehydrogenase activity was not affected in *Arabidopsis thaliana* transgenic cell lines with different SNO/NO contents (Frungillo et al. 2013). In the same study, the Complex I and NADH dehydrogenase activities were shown to be responsive to changes in the SNO/NO levels, confirming the importance of these signaling molecules in the control of mitochondrial respiration (Frungillo et al. 2013).

In addition to modulating mitochondrial bioenergetics, NO is involved in the signal transduction pathways of diverse plant hormones, some of which are involved in the plant response to O_2 deficiency (Lamattina et al. 2003). Plants subjected to flooding often undergo morphological and anatomical changes, such as the formation of aerenchyma and adventitious roots, to increase gas exchange and minimise O_2 deficiency (Sairam et al. 2008). There is strong evidence regarding the role of ethylene in these adaptive responses to hypoxia (Sairam et al. 2008), and a positive effect of NO in promoting ethylene biosynthesis in hypoxic roots has been suggested (Manac'h-Little et al. 2005; Hebelstrup et al. 2012). Interestingly, genetically modified alfalfa roots with low NO emission present no sign of aerenchyma development under hypoxia (Dordas et al. 2003), and NO-deficient *Arabidopsis* plants show a reduction in the amplitude of hypoxia-induced hyponastic growth (Hebelstrup et al. 2012). Thus, an involvement of NO in the ethylene-dependent response of plants to hypoxia has been suggested (Igamberdiev et al. 2005). Additionally, NO has been shown to act as a mediator of many auxin-controlled rooting processes under normoxic conditions (Pagnussat et al. 2003; Guo et al. 2008; Yadav et al. 2010). Whether NO participates in the auxin-mediated formation of adventitious roots in the specific case of plants subjected to hypoxic stress remains to be verified (Igamberdiev et al. 2005).

4.3 Mechanisms of NO Synthesis During O_2 Deficiency: The Increasing Importance of Mitochondrial Nitrite Reduction

Despite the accumulating evidence regarding the importance of NO for hypoxic signaling, the molecular mechanisms by which NO is synthesised in plant cells remain under debate (Moreau et al. 2010). In mammals, a family of several NO synthase (NOS) enzymes has been well established as an important system for NO synthesis (Stuehr et al. 2004). NOSs catalyse the formation of NO and L-Citrulline through the oxidation of the amino acid L-Arginine in a reaction dependent on O_2, NADPH, heme, tetrahydrobiopterin, calmodulin, FAD and FMN (Stuehr et al. 2004). Many plant tissues have been reported to exhibit NO production that is sensitive to inhibitors of mammalian NOSs or L-Citrulline formation from L-Arginine (reviewed by del Río et al. 2004). However, a gene with homology to NOS of animal origin has not been found to date, and thus, the existence of a NOS-like enzyme in higher plants remains controversial (Fröhlich and Durner 2011). Polyamines and hydroxylamines have also been suggested as potential sources for NO synthesis in plants (Tun et al. 2006; Rümer et al. 2009). However, NO production from these substrates and from L-Arginine occurs via an oxidative pathway, and therefore, it would be unlikely to be stimulated under conditions of low O_2 tension, in which reductive mechanisms would be more operative.

In this scenario, nitrite reduction has been considered to be the main source for NO synthesis in plants (Salgado et al. 2006), especially during O_2 deficiency (Gupta et al. 2011a). Under these conditions, NO synthesis is favoured due to the inhibitory effect of low O_2 tensions on the nitrite reductase enzyme, leading to nitrite accumulation in plant tissues (Oliveira and Sodek 2013). Diverse mechanisms of NO synthesis from nitrite have been proposed, including non-enzymatic processes, such as the reduction of nitrite under the acidic pH conditions of the apoplast (Bethke et al. 2004). Among the enzymatic mechanisms, nitrate reductase (NR) first emerged as a potential source for NO synthesis from nitrite in plants (Yamasaki and Sakihama 2000). NR plays an essential role in nitrogen assimilation, catalysing the NAD(P)H-dependent reduction of nitrate to nitrite that is, in turn, reduced to ammonium by nitrite reductase (NiR). Ammonium is then incorporated into amino acids by the glutamine synthetase/ glutamine-2-oxoglutarate aminotransferase system (Lea 1993). During in vitro assays, NR can reduce nitrite as an alternative substrate, resulting in NO synthesis (Yamasaki and Sakihama 2000). However, the efficiency of NR in NO production in vivo is low and requires high-nitrite levels and low-oxygen tensions (Rockel et al. 2002). Thus, the involvement of NR in the NO emission by plant tissues detected under O_2 deficiency has been suggested (Rockel et al. 2002; Planchet et al. 2005; Horchani et al. 2011). Particularly in roots, the participation of a plasma membrane-bound nitrite: NO reductase enzyme in hypoxic NO synthesis has been proposed (Stöhr and Stremlau 2006). In nitrogen-fixing nodules, bacterial nitrate reductase appears to contribute to NO evolution under both hypoxic and normoxic conditions (Meakin et al. 2007; Horchani et al. 2011).

Alternatively, a mitochondrial nitrite reducing activity has emerged as an important mechanism for NO production in tissues subjected to O_2 deprivation. The role of the eukaryotic respiratory chain in NO synthesis was initially documented for mammalian mitochondria, in which electron leakage from Complex III (Kozlov et al. 1999) or the nitrite-reducing activity of COX (Castello et al. 2006) have been demonstrated to be involved in the NO production. The mitochondrial reduction of nitrite to NO has emerged as an important regulator of hypoxic signaling in mammalian physiology, having a protective role against ischaemia/ reperfusion injury (van Faasen et al. 2009). The involvement of the mitochondria in NO synthesis from nitrite in higher plants was demonstrated for the first time in assays of *Arabidopsis* leaf extracts (Modolo et al. 2005) and in tobacco leaves and cell suspensions (Planchet et al. 2005), in which the NO emission was shown to be stimulated by nitrite and suppressed by inhibitors of respiratory complexes. Since then, diverse studies have demonstrated that mitochondria from various plant species can synthesize NO under low-O_2 conditions using nitrite as an alternative acceptor of the electron transport chain (Gupta et al. 2005; Stoimenova et al. 2007; Wulff et al. 2009). In particular, a role for COX in this mechanism has been suggested, although the involvement of Complex III has also been proposed (Igamberdiev et al. 2010; Gupta et al. 2011b). However, the participation of AOX in mitochondrial NO synthesis remains controversial. Some studies have observed a nitrite-dependent NO generation that is sensitive to AOX inhibitors (Gupta et al. 2005; Planchet et al. 2005), whereas Modolo et al. (2005) have not detected a sensitivity to AOX inhibition in NO evolution. Furthermore, in a recent report concerning tobacco leaves, AOX suppression was associated with increased NO levels, and a role for this alternative respiratory enzyme in decreasing the mitochondrial NO synthesis, at least under normoxic conditions, was suggested

(Cvetkovska and Vanlerberghe 2012). In contrast, AOX expression in *Arabidopsis thaliana* cells was proposed to play a role in maintaining NO levels (Wulff et al. 2009) because its expression reduces electron leakage and, consequently, the generation of the superoxide anion that consumes NO (see below).

In addition to AOX, the plant mitochondrial respiratory chain contains at least four alternative non-phosphorylating NAD(P)H dehydrogenases, which divert the electron flow from Complex I. Two of them face the intermembrane space and present distinct properties, separately oxidising cytosolic NADH and NADPH (Millar et al. 2011). A role for these enzymes in nitrite reduction during O_2 deficiency has been demonstrated. Mitochondria isolated from barley and rice roots could oxidise external NAD(P)H in the presence of nitrite under strict anoxia, leading to the synthesis of NO (Stoimenova et al. 2007). This process would allow the regeneration of cytosolic $NAD(P)^+$, which is essential for sustaining ATP synthesis by the glycolytic pathway (Igamberdiev et al. 2010). External NAD(P)H oxidation was also shown to be involved in hypoxic seed germination (Logan et al. 2001). Notably, the activity of external NAD(P)H dehydrogenases would be favoured during O_2 deficiency due to the increased extramitochondrial $NAD(P)H/NAD(P)^+$ ratio, elevated cytosolic calcium levels and lowered pH observed under these conditions (Igamberdiev and Hill 2009; Ramírez-Aguilar et al. 2011). Moreover, hypoxia was recently demonstrated to induce the dissociation of Complex I from the respiratory supercomplexes of plant mitochondria, which would further favour the electron flow from alternative NAD(P)H dehydrogenases under these conditions (Ramírez-Aguilar et al. 2011). Furthermore, the electron transport from NAD(P)H to nitrite by isolated mitochondria from barley and rice roots was associated with ATP production under anaerobiosis, suggesting the generation of the proton motive force in the inner mitochondrial membrane by this pathway (Stoimenova et al. 2007). Although anaerobic ATP production by mitochondria is much lower than that observed under normoxia, it would complement glycolytic ATP synthesis, which might be relevant in situations of low O_2 availability (Stoimenova et al. 2007). Indeed, an effect of nitrite in increasing the ATP/ADP ratio in root nodules submitted to hypoxia has been reported (Horchani et al. 2011).

Overall, these results have highlighted the importance of mitochondrial NO metabolism, especially the mitochondrial reduction of nitrite to NO, in the control of the redox state and energetic status of plant cells for hypoxic survival.

4.4 Mechanisms of NO Degradation During O_2 Deficiency: The Involvement of Respiratory Proteins and Non-symbiotic Hemoglobins

Although, NO exerts various functions in biological systems, it may be toxic to the cells under certain conditions, leading, for example, to apoptotic or necrotic cell death (Murphy 1999). Thus, the control of NO homeostasis is essential to allow this molecule to exert its signaling functions and maintain endogenous NO concentrations at adequate levels for distinct physiological or stress situations. In addition to the mechanisms of synthesis, the processes of degradation are essential for maintaining NO homeostasis. In aerobic aqueous solutions, NO decay may occur due to its auto-oxidation to nitrite (Kharitonov et al. 1994). However, this spontaneous reaction would be rather slow in conditions where O_2 availability is low (de Oliveira et al. 2008). The existence of further mechanisms is, therefore, essential for controlling NO levels under this stress condition.

One of the most studied mechanisms of NO degradation during O_2 deficiency is the oxidation of NO by class 1 non-symbiotic hemoglobins (Igamberdiev et al. 2010). Although these proteins were identified early in plants, their involvement in hypoxic NO metabolism was first demonstrated by Dordas et al. (2003). In this study of transgenic alfalfa root cultures, the expression of sense or antisense barley haemoglobin transcripts led to decreased or increased NO emissions under hypoxia, respectively. Further studies have corroborated the negative correlation between class 1 non-symbiotic haemoglobin expression and NO emission in different experimental systems (Dordas et al. 2004; Perazzolli et al. 2004). These proteins, when oxygenated, catalyse the oxidation of NO to nitrate. The resulting methaemoglobin is then reduced back in a NAD(P)H-dependent enzymatic reaction (Igamberdiev et al. 2005). As the affinity of class 1 non-symbiotic haemoglobin for O_2 is very high, this mechanism of NO degradation may operate even at low O_2 concentrations (Igamberdiev et al. 2005). The involvement of haemoglobin expression in NO-mediated interplay with hormone signaling during O_2 deficiency has also been proposed (reviewed by Hill 2012). Furthermore, a role of NAD(P)H-dependent NO degradation by haemoglobin in regenerating cytosolic NAD(P)$^+$ under hypoxia has been demonstrated (Igamberdiev et al. 2004). This reaction would constitute an alternative to classic fermentative pathways, thereby contributing to the modulation of the redox and energy status of hypoxic cells (Igamberdiev and Hill 2004). Consistent with this hypothesis, the overexpression of barley non-symbiotic haemoglobin led to augmented ATP levels in hypoxic cells, whereas a lower expression of this protein increased the NAD(P)H/NAD(P)$^+$ ratio and decreased hypoxic tolerance (Dordas et al. 2003; Igamberdiev et al. 2004). In symbiotic root nodules, plant leghemoglobins and bacterial proteins have been proposed to be additional players in the modulation of hypoxic NO levels (Sánchez et al. 2011).

The role of plant mitochondria in NO degradation has also emerged, as demonstrated by studies of mitochondria isolated from potato tubers and *Arabidopsis* cells (de Oliveira et al. 2008; Wulff et al. 2009). NO consumption in the presence of plant mitochondria occurred mainly through the non-enzymatic reaction of NO with the superoxide anion (O_2^-) generated by the electron leakage from the respiratory chain (de Oliveira et al. 2008; Wulff et al. 2009). The spontaneous reaction of NO with superoxide results in the production of peroxynitrite ($ONOO^-$) (Radi et al. 2002), which can be further metabolised to nitrite by the peroxynitrite reductase activity of COX (Pearce et al. 2002) or peroxiredoxin (Romero-Puertas et al. 2007). Using substrates and inhibitors of different respiratory proteins, the involvement of external NAD(P)H dehydrogenases, in addition to Complex III, in superoxide generation for NO degradation has been demonstrated (de Oliveira et al. 2008). Inversely, by reducing the superoxide generation, AOX expression was associated with increased NO half-life without interrupting O_2 consumption (Wulff et al. 2009). Therefore, a role for the alternative proteins of plant mitochondria in the control of NO levels has been proposed (de Oliveira et al. 2008; Wulff et al. 2009). Although these studies have been conducted mainly under normoxic conditions, this mechanism of NO degradation would potentially also occur under hypoxia, especially if we consider that electron leakage from the respiratory chain is favoured at low O_2 tensions when the electron flow is lowered and the reduction state of the respiratory components is increased (Blokhina and Fagerstedt 2010). Additionally, the activity of the external NAD(P)H dehydrogenases may be promoted under hypoxia (Igamberdiev and Hill 2009; Ramírez-Aguilar et al. 2011), which would further favour the superoxide-dependent NO degradation by plant mitochondria. In addition to controlling the NO levels, external NAD(P)H may also be relevant for cytosolic $NAD(P)^+$ regeneration for glycolysis and redox status control during hypoxia, as discussed in the previous section.

Notably, this mitochondrial mechanism of NO degradation and NO oxidation by non-symbiotic hemoglobins would not operate under strict anoxia (Igamberdiev et al. 2005; de Oliveira et al. 2008). Indeed, in vitro assays have demonstrated that plant hemoglobins, instead of scavenging NO, catalyse NO synthesis from nitrite under anaerobic conditions (Sturms et al. 2011; Tiso et al. 2012). Therefore, other mechanisms of NO degradation should be active in the complete absence of O_2. One striking possibility is *S*-nitrosoglutathione reductase (GSNOR). This enzyme catalyses the breakdown of *S*-nitrosoglutathione (GSNO), an important intracellular NO reservoir, to oxidised glutathione and ammonia (Leterrier et al. 2011). Despite the growing evidence concerning the role of GSNOR in the response of plants to various stresses (Leterrier et al. 2011), the involvement of this enzyme in the modulation of NO and related species under O_2 deficiency has not been addressed. As GSNOR is a class III alcohol dehydrogenase enzyme (Sakamoto et al. 2002), its potential role in the plant hypoxic response is promising.

4.5 Nitrogen Nutrition and Plant Tolerance to O_2 Deficiency

Early agronomic and experimental observations demonstrated that exogenous supply of nitrate improves the tolerance and survival of diverse plant species subjected to root O_2 deficiency (Arnon 1937; Malavolta 1954; Trought and Drew 1981; Allegre et al. 2004; Thomas and Sodek 2005; Horchani et al. 2010). Nevertheless, this phenomenon is not observed when ammonium is used as an alternative nitrogen source, suggesting that the beneficial effect of nitrate is not solely related to a nitrogen supply for plant growth (Thomas and Sodek 2005). Diverse studies attempted to analyse the mechanisms by which nitrate exerts its ameliorating effect during O_2 deficiency (Sousa and Sodek 2002; Stoimenova et al. 2003), and an involvement of NO homeostasis in the hypoxic response of nitrate-cultivated plants has been proposed (Igamberdiev et al. 2010). Recently, studies of soybean root segments and intact roots indicated the involvement of mitochondrial nitrite reduction to NO in the nitrate-mediated response to hypoxia (Oliveira et al. 2013a, b). Roots from nitrate-cultivated plants presented a less intense hypoxic accumulation of the fermentative products lactate and ethanol compared with ammonium-grown roots. Interestingly, the intense fermentation of ammonium-grown roots was decreased when they were incubated with nitrite, a treatment that induced NO emission to levels similar to those of nitrate-cultivated roots. Furthermore, the nitrite-induced NO emission was sensitive to cyanide, a potent COX inhibitor, corroborating the involvement of the mitochondrial respiratory chain in this mechanism of nitrite reduction to NO (Oliveira et al. 2013a), as previously suggested (Modolo et al. 2005; Planchet et al. 2005; Gupta et al. 2005).

The effect of nitrate cultivation on increasing the NO levels in roots and other plant tissues has been demonstrated in diverse studies (Dordas et al. 2004; Jin et al. 2009; Santos-Filho et al. 2012; Oliveira et al. 2013a, b, c). These observations may be explained by the conversion of nitrate, through NR activity, to nitrite, which in turn can be further reduced to NO by the mitochondrial respiratory chain (Modolo et al. 2005; Oliveira et al. 2013a). Therefore, NR would have a fundamental role in providing the substrate nitrite for NO synthesis, which is in accordance with the low NO emission rate by plants with low NR activity (Modolo et al. 2005; Horchani et al. 2011; Gupta et al. 2012; Oliveira et al. 2013a). Furthermore, the reaction catalysed by NR consumes NAD(P)H and protons and thus acts as an alternative mechanism of NAD(P)H oxidation and as a proton sink during hypoxia (Garcia-Novo and Crawford 1973; Roberts et al. 1985).

The use of nitrite as an alternative electron acceptor by hypoxic plant mitochondria, in addition to NO synthesis, would allow cytosolic $NAD(P)^+$ regeneration and anaerobic ATP production (Stoimenova et al. 2007), explaining the observed effect of nitrite on reducing the accumulation of fermentation end products (Oliveira et al. 2013a, b). NO can be then metabolised by class 1 non-symbiotic hemoglobins via a mechanism that can further contribute to NAD(P)H oxidation in the cytosol (Igamberdiev and Hill 2004). A contribution of external NAD(P)H dehydrogenases to NO degradation through the reaction with superoxide may also be considered (de Oliveira et al. 2008). Finally, the products of NO metabolism (directly, nitrite and indirectly, nitrate and peroxynitrite) can be used to maintain NO synthesis. Therefore, the conjunction of the mechanisms for the control of NO homeostasis constitutes a cycle that could be important to the survival of cells exposed to hypoxia. In this scenario, nitrate nutrition would be essential in primarily feeding this cycle of NO turnover. This model (depicted in Fig. 4.1) is supported by the altered response to hypoxia observed in plants with a disruption in some step of this cycle, such as those observed in plants deficient in NR or in non-symbiotic hemoglobins (Stoimenova et al. 2003; Dordas et al. 2003; Allegre et al. 2004; Igamberdiev et al. 2004).

4.6 Conclusion

Mitochondria, as the main site of O_2 consumption in the cell, have a central role in plant adaptation to low O_2 tension. Recently, the influence of NO on mitochondrial respiration and energy production for the control of plant tolerance to O_2 deprivation has been increasingly recognised. Although many aspects of the mitochondrial NO metabolism have been recently uncovered, as reviewed here, many aspects of this mechanism have not yet been completely revealed. The molecular mechanisms of mitochondrial NO synthesis and degradation and the contribution of the different respiratory complexes to regulate cell metabolism under O_2 deprivation remain to be more thoroughly elucidated. Studies regarding the root-to-shoot communication during O_2 deficiency and the importance of nitrogen leaf metabolism to plant survival during O_2 deprivation are also required areas for further investigation (Oliveira et al. 2013c). Understanding the signaling functions of NO and nitrate in the adaptation of plants to O_2 deprivation and the interactions of these responses with plant hormones remain to be achieved. These studies would allow a more profound understanding of the role of NO in plant adaptation to O_2 deprivation and could contribute to the production of plants more adapted to environmental changes and with a consequently increased productivity under these stressful conditions.

Acknowledgment We thank Conselho Nacional de Desenvolvimento Científico e Tecnológico (CNPq) for financial support (grant 473090/2011–2).

References

Ahsan N, Lee DG, Lee SH et al (2007) A proteomic screen and identification of waterlogging regulated proteins in tomato roots. Plant Soil 295:37–51

Alam I, Lee DG, Kim KH et al (2010) Proteome analysis of soybean roots under waterlogging stress at an early vegetative stage. J Biosci 35:49–62

Allegre A, Silvestre J, Morard P et al (2004) Nitrate reductase regulation in tomato roots by exogenous nitrate: a possible role in tolerance to long-term root anoxia. J Exp Bot 55:2625–2634

Amora Y, Chevionb M, Levinea A (2000) Anoxia pretreatment protects soybean cells against H_2O_2-induced cell death: possible involvement of peroxidases and of alternative oxidase. FEBS Lett 477:175–180

Armstrong W (1979) Aeration in higher plants. Adv Bot Res 7:225–232

Arnon DI (1937) Ammonium and nitrate nitrogen nutrition of barley at different seasons in relation to hydrogen-ion concentration, manganese, copper, and oxygen supply. Soil Sci 44:91–122

Astier J, Rasul S, Koen E et al (2011) S-nitrosylation: an emerging post-translational protein modification in plants. Plant Sci 181:527–533

Bailey-Serres J, Voesenek LACJ (2008) Flooding stress: acclimations and genetic diversity. Annu Rev Plant Biol 59:313–339

Benamar A, Rolletschek H, Borisjuk L et al (2008) Nitrite-nitric oxide control of mitochondrial respiration at the frontier of anoxia. Biochim Biophys Acta 1777:1268–1275

Besson-Bard A, Astier J, Rasul S et al (2009) Current view of nitric oxide-responsive genes in plants. Plant Sci 177:302–309
Bethke PC, Badger MR, Jones RL (2004) Apoplastic synthesis of nitric oxide by plant tissues. Plant Cell 16:332–341
Blokhina O, Fagerstedt KV (2010) Oxidative metabolism, ROS and NO under oxygen deprivation. Plant Physiol Biochem 48:359–373
Borisjuk L, Macherel D, Benamar A et al (2007) Low oxygen sensing and balancing in plant seeds: a role for nitric oxide. New Phytol 176:813–823
Borisjuk L, Rolletschek H (2009) The oxygen status of the developing seed. New Phytol 182:17–30
Brown GC, Cooper CE (1994) Nanomolar concentrations of nitric oxide reversibly inhibit synaptosomal respiration by competing with oxygen at cytochrome oxidase. FEBS Lett 356:295–298
Castello PR, David PS, McClure T et al (2006) Mitochondrial cytochrome oxidase produces nitric oxide under hypoxic conditions: implications for oxygen sensing and hypoxic signaling in eukaryotes. Cell Metab 3:277–287
Cleeter MW, Cooper JM, Darley-Usmar VM et al (1994) Reversible inhibition of cytochrome *c* oxidase, the terminal enzyme of the mitochondrial respiratory chain, by nitric oxide, implications for neurodegenerative diseases. FEBS Lett 345:50–54
Cooper CE, Giulivi C (2007) Nitric oxide regulation of mitochondrial oxygen consumption II: molecular mechanism and tissue physiology. Am J Physiol Cell Physiol 292:C1993–C2003
Corpas FJ, Palma JM, Del Río LA, Barroso JB (2013) Protein tyrosine nitration in higher plants grown under natural and stress conditions. Front Plant Sci 4:29
Cvetkovska M, Vanlerberghe GC (2012) Alternative oxidase modulates leaf mitochondrial concentrations of superoxide and nitric oxide. New Phytol 195:32–39
de Oliveira HC, Wulff A, Saviani EE, Salgado I (2008) Nitric oxide degradation by potato tuber mitochondria: evidence for the involvement of external NAD(P)H dehydrogenases. Biochim Biophys Acta 1777:470–476
del Río LA, Corpas FJ, Barroso JB (2004) Nitric oxide and nitric oxide synthase activity in plants. Phytochemistry 65:783–792
Dordas C, Hasinoff BB, Igamberdiev AU et al (2003) Expression of a stress-induced hemoglobin affects NO levels produced by alfalfa root cultures under hypoxic stress. Plant J 35:763–770
Dordas C, Hasinoff BB, Rivoal J, Hill RD (2004) Class-1 hemoglobins, nitrate and NO levels in anoxic maize cell-suspension cultures. Planta 219:66–72
Drew MC (1997) Oxygen deficiency and root metabolism: injury and acclimation under hypoxia and anoxia. Annu Rev Plant Physiol Plant Mol Biol 487:223–250
Fröhlich A, Durner J (2011) The hunt for plant nitric oxide synthase (NOS): is one really needed? Plant Sci 181:401–404
Frungillo L, de Oliveira JF, Saviani EE et al (2013) Modulation of mitochondrial activity by *S*-nitrosoglutathione reductase in *Arabidopsis thaliana* cell lines. Biochim Biophys Acta 1827:239–247
Garcia-Novo F, Crawford RMM (1973) Soil aeration, nitrate reduction and flooding tolerance in higher plants. New Phytol 72:1031–1039
Geigenberger P (2003) Response of plant metabolism to too little oxygen. Curr Opin Plant Biol 6:247–256
Geigenberger P, Fernie AR, Gibon Y et al (2000) Metabolic activity decreases as an adaptive response to low internal oxygen in growing potato tubers. Biol Chem 381:723–740
Gibbs J, Greenway H (2003) Mechanisms of anoxia tolerance in plants. I. growth, survival and anaerobic catabolism. Funct Plant Biol 30:1–47
Gout E, Boisson AM, Aubert S et al (2001) Origin of the cytoplasmic pH changes during anaerobic stress in higher plant cells. Carbon-13 and phosphorous-31 nuclear magnetic resonance studies. Plant Physiol 125:921–925
Guo K, Xia K, Yang ZM (2008) Regulation of tomato lateral root development by carbon monoxide and involvement in auxin and nitric oxide. J Exp Bot 59:3443–3452

Gupta KJ, Stoimenova M, Kaiser WM (2005) In higher plants, only root mitochondria, but not leaf mitochondria reduce nitrite to NO, *in vitro* and *in situ*. J Exp Bot 56:2601–2609
Gupta KJ, Fernie AR, Kaiser WM, van Dongen JT (2011a) On the origins of nitric oxide. Trends Plant Sci 16:160–168
Gupta KJ, Igamberdiev AU, Manjunatha G et al (2011b) The emerging roles of nitric oxide (NO) in plant mitochondria. Plant Sci 181:520–526
Gupta KJ, Shah JK, Brotman Y et al (2012) Inhibition of aconitase by nitric oxide leads to induction of the alternative oxidase and to a shift of metabolism towards biosynthesis of amino acids. J Exp Bot 63:1773–1784
Hebelstrup KH, van Zanten M, Mandon J et al (2012) Haemoglobin modulates NO emission and hyponasty under hypoxia-related stress in *Arabidopsis thaliana*. J Exp Bot 63:5581–5591
Hill RD (2012) Non-symbiotic haemoglobins-what's happening beyond nitric oxide scavenging? AoB Plants 2012:pls004
Horchani F, Aschi-Smiti S, Brouquisse R (2010) Involvement of nitrate reduction in the tolerance of tomato (*Solanum lycopersicum* L.) to prolonged root hypoxia. Acta Physiol Plant 32:1113–1123
Horchani F, Prévot M, Boscari A et al (2011) Both plant and bacterial nitrate reductases contribute to nitric oxide production in *Medicago truncatula* nitrogen-fixing nodules. Plant Physiol 155:1023–1036
Huang X, von Rad U, Durner J (2002) Nitric oxide induces transcriptional activation of the nitric oxide-tolerant alternative oxidase in Arabidopsis suspension cells. Planta 215:914–923
Igamberdiev AU, Hill RD (2004) Nitrate, NO and haemoglobin in plant adaptation to hypoxia: an alternative to classic fermentation pathways. J Exp Bot 55:2473–2482
Igamberdiev AU, Hill RD (2009) Plant mitochondrial function during anaerobiosis. Ann Bot 103:259–268
Igamberdiev AU, Seregélyes C, Manac'h N, Hill RD (2004) NADH-dependent metabolism of nitric oxide in alfalfa root cultures expressing barley hemoglobin. Planta 219:95–102
Igamberdiev AU, Baron K, Manac'h-Little N et al (2005) The haemoglobin/nitric oxide cycle: involvement in flooding stress and effects on hormone signalling. Ann Bot 96:557–564
Igamberdiev AU, Bykova NV, Shah JK, Hill RD (2010) Anoxic nitric oxide cycling in plants: participating reactions and possible mechanisms. Physiol Plant 138:393–404
Jin CW, Du ST, Zhang YS et al (2009) Differential regulatory role of nitric oxide in mediating nitrate reductase activity in roots of tomato (*Solanum lycocarpum*). Ann Bot 104:9–17
Kharitonov VG, Sundquist AR, Sharma VS (1994) Kinetics of nitric oxide autoxidation in aqueous solutions. J Biol Chem 269:5881–5883
Kozlov AV, Staniek K, Nohl H (1999) Nitrite reductase activity is a novel function of mammalian mitochondria. FEBS Lett 454:127–130
Kreuzwieser J, Hauberg J, Howell KA et al (2009) Differential response of gray poplar leaves and roots underpins stress adaptation during hypoxia. Plant Physiol 149:461–473
Lamattina L, Garcia-Mata C, Graziano M, Pagnussat G (2003) Nitric oxide: the versatility of an extensive signal molecule. Annu Rev Plant Biol 54:109–136
Lasanthi-Kudahettige R, Magneschi L, Loreti E et al (2007) Transcript profiling of the anoxic rice coleoptile. Plant Physiol 144:218–231
Lea PJ (1993) Nitrogen metabolism. In: Lea PJ, Leegood RC (eds) plant biochemistry and molecular biology. Wiley, New York, pp 155–180
Leitner M, Vandelle E, Gaupels F et al (2009) NO signals in the haze: Nitric oxide signaling in plant defence. Curr Opin Plant Biol 12:451–458
Leterrier M, Chaki M, Airaki M et al (2011) Function of S-nitrosoglutathione reductase (GSNOR) in plant development and under biotic/abiotic stress. Plant Signal Behav 6:789–793
Logan DC, Millar AH, Sweetlove LJ et al (2001) Mitochondrial biogenesis during germination in maize embryos. Plant Physiol 125:662–672
Malavolta E (1954) Studies on the nitrogenous nutrition of rice. Plant Physiol 29:98–99
Manac'h-Little N, Igamberdiev AU, Hill RD (2005) Hemoglobin expression affects ethylene production in maize cell cultures. Plant Physiol Biochem 43:485–489

Martí MC, Florez-Sarasa I, Camejo D et al (2013) Response of mitochondrial antioxidant system and respiratory pathways to reactive nitrogen species in pea leaves. Physiol Plant 147:194–206

Maxwell DP, Wang Y, McIntosh L (1999) The alternative oxidase lowers mitochondrial reactive oxygen production in plant cells. Proc Natl Acad Sci USA 96:8271–8276

Meakin GE, Bueno E, Jepson B et al (2007) The contribution of bacteroidal nitrate and nitrite reduction to the formation of nitrosylleghaemoglobin complexes in soybean root nodules. Microbiology 153:411–449

Millar AH, Day DA (1996) Nitric oxide inhibits the cytochrome oxidase but not the alternative oxidase of plant mitochondria. FEBS Lett 398:155–158

Millar AH, Whelan J, Soole KL, Day DA (2011) Organization and regulation of mitochondrial respiration in plants. Annu Rev Plant Biol 62:79–104

Modolo LV, Augusto O, Almeida IM et al (2005) Nitrite as the major source of nitric oxide production by *Arabidopsis thaliana* in response to *Pseudomonas syringae*. FEBS Lett 579:3814–3820

Moreau M, Lindermayr C, Durner J, Klessig DF (2010) NO synthesis and signaling in plants-where do we stand? Physiol Plant 138:372–383

Mugnai S, Azzarello E, Baluska F, Mancuso S (2012) Local root apex hypoxia induces NO-mediated hypoxic acclimation of the entire root. Plant Cell Physiol 53:912–920

Murphy MP (1999) Nitric oxide and cell death. Biochim Biophys Acta 1411:401–414

Narsai R, Rocha M, Geigenberger P, Whelan J, van Dongen JT (2011) Comparative analysis between plant species of transcriptional and metabolic responses to hypoxia. New Phytol 190:472–487

Navarre DA, Wendehenne D, Durner J et al (2000) Nitric oxide modulates the activity of tobacco aconitase. Plant Physiol 122:573–582

Ober ES, Sharp RE (1996) A microsensor for direct measurement of O_2 partial pressure within plant tissues. J Exp Bot 47:447–457

Oliveira HC, Sodek L (2013) Effect of oxygen deficiency on nitrogen assimilation and amino acid metabolism by soybean root segments. Amino Acids 44:743–755

Oliveira HC, Salgado I, Sodek L (2013a) Involvement of nitrite in the nitrate-mediated modulation of fermentative metabolism and nitric oxide production of soybean roots during hypoxia. Planta 237:255–264

Oliveira HC, Salgado I, Sodek L (2013b) Nitrite decreases ethanol production by intact soybean roots submitted to oxygen deficiency: a role for mitochondrial nitric oxide synthesis? Plant Signal Behav 8:e23578

Oliveira HC, Freschi L, Sodek L (2013c) Nitrogen metabolism and translocation in soybean plants subjected to root oxygen deficiency. Plant Physiol Biochem 66:141–149

Pagnussat GC, Lanteri ML, Lamattina L (2003) Nitric oxide and cyclic GMP are messengers in the indole acetic acid-induced adventitious rooting process. Plant Physiol 132:1241–1248

Palmieri MC, Sell S, Huang X et al (2008) Nitric oxide-responsive genes and promoters in *Arabidopsis thaliana*: a bioinformatics approach. J Exp Bot 59:177–186

Parani MR, Myers R, Weirich H et al (2004) Microarray analysis of nitric oxide responsive transcripts in *Arabidopsis*. Plant Biotechnol J 2:359–366

Pearce LL, Kanai AJ, Birder LA et al (2002) The catabolic fate of nitric oxide: the nitric oxide oxidase and peroxynitrite reductase activities of cytochrome oxidase. J Biol Chem 277:13556–13562

Perazzolli M, Dominici P, Romero-Puertas MC et al (2004) Arabidopsis nonsymbiotic hemoglobin AHb1 modulates nitric oxide bioactivity. Plant Cell 16:2785–2794

Planchet E, Jagadis Gupta K et al (2005) Nitric oxide emission from tobacco leaves and cell suspensions: rate limiting factors and evidence for the involvement of mitochondrial electron transport. Plant J 41:732–743

Radi R, Cassina A, Hodara R (2002) Nitric oxide and peroxynitrite interactions with mitochondria. Biol Chem 383:401–409

Ramirez L, Simontacchi M, Murgia I et al (2011) Nitric oxide, nitrosyl iron complexes, ferritin and frataxin: a well equipped team to preserve plant iron homeostasis. Plant Sci 181:582–592
Ramírez-Aguilar SJ, Keuthe M, Rocha M et al (2011) The composition of plant mitochondrial supercomplexes changes with oxygen availability. J Biol Chem 286:43045–43053
Roberts JKM, Andrade FH, Anderson IC (1985) Further evidence that cytoplasmic acidosis is a determinant of flooding intolerance in plants. Plant Physiol 77:492–494
Rocha M, Licausi F, Araújo WL et al (2010) Glycolysis and the TCA-cycle are linked by alanine aminotransferase during hypoxia induced by waterlogging of *Lotus japonicus*. Plant Physiol 152:1501–1513
Rockel P, Strube F, Rockel A et al (2002) Regulation of nitric oxide (NO) production by plant nitrate reductase in vivo and in vitro. J Exp Bot 53:103–110
Romero-Puertas MC, Laxa M, Mattè A et al (2007) *S*-nitrosylation of peroxiredoxin II E promotes peroxynitrite-mediated tyrosine nitration. Plant Cell 19:4120–4130
Rümer S, Gupta KJ, Kaiser WM (2009) Plant cells oxidize hydroxylamines to NO. J Exp Bot 60:2065–2072
Sachs MM, Freeling M, Okimoto R (1980) The anaerobic proteins of maize. Cell 20:761–767
Sairam RK, Kumutha D, Ezhilmathi K et al (2008) Physiology and biochemistry of waterlogging tolerance in plants. Biol Plant 52:401–412
Sakamoto A, Ueda M, Morikawa H (2002) Arabidopsis glutathione-dependent formaldehyde dehydrogenase is an S-nitrosoglutathione reductase. FEBS Lett 515:20–24
Sánchez C, Cabrera JJ, Gates AJ et al (2011) Nitric oxide detoxification in the rhizobia-legume symbiosis. Biochem Soc Trans 39:184–188
Santos-Filho PR, Vitor SC, Frungillo L et al (2012) Nitrate reductase- and nitric oxide-dependent activation of sinapoylglucose:malate sinapoyltransferase in leaves of Arabidopsis thaliana. Plant Cell Physiol 53:1607–1616
Salgado I, Modolo LV, Augusto O et al (2006) Mitochondrial nitric oxide synthesis during plant-pathogen interactions: role of nitrate reductase in providing substrates. In: Lamattina L, Polacco JC (eds) Plant cell monographs, nitric oxide in plant growth, development and stress physiology, vol 6. Springer, Berlin, pp. 239–254
Saviani EE, Orsi CH, Oliveira JFP et al (2002) Participation of the mitochondrial permeability transition pore in nitric oxide-induced plant cell death. FEBS Lett 510:136–140
Shingaki-Wells RN, Huang S, Taylor NL et al (2011) Differential molecular responses of rice and wheat coleoptiles to anoxia reveal novel metabolic adaptations in amino acid metabolism for tissue tolerance. Plant Physiol 156:1706–1724
Siddiqui MH, Al-Whaibi MH, Basalah MO (2011) Role of nitric oxide in tolerance of plants to abiotic stress. Protoplasma 248:447–455
Simonin V, Galina A (2013) Nitric oxide inhibits succinate dehydrogenase-driven oxygen consumption in potato tuber mitochondria in an oxygen tension-independent manner. Biochem J 449:263–273
Skutnik M, Rychter AM (2009) Differential response of antioxidant systems in leaves and roots of barley subjected to anoxia and post-anoxia. J Plant Physiol 166:926–937
Sousa CAF, Sodek L (2002) The metabolic response of plants to oxygen deficiency. Braz J Plant Physiol 14:83–94
Stamler JS, Singel DJ, Loscalzo J (1992) Biochemistry of nitric oxide and its redox-activated forms. Science 258:1898–1902
Stöhr C, Stremlau S (2006) Formation and possible roles of nitric oxide in plant roots. J Exp Bot 57:463–470
Stoimenova M, Libourel IGL, Ratcliffe RG, Kaiser WM (2003) The role of nitrate reduction in the anoxic metabolism of roots II. Anoxic metabolism of tobacco roots with or without nitrate reductase activity. Plant Soil 253:155–167
Stoimenova M, Igamberdiev AU, Gupta KJ, Hill RD (2007) Nitrite-driven anaerobic ATP synthesis in barley and rice root mitochondria. Planta 226:465–474
Stuehr DJ, Santolini J, Wang ZQ et al (2004) Update on mechanism and catalytic regulation in the NO synthases. J Biol Chem 279:36167–36170

Sturms R, DiSpirito AA, Hargrove MS (2011) Plant and cyanobacterial hemoglobins reduce nitrite to nitric oxide under anoxic conditions. Biochemistry 50:3873–3878

Thomas AL, Sodek L (2005) Development of nodulated soybean plant after flooding of the root system with different sources of nitrogen. Braz J Plant Physiol 17:291–297

Tiso M, Tejero J, Kenney C et al (2012) Nitrite reductase activity of nonsymbiotic hemoglobins from *Arabidopsis thaliana*. Biochemistry 51:5285–5292

Trought MCT, Drew MC (1981) Alleviation of injury to young wheat plants in anaerobic solution cultures in relation to the supply of nitrate and other inorganic nutrients. J Exp Bot 32:509–522

Tun NN, Santa-Catarina C, Begum T et al (2006) Polyamines induce rapid biosynthesis of nitric oxide (NO) in *Arabidopsis thaliana* seedlings. Plant Cell Physiol 47:346–354

van Dongen JT, Schurr U, Pfister M, Geigenberger P (2003) Phloem metabolism and function have to cope with low internal oxygen. Plant Physiol 131:1529–1543

van Dongen JT, Fröhlich A, Ramírez-Aguilar SJ et al (2009) Transcript and metabolite profiling of the adaptive response to mild decreases in oxygen concentration in the roots of *Arabidopsis* plants. Ann Bot 103:269–280

van Faassen EE, Bahrami S, Feelisch M et al (2009) Nitrite as regulator of hypoxic signaling in mammalian physiology. Med Res Rev 29:683–741

Vanlerberghe GC, Cvetkovska M, Wang J (2009) Is the maintenance of homeostatic mitochondrial signaling during stress a physiological role for alternative oxidase? Physiol Plant 137:392–406

Wendehenne D, Hancock JT (2011) New frontiers in nitric oxide biology in plant. Plant Sci 181:507–508

Wulff A, Oliveira HC, Saviani EE, Salgado I (2009) Nitrite reduction and superoxide-dependent nitric oxide degradation by *Arabidopsis* mitochondria: influence of external NAD(P)H dehydrogenases and alternative oxidase in the control of nitric oxide levels. Nitric Oxide 21:132–139

Yadav S, David A, Bhatla SC (2010) Nitric oxide modulates specific steps of auxin-induced adventitious rooting in sunflower. Plant Signal Behav 5:1163–1166

Yamasaki H, Sakihama Y (2000) Simultaneous production of nitric oxide and peroxynitrite by plant nitrate reductase: in vitro evidence for the NR-dependent formation of active nitrogen species. FEBS Lett 468:89–92

Yamasaki H, Shimoji H, Ohshiro Y, Sakihama Y (2001) Inhibitory effects of nitric oxide on oxidative phosphorylation in plant mitochondria. Nitric Oxide: Biol Chem 5:261–270

Zabalza A, van Dongen JT, Froehlich A et al (2009) Regulation of respiration and fermentation to control the plant internal oxygen concentration. Plant Physiol 149:1087–1098

Zottini M, Formentin E, Scattolin M et al (2002) Nitric oxide affects plant mitochondrial functionality in vivo. FEBS Lett 515:75–78

Chapter 5
Production of Nitric Oxide by Marine Unicellular Red Tide Phytoplankton, *Chattonella marina*

Daekyung Kim and Tatsuya Oda

Abstract In this chapter, we show several lines of evidence for the production of nitric oxide (NO) by unicellular red tide phytoplankton *Chattonella marina* under the normal growth conditions. Chemiluminescence (CL) assay suggested that *C. marina* produced NO in a cell-number-dependent manner, and the level of NO decreased by the addition of carboxy-PTIO, a specific NO scavenger. NO generation by *C. marina* was also confirmed by a spectrophotometric assay based on the measurement of the diazo-reaction positive substances (NO_x) and by fluorometric assay using highly specific fluorescent indicator of NO. Furthermore, the NO level in *C. marina* was significantly decreased by L-NAME, a specific NO synthase (NOS) inhibitor. The addition of L-arginine increased the NO level, whereas $NaNO_2$ had no effect. These results suggest that a NOS-like enzyme is mainly responsible for NO generation in *C. marina*.

Keywords *Chattonella marina* · Chemiluminescence · Fluorescence · Nitric oxide · Nitrite · Peroxynitrite

5.1 Introduction

Chattonella marina (*C. marina*), a raphidophycean flagellate, is one of the noxious red tide phytoplankton species, which causes massive mortality of cultured and natural fish, especially the yellow tail *Seriola quinqueradiata* in Japan

D. Kim
Jeju Center, Korea Basic Science Institute (KBSI), Smart Building 1F, 2170-2 Jeju Science Park, Ara-dong, Jeju-Si, Jeju Special Self-Governing Province 690-756, Korea

T. Oda (✉)
Faculty of Fisheries, Division of Biochemistry, Nagasaki University, Bunkyo-machi 1-14, Nagasaki 852-8521, Japan
e-mail: t-oda@nagasaki-u.ac.jp

M. N. Khan et al. (eds.), *Nitric Oxide in Plants: Metabolism and Role in Stress Physiology*, DOI: 10.1007/978-3-319-06710-0_5,

(Okaichi 1989). Previous studies have demonstrated that *Chattonella* spp. generate reactive oxygen species (ROS) such as O_2^- and H_2O_2 (Kawano et al. 1996; Oda et al. 1997). Furthermore, our studies using electron spin resonance (ESR) spectroscopy with the spin traps 5,5-dimethyl-1-pyrroline-*N*-oxide and *N*-t-butyl-a-phenylnitrone showed that *C. marina* generates hydroxyl radical (OH) (Oda et al. 1992a). Since ROS are generally considered to be toxic against living organisms (Oda et al. 1989), the ROS generated by *Chattonella* spp. may at least be partly involved in the fish-kill mechanism. Furthermore, we have found that *C. marina* exhibited ROS-mediated toxic effect on a marine bacterium, *Vibrio alginolyticus* (Oda et al. 1992b). In addition to *Chattonella* spp., it has been reported that another raphidophycean flagellate, *Heterosigma akashiwo* also showed ROS-mediated toxicity on rainbow trout (Yang et al. 1995). Thus, it seems that the production of ROS is a common feature of raphidophycean flagellates (Kim et al. 1999).

During our search for toxic factors apart from ROS, we presumed that *C. marina* produces nitric oxide (NO) under normal growth conditions. Nitric oxide (NO), a gaseous free radical, plays many significant signaling roles not only in animals but also in plants. This membrane permeable compound is involved in the regulation of plant metabolism, gene expression (Lamattina et al. 2003), and plant-pathogen interaction (Zhang et al. 2003; Yamamoto et al. 2004). In addition to various higher plants, NO producing activities have been observed in green algae and cyanobacteria (Mallick et al. 1999, 2000; Sakihama et al. 2002). Evidence for the existence of an endogenous pathway for NO synthesis in the plant kingdom has been reported (Ninnemann and Maier 1996; Cueto et al. 1996; Durner et al. 1998). Since *C. marina* is a phototrophic, these findings prompted us to measure the generation of NO in *C. marina*. In this study, we employed the chemiluminescence reaction of NO with luminol-H_2O_2 that has been reported as a sensitive and specific NO-detection method (Kikuchi et al. 1993). In this assay system, *C. marina* induced potent chemiluminescence emission that was abolished by carboxy-PTIO, a NO-specific scavenger (Pfeiffer et al. 1997). The results obtained by other NO detection systems such as a spectrophotometric assay which measures the product of a nitrite azo-coupling reaction, and fluorometric assay using NO-specific fluorescent probe also supported the generation of NO by *C. marina*.

5.2 Synthesis of NO in *C. marina* Cell Suspension

We measured NO in *C. marina* cell suspension by adopting three different methods:

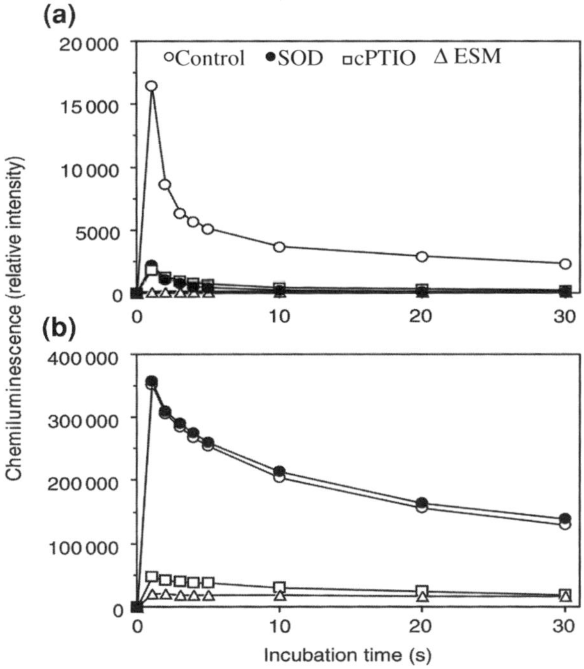

Fig. 5.1 Effects of superoxide dismutase (SOD) and carboxy-PTIO on luminol-dependent CL responses in *C. marina* in the absence (**a**) or presence (**b**) of 2 mM of H_2O_2. **a** After the addition of luminol (final 50 μM) alone (○) or with SOD (final 100 units/mL) (•) or with cPTIO (final 2 mM) (□) simultaneously to *C. marina* cell suspension (1.5–2.0 $\times 10^4$ cells/mL), or to ESM medium without *C. marina* cells (Δ), the CL responses were measured at 27 °C immediately. **b** After the addition of luminol/H_2O_2 (final 50 μM /2 mM) alone (○) or with SOD (•) or with carboxy-PTIO (□) simultaneously to *C. marina* cell suspension (1.5–2.0 $\times 10^4$ cells/mL), or to ESM medium without *C. marina* cells (Δ), the CL responses were measured at 27 °C immediately

5.2.1 Chemiluminescence (CL) Reaction

This method is based on the formation of luminol and peroxynitrite in the reaction of luminol with exogenously added H_2O_2 as described previously (Kikuchi et al. 1993). After addition of H_2O_2 (2 mM) and luminol (50 μM) to *C. marina* cell suspension in Erd Schreiber modified (ESM) medium, the CL emission was recorded immediately for the designated periods of time with 1,254 lumino meter. The reaction mixtures (1 mL) typically consisted of 0.8 mL flagellate cell suspension (0.3–3.0 × 10^4 cells/mL), 0.1 mL H_2O_2 solution, and 0.05 mL luminol solution, and 0.05 mL ESM medium or other reagent solution. The mixtures with ESM medium but without flagellate cells were used as control. All CL analyses were carried out at 27 °C. Since the maximal CL response of *C. marina* was attained within the first 10 s, quantitative analysis for the activity of *C. marina* to

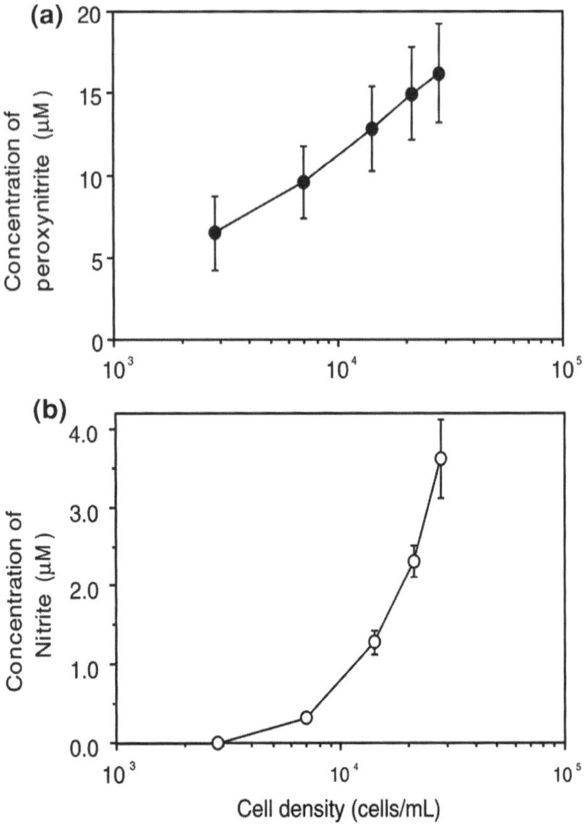

Fig. 5.2 Cell density-dependent NO (**a**) and nitrite (**b**) production in *C. marina* cell suspension. **a** After the addition of luminol/H_2O_2 to varying concentrations of *C. marina* cell suspension, CL intensity of integrated emission during the initial 10 s in each cell suspension was measured. Based on the calibration curve for peroxynitrite, the level of NO equivalent to peroxynitrite in each cell suspension was estimated. **b** Nitrite level in each *C. marina* cell suspension was measured spectorophotometrically using the Griess reaction as described in the text. Each value represents an average of triplicate measurements and each bar indicates the standard deviation

induce CL was expressed in terms of relative intensity of integrated emission during the first 10 s. A standard calibration curve for peroxynitrite was made from relationship between the CL intensity of integrated emission during the first 10 s and concentration of peroxynitrite.

As shown in Fig. 5.1a, luminol-mediated CL emission was observed in *C. marina* cell suspension immediately after addition of luminol. The CL was inhibited by superoxide dismutase (SOD) (100 units/mL), suggesting the involvement of O_2^- in the CL response. Interestingly, cPTIO (final 2 mM), a specific NO scavenger, also suppressed the CL. It has been reported that O_2^- reacts with NO to form peroxynitrite ($ONOO^-$), a strong oxidizing agent, which can induce luminol-mediated CL (Kikuchi et al. 1993). $ONOO^-$ is also reported to be generated from the reaction of NO and H_2O_2, and based on this reaction, sensitive and specific NO detection assay method has been developed (Kikuchi et al. 1993). When 2 mM of H_2O_2 was added with luminol to *C. marina*, a dramatic increase in CL was observed (Fig. 5.1b). The luminol-H_2O_2 mediated CL was not affected by SOD, but suppressed by cPTIO. These results suggest that for the most part, CL in the luminol-H_2O_2 system is due to NO produced by

C. marina, and the level of NO is considerably higher than that of O_2^-. As shown in Fig. 5.2, the NO production by *C. marina* was cell density-dependent, and based on the calibration using $ONOO^-$ solution as standard, the level of NO for 10^4 cells /mL was estimated to be nearly 10 μM.

5.2.2 Nitrite Determination

To further confirm the NO formation by *C. marina*, we measured the accumulation of nitrite, a stable reaction product derived from NO, in cell suspension by the Griess reaction (Green et al. 1982). Griess reagent (100 μL: 1 % sulfanilamide and 0.1 % naphthyl-ethylene diamine dihydrochloride in 5 % phosphoric acid) was added to 100 μL of *C. marina* cell suspension. After incubation at room temperature for 10 min, the optical density was measured at 540 nm using a micro-plate reader (TOSOH, MPR-A4i, Tosoh Corp., Tokyo, Japan).

As shown in Fig. 5.2b, nitrite was detected in *C. marina* cell suspension in a cell density-dependent manner, but the nitrite level was lower than that of NO estimated by luminol-H_2O_2 assay (Fig. 5.2a). This may be due to the conversion of nitrite to nitrate by further oxidation. Since ESM medium originally contains high concentration of nitrate as a nutrient, we could not measure the entire NO_x level (NO_2/NO_3) by the addition of nitrate reductase that convert NO_3 to NO_2 as is the usual procedure.

5.2.3 Fluorescent Probe Detection

As another independent measurement technique for NO formation, we also used the NO-reactive fluorescent probe diaminofluorescein-FM diacetate (DAF-FM DA) (Kojima et al. 1999) from *C. marina*. DAF-FM DA is membrane-permeable ester derivative of DAF-FM, which has been used to load living cells, in which DAF-FM DA is hydrolyzed by intracellular esterases to release DAF-FM, which is converted to the fluorescent triazole derivative DAF-FM T after reaction with NO.

After addition of DAF-FM DA into *C. marina* cell suspension, gradual increase in the fluorescence intensity was observed, however, it was completely inhibited in the presence of cPTIO. Fluorescence microscopic observation also suggested that NO was produced inside *C. marina* cells, and the emission of the bright fluorescence from *C. marina* was prevented by cPTIO (Fig. 5.3a, b).

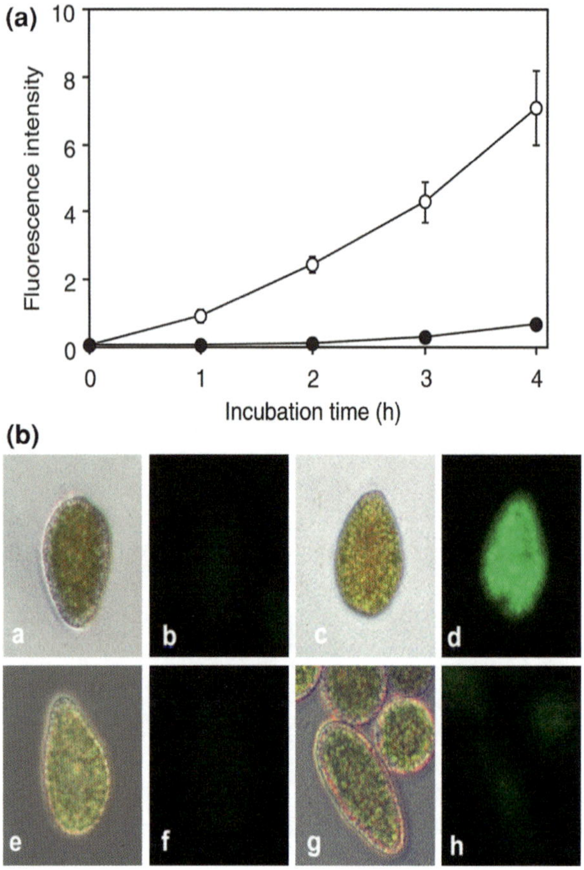

Fig. 5.3 Kinetics of NO production estimated by the fluorescence indicator DAF-FM DA. **a** *C. marina* cell suspension was incubated with (•) or without (○) 2 mM of cPTIO for 5 min at 27 °C. DAF-FM DA (final 10 μM) was then added to each cell suspension and the fluorescence intensities were measured at the indicated periods of time. Each value represents an average of triplicate measurements and each bar indicates the standard deviation. **b** Microscopic observation of NO production in *C. marina*. Phase-contrast micrographs (*a*, *c*, *e*, *g*) and fluorescence micrographs (*b*, *d*, *f*, *h*) of *C. marina*. *C. marina* cells were incubated without (*a*, *b*, *c*, *d*) or with (*e*, *f*, *g*, *h*) 2 mM of carboxy-PTIO for 5 min at 27 °C. Each cell suspension was observed at 0 (*a*, *b*, *e*, *f*) and 3 h (*c*, *d*, *g*, *h*) after addition of DAF- FM DA

5.3 Involvement of NO Synthase (NOS) and Nitrate Reductase (NR) in NO Production by *C. marina*

To gain insight into the source of NO in *C. marina*, we examined the effect of N^G-Nitro-L-arginine methyl ester (L-NAME). This is an inhibitor of NOS that blocked NO production in the mouse macrophage cell line RAW264.7 cells (Sanzen et al. 2001). As shown in Fig. 5.4a, L-NAME showed inhibitory effect on

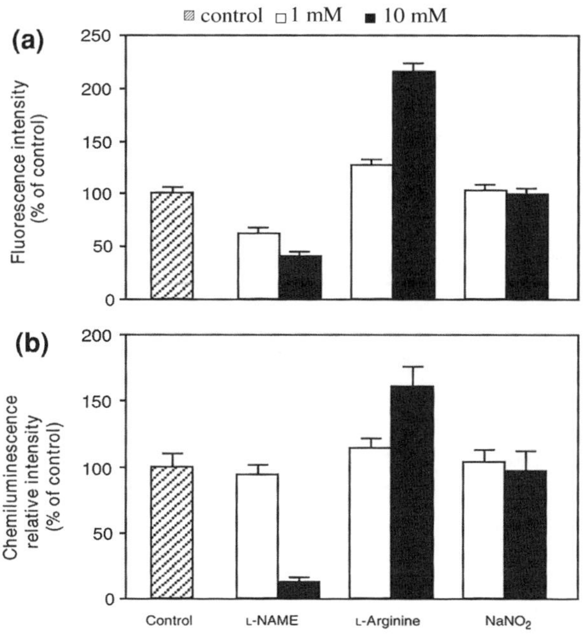

Fig. 5.4 Effects of various agents on NO production in *C. marina* as measured by fluorometric (**a**) and CL assay (**b**). **a** Cells (1.5–2.0 $\times 10^4$ cells/mL) were treated with NOS inhibitor (L-NAME), L-arginine, or $NaNO_2$ at 1 mM (□) or 10 mM (■) for 5 min at 27 °C. After the addition of DAF-FM DA (final 10 μM) to each treated cells, the fluorescence intensities were measured immediately. (▨), untreated control. **b** Cells (1.5–2.0 $\times 10^4$ cells/mL) were treated with NOS inhibitor (L-NAME), L-arginine, or $NaNO_2$ at 1 mM (□) or 10 mM (■) for 5 min at 27 °C. After the addition of luminol/H_2O_2 (final 50 μM /2 mM) to each treated cells, the CL responses were measured immediately. Each value represents an average of triplicate measurements and each bar indicates the standard deviation. (▨), untreated control

NO production by *C. marina* as measured by fluorometric assay using DAF-FM DA. Furthermore, the addition of L-arginine, a substrate for NO synthase (NOS), resulted in the increased NO production. In the luminol-H_2O_2 assay, basically similar results were obtained (Fig. 5.4b). These observations suggest that a NOS-like enzyme is mainly responsible for NO generation in *C. marina*. It has been reported that unicellular green alga *Chlorella sorokiniana* cells produced NO only under anaerobic conditions, which was strictly dependent on nitrite supply (Tischner et al. 2004). Nitrite-dependent NO production has been observed in some higher plants (Meyer et al. 2005), and in some cases, NO is likely to be produced by nitrate reductase (NR) (Yamasaki et al. 1999), that reduces nitrate to nitrite and can further reduce nitrite to NO. However, nitrite had no effect on NO production in *C. marina* (Fig. 5.4).

Our results demonstrated that unicellular marine phytoplankton, *C. marina* produced a considerable amount of NO under normal growth conditions. In mammalian cells, NO is mainly produced from L-arginine by the enzyme NO synthase which yields L-citruline and NO (Stuehr 1999). In plants, apart from NO synthesis from L-arginine through NOS-like enzyme, NO production from nitrite through nitrate reductase has been demonstrated as a major NO generation system (Yamasaki et al. 1999). Moreover, NO can also be synthesized by nonenzymatic spontaneous nitrite reduction under acidic conditions (Bethke et al. 2004). Since the NO production in *C. marina* was inhibited by NOS-inhibitor L-NAME, it was suggested that main source of NO production in this alga is NOS-like enzyme. In agreement with this notion, when the NOS substrate L-arginine (10 mM) was added to the *C. marina* cell suspension, the production of NO increased significantly. In contrast to *C. marina*, previous reports have indicated that in the unicellular alga *Chlamydomonas reinhardtii*, NO production was exclusively due to NR activity (Sakihama et al. 2002), and a NOS-like activity was not involved. Similar to *C. reinhardtii*, it has been shown that NO production in *Chlorella sorokiniana* in dark was strictly dependent on nitrite supply (Tischner et al. 2004). Furthermore, it has been suggested that NR-mediated NO production activities of these micro algae were linked with photosynthetic-electron transport system since illumination of the algae cells deppresses NO production, and the suppressive effect was reversed by 3,4-dichlorophenyl-1,1-dimethylurea, a photosynthesis inhibitor (Sakihama et al. 2002).

5.4 Conclusion

Our findings suggest that these unicellular algae may have a common NO production mechanism that can be affected by various growth conditions. In the case of *C. marina*, however, continuous NO production was observed under normal growth conditions even under illumination, and exogenously added nitrite had no effect, suggesting NR may not be involved in NO generation. Although the biological significance of NO production in *C. marina* is still debatable, it may differ from such inducible enzyme system that requires certain stimuli for the expression of enzymatic activity. The specific stimuli or culture conditions that influence the level of NO in *C. marina* have not been discovered so far. At present, we cannot completely rule out the possibility of NO production for defense purpose in *C. marina*, but it seems most probable that *C. marina* produces NO during the normal course of nitrogen assimilation.

References

Bethke PC, Badger MR, Jones RL (2004) Apoplastic synthesis of nitric oxide by plant tissues. Plant Cell 16:332–341

Cueto M, Hernandez-Perera O, Martin R et al (1996) Presence of nitric oxide synthase activity in roots and nodules of *Lupinusalbus*. FEBS Lett 398:159–164

Durner J, Wendehenne D, Klessig DF (1998) Defense gene induction in tobacco by nitric oxide, cyclic GMP and cyclic ADP-ribose. Proc Natl Acad Sci USA 95:10328–10333
Green LC, Wagner DA, Glogowski J et al (1982) Analysis of nitrate, nitrite, and [^{15}N] nitrate in biological fluids. Anal Biochem 126:131–136
Kawano I, Oda T, Ishimatsu A, Muramatsu T (1996) Inhibitory effect of the iron chelator Desferrioxamine (Desferal) on the generation of activated oxygen species by *Chattonella marina*. Mar Biol 126:765–771
Kikuchi K, Nagano T, Hayakawa H et al (1993) Detection of nitric oxide production from perfused organ by luminol-H_2O_2 system. Anal Chem 65:1794–1799
Kim D, Nakamura A, Okamoto T et al (1999) Toxic potential of the raphidophyte *Olisthodiscus luteus*: mediation by reactive oxygen species. J Plankton Res 21:1017–1027
Kojima H, Urano Y, Kikuchi K et al (1999) Fluorescent indicators for imaging nitric oxide production. Angew Chem Int Ed Engl 38:3209–3212
Lamattina L, García-Mata C, Graziano M, Pagnussat G (2003) Nitric oxide: the versatility of an extensive signal molecule. Annu Rev Plant Biol 54:109–136
Mallick N, Rai LC, Mohn FH, Soeder CJ (1999) Studies on nitric oxide (NO) formation by the green alga *Scenedesmus obliquus* and the diazotrophic cyanobacterium *Anabaena doliolum*. Chemosphere 39:1601–1610
Mallick N, Mohn FH, Rai LC, Soeder CJ (2000) Evidence for the non-involvement of nitric oxide synthase in nitric oxide production by the green alga *Scenedesmus obliquus*. J Plant Physiol 156:423–426
Meyer C, Lea US, Provan F et al (2005) Is nitrate reductase a major player in the plant NO (nitric oxide) game? Photosyn Res 83:181–189
Ninnemann H, Maier J (1996) Indications for the occurrence of nitric oxide synthases in fungi and plants and the involvement in photoconidiation of *Neurospora crassa*. Photochem Photobiol 64:393–398
Oda T, Akaike T, Hamamoto T et al (1989) Oxygen radicals in influenza-induced pathogenesis and treatment with pyran polymer conjugated SOD. Science 244:974–976
Oda T, Akaike T, Sato K et al (1992a) Hydroxyl radical generation by red tide algae. Arch Biochem Biophys 294:38–43
Oda T, Ishimatsu A, Shimada M et al (1992b) Oxygen-radical-mediated toxic effects of the red tide flagellate *Chattonella marina* on *Vibrio alginolyticus*. Mar Biol 112:505–509
Oda T, Nakamura A, Shikayama M et al (1997) Generation of reactive oxygen species by raphidophycean phytoplankton. Biosci Biotechnol Biochem 61:1658–1662
Okaichi T (1989) Red tide problems in the Seto Inland Sea, Japan. In: Okaichi T, Anderson DM, Nemoto T (eds) Red tides: biology, environmental science, and toxicology. Elsevier, New York, pp 137–142
Pfeiffer S, Leopold E, Hemmens B et al (1997) Interference of carboxy-PTIO with nitric oxide- and peroxynitrite-mediated reactions. Free Radic Biol Med 22:787–794
Sakihama Y, Nakamura S, Yamasaki H (2002) Nitric oxide production mediated by nitrate reductase in the green alga *Chlamydomonas reinhardtii*: an alternative NO production pathway in photosynthetic organisms. Plant Cell Physiol 43:290–297
Sanzen I, Imanishi N, Takamatsu N et al (2001) Nitric oxide-mediated antitumor activity induced by the extract from *Grifola frondosa* (Maitake mushroom) in a macrophage cell line, RAW264.7. J Exp Clin Caner Res 20:591–597
Stuehr DJ (1999) Mammalian nitric oxide synthases. Biochim Biophys Acta 1411:217–230
Tischner R, Planchet E, Kaiser WM (2004) Mitochondrial electron transport as a source for nitric oxide in the unicellular green alga *Chlorella sorokiniana*. FEBS Lett 576:151–155
Yamamoto A, Katou S, Yoshioka H et al (2004) Involvement of nitric oxide generation in hypersensitive cell death induce by elicitin in tobacco cell suspension culture. J Gen Plant Pathol 70:85–92
Yamasaki H, Sakihama Y, Takahashi S (1999) An alternative pathway for nitric oxide production in plants: new features of an old enzyme. Trends Plant Sci 4:128–129

Yang CZ, Albright LJ, Yousif AN (1995) Oxygen-radical-mediated effects of the toxic phytoplankter *Heterosigma carterae* on juvenile rainbow trout *Oncorhynchus mykiss*. Dis Aquat Org 23:101–108

Zhang C, Czymmek KJ, Shapiro AD (2003) Nitric oxide does not trigger early programmed cell death events but may contribute to cell-to-cell signaling governing progression of the *Arabidopsis* hypersensitive response. Mol Plant Microbe Interact 16:962–972

Chapter 6
Identification of Nitrosylated Proteins (SNO) and Applications in Plants

Jean-Benoît Peltier, Abasse Fares and Michel Rossignol

Abstract Over the last decade, due to its broad biological effects, nitric oxide (NO) has triggered a huge interest in the plant science too. Nitrosylation of cystein thiol residues (SNO) in proteins has been shown to be the main target of endogenously produced NO or in a biological sample exposed to this gas. This chapter summarizes the hitherto 18 different methods developed to identify and quantify nitrosylated proteins. These methods derive mostly from the original "Biotin-Switch" technique (BS) published in 2001 but new approaches try to circumvent BS weaknesses. Surprisingly, out of this bloomy panel only a couple of methods have been used in plants. By collecting all the plant published data up to now, we "blasted" them against the proteome of the plant model *Arabidopsis* and identified 373 nonredundant nitrosylated proteins. We then provide the first overview of plant nitrosylated proteome showing a wide range of functions and cellular compartments involved in NO signaling/targeting. This plant nitrosylated proteome resource expands our current understanding on NO-targeted proteins and facilitates comparisons with new nitrosylated protein data.

Keywords *Arabidopsis* · Biotin-Switch · Nitric oxide · Plant nitrosylome · *S*-nitrosylation

---

J.-B. Peltier (✉)
Campus SupAgro-INRA, Bât 7 Equipe canaux, 34060 Montpellier Cedex 1, France
e-mail: peltier@supagro.inra.fr

A. Fares
CEA Saclay DSV/iBiTec-S/SIMOPRO/LCV, Bât 152, 91191 Gif-sur-Yvette, France

M. Rossignol
Campus SupAgro-INRA, Bât 13, 34060 Montpellier Cedex 1, France

M. N. Khan et al. (eds.), *Nitric Oxide in Plants: Metabolism and Role in Stress Physiology*, DOI: 10.1007/978-3-319-06710-0_6,

6.1 Introduction

Nitric oxide (NO) was considered for a long time as a toxic by-product of oxidative metabolism and its direct link with the vasodilatory effect of nitroglycerin used in medication since the mid-nineteenth century was only deciphered in the 70s and 80s (Arnold et al. 1977; Furchgott and Zawadzki 1980; Ignarro et al. 1987; Palmer et al. 1987). NO is a lipophilic molecule that diffuses through membranes, and it is also one of very few gases involved in signaling in both prokaryotes and eukaryotes. Indeed, NO functions as a signal molecule in plants and its implication, firstly discovered in 1998 in plant defense (Delledonne et al. 1998; Durner et al. 1998) continues to manifest in many parts of plant growth and development (Astier and Lindermayr 2012; Mur et al. 2013). NO has a very short biological half life (5–15 s) and its lability occurs through enzymatic (GSNO reductase, thioredoxin) and nonenzymatic (ascorbate, trans-nitrosylation, UV) mechanisms (Hess et al. 2001). It modifies cysteine thiols (*S*-nitrosylation), the hydroxyphenyl group of tyrosines (Tyrosine nitration) and transition metal centers (metal nitrosylation) of a wide spectrum of functional proteins. *S*-nitrosylation is now established as a key post-translational modification (PTM) and identification of *S*-nitrosothiols (SNO) has become a priority despite their lability and redox sensitivity. A breakthrough in 2001 (Jaffrey and Snyder 2001) enables detection/identification of *S*-nitrosylated proteins with the so-called "Biotin-Switch" method (BS). Further developed methods allowed not only to identify *S*-nitrosylated proteins but also the nitrosylated sites and the quantitation of these sites. In this review, we provide the so far most complete overview of the multiplicity of *S*-nitrosothiols identification methods and a few ones experienced in plants. Furthermore, we summarize all the *S*-nitrosylated proteins found up to now in plants and discussed the need of improved methods to detect endogenous *S*-nitrosothiols.

6.2 Biotin-Switch and Relatives

The Biotin-Switch was introduced in 2001 by (Jaffrey and Snyder 2001), and still constitutes the most popular method to detect and isolate SNO. This method implies three chemical steps which select and convert unstable nitrosothiols to stable biotin conjugates. In a first step, free thiols on proteins are blocked with molecules like Methyl MethaneThioSulfonate (MMTS) or alkylating agent like *N*-EthylMaleimide (NEM), acrylamide, or iodoacetamide. At this stage, proteins are usually denaturated to provide a better accessibility to the blocking/alkylating agent. In a second step, SNO groups are reduced by ascorbate forming new free thiols. Eventually, in a third step, new free thiols react with Biotin-HPDP (*N*-[6-(biotinamido) hexyl]-30-(20-pyridyldithio)-propionamide) a reversible thiol reacting agent linking biotin to thiol groups. Biotin is then used to select biotin-labeled proteins on an avidin column. After elution of the biotin-tagged proteins, tagged proteins are usually

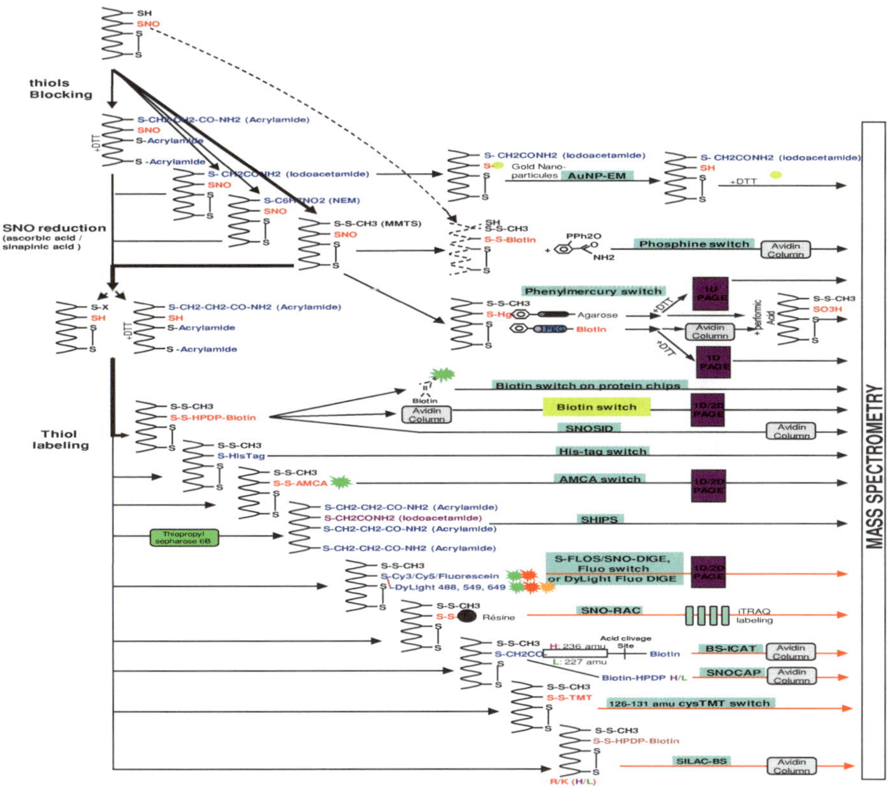

Fig. 6.1 Methods used to detect nitrosothiols. Generally, cysteins are under three different states (out of 11 possible): free thiols, disulphide bridges, or nitrosothiol. Most methods to detect nitrosothiols are indirect because they require a chemical substitution of the nitrosothiol to a more stable bond. Three methods allow a direct reaction with the nitrosothiol (Au–NP–EM, phosphine switch, and phenylmercury switch). Phosphine switch is the only method allowing in theory to avoid previously block free thiols because phosphine reacts specifically with nitrosothiols. All the other methods can be considered as BS derivatives with three steps: free thiol blocking, reduction of SNO, and labeling of the new free thiols. Three methods allow nitrosothiol quantitations for pair comparaisons (BS-ICAT/SNOCAP) or multiple comparaisons (SNORAC, CysTMT). In SNORAC free thiols providing from SNO reduction of each sample are selected on a column meaning that each sample passes on its own column before protein digestion and labeling. Cys-TMT avoids the possible bias induced by the columns. BS-ICAT uses labeling markers (H/L) with a mass difference of 9 amu between the two labeling forms accepting a mass precision less critical than for iTRAQ et cys-TMT labeling where the mass differences are only 1 amu between the different markers

trypsin digested and resulting biotin-conjugated peptides are analyzed via tandem mass spectrometry or detected on Western blots. This BS method aims at enriching SNO in a complex biological sample through biotin derivatization. Many versions of the original method have been developed increasing sometimes its potential as the use of different isotope form of the biotin-alkylating agent allowing quantitation of SNO (Fig. 6.1).

6.2.1 SNOSID (S–NO Site Identification)

Hao et al. (2006) introduced a slight but interesting modification in the original BS method. In SNOSID, proteins are digested before the enrichment on the avidin column allowing to select peptides and *de facto* identify nitrosylation sites. A direct and crucial consequence of the selection of biotin conjugated peptides instead of biotin conjugated proteins is the important decrease of false positives. Indeed, selected peptides that do not contain a modified cystein are discarded in SNOSID (Fig. 6.1).

6.2.2 His-Tag Switch

Camerini et al. (2007) used a modified version on the alkylating agent iodoacetamide (I–CH_2–CHO–NH_2 vs. I–CH_2–CO–Gly–Arg–Ala–$(His)_6$) instead of Biotin-HPDP to label nitrosothiols after a first alkylation of the free thiols by NEM. Labeled proteins are then selected on a nickel column and, after elution, are separated on 1D SDS-PAGE. Proteins are stained with Coomassie blue or transfered on a blot and revealed with an anti-His antibody (Western). Proteins are in gel digested and generated peptides are analyzed by mass spectrometry (MALDI-TOF and/or Q-TOF). The tag is not removed but reduced by the trypsic digestion (the final mass shift of the modified cysteine is 271.12 amu) and do not hamper the peptide fragmentation/flight process during MS analysis (Fig. 6.1).

6.2.3 DyLight Fluor DIGE, S-FLOS/SNO-DIGE, AMCA Switch and "Fluorescent Switch"

A quantitative dimension was introduced by Sun et al. (2007). Santhanam et al. (2008) using newly developed fluorescent probes (DyLight maleimide sulfhydryl-reactive fluors from Pierce) or diverting fluorescent probe (CyDye-maleimide) commonly used in 2D-DIGE (Differential In Gel-Electrophoresis), respectively. Fluorescent probes were introduced to compensate the lack of sensitivity observed with biotin-HPDP labeling. Excepting the labeling switch (Fluo vs. biotin), the primary steps in the protocol do not change much. After free thiol blocking by MMTS and reduction of SNO via ascorbate, new free thiol generated are differentially labeled with DyLight 488, 549 and 649 (Sun et al. 2007) or Cy3 or Cy5-

maleimide (Santhanam et al. 2008). Labeled proteins are then quantified on 1D- or 2D-PAGE and no secondary methods of detection via avidin binding or western blot analysis are needed. Initially, the method developed by Santhanam et al. (2008) was named Selective Fluorescent Labeling Of *S*-nitrosothiols (*S*-FLOS), but later on it was renamed *S*-NitrOsothiol Difference In Gel Electrophoresis (SNO-DIGE) by Chouchani et al. 2010. Additionally, *S*-FLOS was shown to be compatible with in situ tissue staining.

Han et al. (2008) used ACMA (7-amino-4-methylcoumarin-3-acetic acid) another type of fluorescent probe to label free thiols. In this ACMA switch, the use of a single probe did not allow SNO quantitation as described with DyLight Fluor DIGE or *S*-FLOS/SNO-DIGE. Finally, in the same seam Tello et al. (2009) used Fluorescein-5-maleimide after reduction of SNO with ascorbate (100 mM) to label nitrosylated proteins (Fig. 6.1).

6.2.4 BS-ICAT and SNOCAP

Chen et al. (2008) replaced biotin-HPDP with ICAT (Isotope-coded affinity tag) reagents to study nitrosylation of a particular protein (tyrosine phosphatase 1B). ICAT reagents have been used for a decade in quantitative proteomics and they are present in two isotopic forms (light/heavy). Wu et al. (2011) used BS-ICAT as a more general method to investigate protein nitrosylation. ICAT reagents react with free thiol and are coupled to biotin. They are actually very similar to biotin-HPDP except that they contain a linker composed of nine isotopic carbons between the HPDP and biotin groups. After labeling and selection on avidin column, peptides are eluted in acidic conditions and the biotin tag is cleaved off after an incubation in a strong acid. The remnant tag differentiates the heavy form from the light one by 9 amu. This method allows to let go of gel step and *ipso facto* reduces bias linked to this step as seen with *S*-FLOS or SNO-DIGE. BS-ICAT permits to combine samples before digestion, reducing part of the technical variability. BS-ICAT is limited to pair comparisons. In the case of SNOCAP (Paige et al. 2008), two isotopes of the biotin-HPDP (L/H) were synthetized and used like ICAT reagents. The two isotopes can be distinguished from each other by 4 amu. After a GSNO/GSH treatment, more than 100 proteins were identified in the GSNO treated sample and none in the GSH allowing authors to conclude that the first step in BS consisting of free thiol blocking is complete (Fig. 6.1).

6.2.5 SNO-RAC

The use of resins, able to interact with free thiols, has been introduced many years ago to facilitate the study of redox proteins. After blocking free thiols with MMTS and reducing SNO with ascorbate, Forrester et al. (2009) used this type of resins

(resins linked to 2- or 4-pyridyl disulfide) to fix SNO (RAC pour Resin Assisted Capture). The switch between biotin-HPDP to resin allows to potentially digest the fixed proteins on the column directly. The isobaric labeling (iTRAQ) of peptides afterwards bring a quantitative dimension of this approach and open the possibility for kinetic studies but do not allow to select the isobaric-labeled peptides on the same column. SNO-RAC seems to be more sensitive than classical BS for high molecular weight proteins probably because of the lower number of precipitation steps (Fig. 6.1).

6.2.6 BS on Protein Microarrays

In order to increase the depth of the analysis, Foster et al. (2009) used BS on protein microarrays after GSNO treatment. On these microarrays yeast proteins (4000 ORF), coupled to GST, were fixed on the plate surface. This strategy helps to solve the problem of the natural dynamic range in protein expression in any proteome. After a classical labeling with biotin-HPDP, SNO are detected with a primary anti-biotin antibody coupled with a fluorescent secondary antibody. Moreover, this method allows to test reagents, co-factors or enzymes involved in nitrosylation regulation (Fig. 6.1).

6.2.7 SHIPS

Liu et al. (2010) used two different alkylating agents to differentiate free thiols or those involved in disulphide bridges to nitrosothiols. Denaturated proteins were treated with DTT to open all disulphide bridges following by an acrylamide treatment to this time alkylate all free thiols. SNO are then reduced with ascorbate (50 mM). Proteins are digested in solution and free thiols are captured on a resin (TS6B: Thiopropyl Sepharose 6B) in the same way as in SNO-RAC. The disulphide bridges between peptides and resin are cleaved off and eluted with DTT (10 mM) and free thiols are eventually alkylated with iodoacetamide (55 mM). Peptides were analyzed by ESI-MS/MS. If one peptide contains more than one cystein, one can distinguish the original status of different cysteins due to the mass shift induced by the two different alkylating agents (acrylamide (+71 amu) vs. iodoacetamide (+57 amu). This method named (SHIPS) for "Site-specific High-throughput Identification of Protein *S*-nitrosylation" is an elegant way to study the redox status of different cysteins but is based on an a priori strong stability of SNO during DTT treatment (Fig. 6.1).

6.2.8 Biotin/Cys-TMT Switch and SILAC-BS

Murray et al. (2012), used a new thiol reactive agent, CysTMT6 (cys tandem Mass Tag), present in six isobaric forms replacing biotin-HPDP for quantitative nitrosylation studies. With the help of these isobaric forms, multiple comparisons are possible and SNO can be distinguished from cystein glutathionylation and in theory putative leakage in thiol blocking by NEM. A short range of GSNO concentration (2, 10, 20 μM) allowing the authors to conclude that about 28 % of all cysteins can be nitrosylated. A resin able to trap TMT is used to select labeled peptides and in this study 691 nitrosylated proteins were identified. Benhar et al. (2010), employed Stable Isotope Labeling by Amino acids in Cell culture (SILAC) coupled to BS to identify *S*-nitrosylated targets of thioredoxin. Two populations of cells were grown in parallel for six passages on light (Arg0/Lys0) or heavy (Arg10/Lys8) media before treatment with NO donor and BS labeling. Samples were then subjected to streptavidin pull-down before analysis. The use of SILAC labeling allows to combine samples directly after cell growth reducing technical variability of all further steps (Fig. 6.1).

6.3 Methods Using a Direct SNO Reduction

BS methods are based on an a priori fairly complete free thiol blocking and crucial additional time-consuming controls have to be done as omission of ascorbate reduction or NO donors (not possible when studying endogenous nitrosylation). One usual control is the simple omission of biotin-HPDP but it cannot reveal incomplete thiol blocking and it is then not a sufficient and suitable control. Direct SNO reduction seems to be a promising but still challenging way to improve identification of SNO.

6.3.1 Phenylmercury Reduction

SNO reaction with phenylmercury compound leads to the formation of a very stable thiol mercury bond. Doulias et al. (2010), used an organomercury resin (MRC) synthetized by conjugation of ρ-amino-phenylmercury acetate on Affigel10 agarose beads and activated with *N*-hydroxysuccinimide. They used also phenylmercury-polyethyleneglycol-biotin (mPEGb) to capture nitrosylated peptides or proteins. The overall method can be divided into three steps: (i) free thiol blocking with MMTS (ii) capture on MRC or mPEGb and elution (β-Mercaptoethanol or performic acid) of the nitrosylated peptides/proteins, (iii) analysis of the selected peptides by ESI-MS/MS. The reduction step of SNO with ascorbate was omitted because phenylmercury reacts directly with SNO. Mild concentration of

Table 6.1 Complementary approaches to detect in situ nitrosylation

Authors	Year	In situ	Material/tissue
Gow et al.	2002	SNO antibody	Endothelial cells, macrophages, neuronal cells...
Ckless et al.	2004	Fuorophore (MPB)-conjugated streptavidin	Mouse alveolar epithelial cell and mouse lung

Table 6.2 Websites allowing to verify in silico nitrosylation sites

Authors	Year	Bio-info name	Website	Type of studies
Xue et al.	2010	GPS-SNO	http://sno.biocuckoo.org/	*S*-nitrosylation sites
Lee et al.	2011	SNOSite	http://csb.cse.yzu.edu.tw/SNOSite/	*S*-nitrosylation sites
Li et al.	2011	CPR-SNO	http://math.cau.edu.cn/CPR-SNO/CPR-SNO.html	*S*-nitrosylation sites
Liu et al.	2011	GPS-YNO2	http://yno2.biocuckoo.org/	*Y*-nitration sites

performic acid was sufficient to oxidate free thiol to sulfonic acid creating a specific MS signature. Negative control like the use of UV (degrade SNO) for 7 min, DTT (10 mM, 30 min, 37 °C), Ascorbate + Cu^{2+}, $HgCl_2$ were used to validate this approach. A total of 328 peptides held in 192 proteins were identified in this study (Fig. 6.1).

6.3.2 Phosphine Switch

Two groups used phosphine derivatives and showed that they react very (phosphine thioesters) or fairly (triarylphosphine) specifically with SNO. Bechtold et al. (2010) used the water soluble XPTS (tris(4,6-dimethyl-3-sulfonatophenyl)-phosphine). XPTS is able to react with SNO to form a covalent *S*-alkylphosphonium. However, some disulphide bridges (GSSG and cysteine) can be reduced to some extent (3–15 %). So far, this method does not allow to select modified peptides and its use is restricted to model proteins. However, Zhang et al. (2010) used another phosphine derivative coupled to biotin to detect SNO. Unfortunately, this compound is not any longer water soluble but can be easily dissolved in DMSO. The phosphine does not seem to react with free thiol or amines permitting in theory to skip the free thiol blocking step with for instance MMTS. Nevertheless, this blocking step was kept in the protocol when biological samples were used for two reasons (Xian pers. comm.): (i) to block free thiol which might help to stabilize SNO entities; (ii) to limit exchange between disulphide bridges formed between SNO and phosphine in one hand and free thiol in other hand that could lead to false positive identifications (Fig. 6.1).

Table 6.3 Nitrosothiol studies in plants: Biotin-Switch was mostly used to identify nitrosothiols in plants and more than half of the studies focus on a particular protein

Authors (PMID)	Year	Methods	Material/tissue	Protein identities or nb of SNO
Perazzolli et al. (15367716)	2004	Biotin-Switch	*A. thaliana*	Ahb1
Lindermayr et al. (15734904)	2005	Biotin-Switch	*A. thaliana* leaves and cell line	52 proteins
Lindermayr et al. (16365035)	2006	Direct ESI-MS/MS (QTOF-MS)	*A. thaliana*	MAT
Belenghi et al. (17110382)	2007	Biotin-Switch	*A. thaliana* leaves	Metascaspase 9
Romero-Puertas et al. (18165327)	2007	Biotin-Switch	*A. thaliana* leaves	PrxII E
Serpa et al. (17686455)	2007	Biotin-Switch	*A. thaliana*	AtMYB2
Abat et al. (18445036)	2008	Biotin-Switch	*Kalanchoe pinnata* leaves	19 proteins
Romero-Puertas et al. (18297659)	2008	Biotin-Switch	*A. thaliana* leaves	16 proteins
Tada et al. (18635760)	2008	Biotin-Switch	*A. thaliana*	NPR1
Holtgrefe et al. (18298409)	2008	Biotin-Switch	*A. thaliana*	GAPDH
Tanou et al. (19682288)	2009	Biotin-Switch	*Citrus aurantium* L. leaves	49 proteins
Abat and Deswal (19655309)	2009	Biotin-Switch	*Brassica juncea* seeds	20 proteins (15 differentially nitrosylated)
Wang et al. (19017644)	2009	Biotin-Switch	*A. thaliana* leaves	AtSABP3
Elviri et al. (20227082)	2010	RPLC et ESI-MS/MS (LIT-MS)	*A. thaliana* cell line	Phytochelatins PC2, 3, 4, 5
Palmieri et al. (20089767)	2010	Biotin-Switch	Mitochondria of *A. thaliana* leaves	11 proteins
Wawer et al. (20397974)	2010	Biotin-Switch	*Nicotiana tabacum* cell line	NtOSAK & GAPDH
Lindermayr et al. (20716698)	2010	Biotin-Switch	*A. thaliana*	NPR1 & TGA1
van der Linde et al. (21782461)	2011	Biotin-Switch	*A. thaliana*	F6P aldolase
Wimalasekera et al. (21471330)	2011	Biotin-Switch	*A. thaliana*	ND
Yun et al. (21964330)	2011	Biotin-Switch modified	*A. thaliana*	NADPH oxidase
Terrile et al. (22171938)	2012	Biotin-Switch	*A. thaliana*	TIR1/AFB
Fares et al. (22115780)	2011	BS-ICAT	*A. thaliana* cell line	46 proteins
Ortega-Galisteo et al. (22213812)	2012	Biotin-Switch	*Pisum sativum*	6 peroxisomal proteins
Kato et al. (22924747)	2012	Biotin-Switch	*Solanum tuberosum*	80 proteins
Tanou et al. (22780834)	2012	Biotin-Switch	*C. aurantium L.* leaves and roots	141 proteins
Feng et al. (23443557)	2013	Biotin-Switch	*A. thaliana*	AHP1
Camejo et al. (23238061)	2013	Biotin-Switch	*P. sativum*mitochondria	22 proteins

6.3.3 SNO Reduction by Gold Nanoparticules

Faccenda et al. (2010) used gold nanoparticules (AuNP) to select nitrosylated peptides. In a first step comparable to BS, free thiols are alkylated with iodoacetamide and proteins are digested with trypsin. Gold nanoparticules are added, incubated with peptides, and collected by centrifugation. DTT was then added to displace the interactions between gold nanoparticules and nitrosylated peptides through an excess of free thiols. Peptides were analyzed with MALDI-TOF. However, thiols and also thioesters (methionine) have affinity for AuNP, and this affinity is stronger than for any other chemical groups present in proteins. In order to distinguish nitrosothiols to free glutathionylated peptides or peptides with methionine, Faccenda et al. (2010) modified the original protocol. They alkylated free thiols with iodoacetamide and reduced nitrosothiols with ascorbate. The new free thiols generated were alkylated by NEM, and proteins were digested with trypsin. Peptides were incubated with AuNP. By this way, nitrosylated peptides can be identified through NEM signature; glutathionylated peptides and peptides with methionine are enriched with AuNP while original free thiols are carbamidomethylated (Fig. 6.1).

6.3.4 Complementary Approaches to Identify Nitrosothiols

Nitrosothiols can be directly in situ detected, thanks to specific antibodies (Gow et al. 2002). Nevertheless, a BS derivative was used in in situ approaches. Indeed, after BS labeling, streptavidin coupled to a fluorophore permitted to visualize nitrosothiols with confocal microscopy (Ckless et al. 2004) (Table 6.1). These studies got a limited impact on nitrosylation studies because of the different constraints (antibody specificity, high background, etc.). Eventually, different bioinformatic applications trained on data present in the literature allow to verify the "orthodoxy" of the detected sites in a study (Table 6.2).

6.4 Assessment of Protein Nitrosylation in Plants

Nitrosylation data in plants are scarce, and most of them refer to a study of a single protein (Perazzolli et al. 2004; Belenghi et al. 2007; Serpa et al. 2007; Tada et al. 2008). However, several large-scale studies (as Lindermayr et al. 2005; Tanou et al. 2009; Lin et al. 2012, see Table 6.3) contributed to broaden our knowledge on SNO in plants. So far, 28 contributions collected in the literature allowed the overall identification of 373 nonredundant putative nitrosylated proteins (Table 6.4). It represents the hitherto putative plant nitrosylome showing that many cellular compartments are NO target with a very large diversity of protein

Table 6.4 Plant nitrosylome: All the plant nitrosylated proteins identified so far have been collected and blasted against the *Arabidopsis* proteome

Original accession	*Arabidpsis* number	Blast e-value	Undistinguishable blast hits	Definition	MapManBin	TMHMM	TargetP	PMID number
Os08g0459300	AT1G02180	1.0e-33	–	Ferredoxin-related	1.1.5.2 PS.lightreaction.other electron carrier (ox/red).ferredoxin	0	S	22106097
Os05g0310500	AT1G02305	9.6e-95	–	Cysteine proteinases superfamily protein	29.5.3 protein.degradation.cysteine protease	1	S	22106097
NP_849577; Os01g0323600	AT1G02500	1.7e-207	–	*S*-adenosylmethionine synthetase 1	13.1.3.4.11 amino acid metabolism.synthesis.aspartate family.methionine. *S*-adenosylmethionine synthetase	0	–	16365035; 22106097
–	AT1G02560	–	–	Nuclear encoded CLP protease 5	29.5.5 protein.degradation.serine protease	0	C	18297659
–	AT1G03130	–	–	Photosystem I subunit D-2	1.1.2.2 PS.lightreaction.photosystem I.PSI polypeptide subunits	0	C	15734904
–	AT1G04410	–	–	Lactate/malate dehydrogenase family protein	8.2.9 TCA/org.transformation. other organic acid transformaitons.cyt MDH	0	Cytosol	15734904
–	AT1G04710	–	–	Peroxisomal 3-ketoacyl-CoA thiolase 4	11.9.4.5 lipid metabolism.lipid degradation.beta-oxidation	0	Peroxisome	22115780
Os07g0141400; gi|42414435; P93566	AT1G06680	1.7e-81	–	PsbP-1 OEC23 Tat lTP (model with cTP)	1.1.1.2 PS.lightreaction.photosystem II.PSII polypeptide subunits	0	C	22106097; 22780834; 22924747; 15734904
Os03g0295800	AT1G07080	5.9e-63	–	Thioredoxin superfamily protein	35.1 not assigned.no ontology	0	S	22106097
Q43824; Os07g0694700; BAC22953, ABA10743	AT1G07890	2.9e-109	AT3G09640	Ascorbate peroxidase 1	21.2.1 redox.ascorbate and glutathione.ascorbate	0	Cytosol	19655309; 22115780; 22106097; 22924747
gi|226494632; BAC23049	AT1G07920	5.2e-222	AT5G60390, AT1G07930, AT1G07940	Elongation factor 1-alpha/EF-1-alpha	29.2.4 protein.synthesis.elongation	0	Cytosol	19682288; 22924747; 15734904
–	AT1G07930	–	–	Elongation factor 1-alpha (E-Tu)	29.2.4 protein.synthesis.elongation	0	–	22115780; 15734904
–	AT1G07940	–	–	GTP binding Elongation factor Tu family protein	29.2.4 protein.synthesis.elongation	0	–	15734904

(continued)

Table 6.4 (continued)

Original accession	*Arabidpsis* number	Blast e-value	Undistinguishable blast hits	Definition	MapManBin	TMHMM	TargetP	PMID number
ABK54292	AT1G08830	7.2e-42	–	Superoxide dismutase (Cu-Zn) (SODCC)(CSD1) (this is not the chloroplast protein)	21.6 redox.dismutases and catalases	0	–	22924747; 15734904
A9JNV7	AT1G09195	1.8e-54	–	HD subdomain (InterPro:IPR006674)	35.2 not assigned.unknown	0	M	19655309
gi|34418745/ gi|225450527	AT1G09340	1.2e-169	–	HIP1.3 (cytosol–heteroglycan-interaction) or Rap38/ CSP41B (plastid-RNA binding)	27 RNA	0	Cytosol; plastid stroma	22780834; 15734904
gi|55392843, gi|227937359, ABB87108	AT1G09640	4.4e-74	AT1G57720	Translation elongation factor EF1B gamma chain	29.2.4 protein.synthesis.elongation	0	Cytosol	22780834; 22924747; 15734904
–	AT1G09780	–	–	Phosphoglycerate mutase	4.11 glycolysis.phosphoglycerate mutase	0	Cytosol	22115780
Os05g0508300	AT1G09850	8.0e-123	–	Xylem bark cysteine peptidase 3	29.5.3 protein.degradation.cysteine protease	0	S	22106097
gi|29550305	AT1G10370	2.7e-12	AT1G10360	Glutathione *S*-transferase (ATGSTU17/ERD9/GST30/ GST30B)	26.9 misc.glutathione *S* transferases	0	S	22780834
ACC93586	AT1G10760	0	AT4G24450	PEP/pyruvate binding domain	2.2.2.3 major CHO metabolism.degradation. starch.glucan water dikinase	0	C	22924747
gi|19568098	AT1G10940	–	–	Serine/threonine kinase 1 (ASK1)	29.4 protein.postranslational modification	0	–	20397974
gi|3915699; gi|3915699; gi|46208295	AT1G11860	1.7e-193	–	Glycine cleavage T-protein	13.2.5.2 amino acid metabolism.degradation.serine-glycine-cysteine group.glycine	0	M	22213812; 23238061; 22780834; 15734904
Os04g0459500; gi|188293203; gi|77540210	AT1G12900	1.8e-157	AT3G26650	Glyceraldehyde-3-phosphate dehydrogenase A-2 (GAPA-2)	1.3.4 PS.calvin cyle.GAP	0	C	22106097; 22780834; 19682288; 18297659; 15734904
–	AT1G14980	–	–	Chaperonin 10	29.6 protein.folding	0	M	15734904
Os05g0582800	AT1G15000	3.1e-105	–	Serine carboxypeptidase (S10)	29.5.5 protein.degradation.serine protease	0	S	22106097
Os02g0328300	AT1G15140	5.0e-82	–	Oxidoreductase NAD-binding domain-containing protein	26.7 misc.oxidases—copper, flavone etc.	0	C	22106097

(continued)

Table 6.4 (continued)

Original accession	*Arabidpsis* number	Blast e-value	Undistinguishable blast hits	Definition	MapManBin	TMHMM	TargetP	PMID number
Q8H9E7; gi|30315017	AT1G17180	5.5e-83	AT1G78380, AT1G78340, AT1G17170, AT1G17190, AT1G78320, AT1G78360	Glutathione *S*-transferase TAU 25	26.9 misc.glutathione *S* transferases	0	–	19655309; 19682288
gi|119367468; P58515, Q41480, P58518	AT1G17860	1.3e-26	AT1G73260, AT1G73325	Trypsin and protease inhibitor family protein/Kunitz family protein	20.1.7.6 stress.biotic.PR-proteins.proteinase inhibitors	0	S	22780834; 22924747
gi|225456197	AT1G18270	0	–	Ketose-bisphosphate aldolase class-II family protein	35.1 not assigned.no ontology	0	S	19682288
gi|37509041	AT1G18980	1.8e-31	AT1G18970, AT1G09560, AT1G02335, AT3G62020	RmlC-like cupins superfamily protein	20.2.99 stress.abiotic.unspecified	0	S	22780834
AAY47048; Os05g0116100; ABA40439	AT1G19570	1.2e-82	AT1G75270	Dehydroascorbate reductase-1 (DHAR1)—monomeric-cytosol and peroxisome	21.2.1 redox.ascorbate and glutathione.ascorbate	0	Cytosol; peroxisome	22115780; 22924747; 22106097
P00296; Os06g0101600	AT1G20340	5.4e-37	AT1G76100	Plastocyanin-1 (PC-1)	1.1.5.1 PS.lightreaction.other electron carrier (ox/red).plastocyanin	1	C	22924747; 22106097
–	AT1G20620	–	–	Catalase 3	21.6 redox.dismutases and catalases	0	Peroxisome	20089767
gi|225459587	AT1G21750	2.1e-177	AT1G77510	PDI Calsequestrin and thioredoxin domain	21.1 redox.thioredoxin	0	S/ER	19682288
–	AT1G22300	–	–	14-3-3 protein GF14 epsilon (grf10)	30.7 signaling.14-3-3 proteins	0	–	22115780; 15734904
Os03g0390400	AT1G22450	1.2e-39	–	Cytochrome c oxidase subunit 6b	9.7 mitochondrial electron transport/ATP synthesis.cytochrome c oxidase	0	Mitochondria	22106097
gi|255573724, gi|38026260, gi|38026260	AT1G23190	4.3e-273	AT1G70730	Phosphoglucomutase-3 (PGM-3)	4.2 glycolysis.PGM	0	Cytosol	22780834
–	AT1G23310	–	–	Glutamate:glyoxylate aminotransferase	1.2.3 PS.aminotransferases peroxisomal	0	Peroxisome	15734904
gi|21650315	AT1G24020	1.1e-08	AT5G45860, AT5G45870, AT5G28000	Bet v I allergen family protein	20.2.99 stress.abiotic.unspecified	0	–	22780834

(continued)

Table 6.4 (continued)

Original accession	*Arabidpsis* number	Blast e-value	Undistinguishable blast hits	Definition	MapManBin	TMHMM	TargetP	PMID number
–	AT1G26630	–	–	Eukaryotic translation initiation factor 5A-1 (eIF-5A 1) protein	29.2.3 protein.synthesis.initiation	0	–	15734904
Os09g0537700	AT1G26820	1.7e-42	–	Ribonuclease 3 (RNS3)	27.1.19 RNA.processing. ribonucleases	0	S	22106097
gi|28630973	AT1G29910	5.7e-97	AT1G29920, AT2G34430, AT1G29930, AT2G34420	LHCII-1.1-100 % identical to lhcb-1.2	1.1.1.1 PS.lightreaction.photosystem II.LHC-II	0	C	22780834
gi|110862369	AT1G31330	6.3e-75	–	psaF—subunit III—LTP—hydrophobic	1.1.2.2 PS.lightreaction.photosystem I.PSI polypeptide subunits	1	C	22780834
gi|147859917; Os02g0698000; gi|46206611; NP_174486	AT1G32060	9.5e-175	–	Phosphoribulokinase-2 (PRK-2)	1.3.12 PS.calvin cyle.PRK	0	C	19682288; 22106097; 22780834; 22924747; 18297659
gi|737595	AT1G32470	5.8e-56	AT2G35370	Glycine cleavage system H protein	1.2.4.4 PS.photorespiration.glycine cleavage.H protein	0	Mitochondria	22213812; 15734904; 20089767
Os03g0122200	AT1G32990	1.7e-63	–	50S ribosomal protein L11	29.2.1.1.1.2.11 protein.synthesis. ribosomal protein.prokaryotic. chloroplast.50S subunit.L11	0	C	22106097
CAB92956	AT1G35720	6.5e-112	–	Annexin—AnnAt1	31.1 cell.organisation	0	Cytosol	22924747; 22115780; 15734904
gi|188419599	AT1G42970	1.4e-95	–	Glyceraldehyde-3-phosphate dehydrogenase B (GAPB)	1.3.4 PS.calvin cyle.GAP	0	C	22780834; 15734904
Os01g0866400	AT1G43670	3.3e-156	–	D-fructose-1	2.1.1.3 major CHO metabolism. synthesis.sucrose.FBPase	0	–	22106097
gi|21652048	AT1G44575	3.1e-57	–	psbS (NPQ4—null mutant)	1.1.1.2 PS.lightreaction.photosystem II.PSII polypeptide subunits	0	C	22780834
–	AT1G47128	–	–	Cysteine proteinase (RD21A)/thiol protease	29.5.3 protein.degradation.cysteine protease	0	Vacuole	22115780
gi|9955321; AAN23154	AT1G48030	3.1e-215	AT3G17240	E3-1 dihydrolipoamide dehydrogenase 1 (mtlpd1)	8.1.1.3 TCA/org.transformation. TCA.pyruvate DH.E3	0	Mitochondria	22213812; 22924747; 22115780; 20089767
Q9M7C1	AT1G48130	6.2e-107	–	1-Cys peroxiredoxin (1-Cys Prx)	21.5 redox.periredoxins	0	–	19655309

(continued)

Table 6.4 (continued)

Original accession	*Arabidpsis* number	Blast e-value	Undistinguishable blast hits	Definition	MapManBin	TMHMM	TargetP	PMID number
gi|21388550, gi|1346501	AT1G53240	1.6e-131	AT3G15020	Malate dehydrogenase [NAD]	8.1.9 TCA/org.transformation. TCA.malate DH	0	M	23238061; 15734904
CAA64769	AT1G55020	4.3e-266	–	LOX1 (Lipoxygenase 1); lipoxygenase	17.7.1.2 hormone metabolism. jasmonate.synthesis-degradation.lipoxygenase	0	–	22924747
–	AT1G55480	–	–	Protein containing PDZ domain, a K-box domain, and a TPR region	35.2 not assigned.unknown	0	C	15734904
–	AT1G55490	–	–	Chaperonin 60 beta	29.6 protein.folding	0	C	15734904
Q9SGT4; Os02g0519900; gi|28390642	AT1G56070	0	AT3G12915	Elongation factor 2	29.2.4 protein.synthesis.elongation	0	–	19655309; 22115780; 22106097; 22780834; 15734904
AAC26785; gi|110845353	AT1G56190	1.9e-201	AT3G12780	Phosphoglycerate kinase-1 (PGK-2)	1.3.3 PS.calvin cyle.phosphoglycerate kinase	0	C	22924747; 22780834; 15734904
Os07g0246200	AT1G56340	2.5e-151	AT1G09210	Calreticulin 1 (CRT1)	30.3 signaling.calcium	1	S/ER	22106097
–	AT1G60710	–	–	Aldo/keto reductase family protein	17.2.3 hormone metabolism.auxin.induced-regulated-responsive-activated	0	–	22115780
Os08g0104600; CAC38395	AT1G60950	5.7e-26	AT1G10960	Fd2-leaf ferredoxin	1.1.5.2 PS.lightreaction.other electron carrier (ox/red).ferredoxin	0	C	22106097; 22924747
–	AT1G63000	–	–	NAD dependent epimerase/dehydratase family; RmlD substrate binding domain	10.1.11 cell wall.precursor synthesis.UER	0	–	22115780
gi|55935638	AT1G63940	1.8e-68	–	Monodehydroascorbate reductase (MDHAR) (dual mito and chloro)	21.2.1 redox.ascorbate and glutathione.ascorbate	0	Mitochondria; plastid stroma	22780834
gi|215713553	AT1G64260	2.5e-14	AT1G49920, AT1G64255	MuDR family transposase	27.3.99 RNA.regulation of transcription. unclassified	0	–	23238061
–	AT1G64280	–	–	regulatory protein (NPR1)	27.3.61 RNA.regulation of transcription.NPR1/NIM1	0	–	18635760; 20716698
–	AT1G65930	–	–	NADP+ isocitrate dehydrogenase (ICDH)	8.1.4 TCA/org.transformation.TCA.IDH	0	Cytosol	22115780
AAP34571	AT1G65980	1.7e-72	AT1G65970, AT1G60740	Peroxiredoxin IIB (PrxII B)	21.5 redox.periredoxins	0	–	22924747; 15734904

(continued)

Table 6.4 (continued)

Original accession	*Arabidpsis* number	Blast e-value	Undistinguishable blast hits	Definition	MapManBin	TMHMM	TargetP	PMID number
–	AT1G66240	–	–	Homolog of anti-oxidant 1	15.2 metal handling.binding, chelation and storage	0	Cytosol	15734904
–	AT1G67090	–	–	Ribulose bisphosphate carboxylase small chain 1A	1.3.2 PS.calvin cyle.rubisco small subunit	0	C	15734904
gi|223528624; gi|13873334	AT1G68010	1.7e-175	–	Hydroxypyruvate reductase 1 (HPR1)	1.2.6 PS.photorespiration. hydroxypyruvate reductase	0	Peroxisome	22213812; 19682288
Os12g0555500	AT1G69200	0.0011	–	pfkB-type carbohydrate kinase (FLN2) (TAC)	29.4.1 protein.postranslational modification.kinase	0	Plastid	22106097
Os03g0758800; P56335	AT1G69410	1.1e-72	AT1G13950, AT1G26630	Eukaryotic translation initiation factor 5A	29.2.3 protein.synthesis.initiation	0	–	22106097; 22924747
gi|42477931	AT1G70890	2,4	AT1G70840, AT1G70830	Major latex protein (MLP) related); Pathogenesis-related protein Bet v I family	20.2.99 stress.abiotic.unspecified	0	–	
Os01g0962700	AT1G71695	2.1e-92	–	Peroxidase 12 (PER12) (P12) (PRXR6)	26.12 misc.peroxidases	1	S/cell wall	
Q41435; P01080	AT1G72060	0.0061	–	Serine-type endopeptidase inhibitors	35.2 not assigned.unknown	1	S	22924747
AAL60248; AAB28594	AT1G72290	2.3e-05	AT1G73325	Kunitz family trypsin and protease inhibitor protein	20.1.7.6 stress.biotic.PR-proteins.proteinase inhibitors	0	S	22924747
–	AT1G73010	–	–	Phosphate starvation-induced gene 2	26.13 misc.acid and other phosphatases	0	–	22115780
AAB53203	AT1G76180	1.4e-20	AT1G20450	Dehydrin -ERD14	20.2.99 stress.abiotic.unspecified	0	Peroxisome	22924747
P14673	AT1G77120	3.9e-162	–	ADH1 (ALCOHOL DEHYDROGENASE 1)	5.3 fermentation.ADH	0	–	22924747; 22115780
gi|189014940, gi|38052125, gi|45451793	AT1G78300	4.9e-107	AT1G35160, AT4G09000, AT3G02520, AT5G38480	14-3-3 protein GF14 omega (grf2)—cytosol and nucleus	30.7 signaling.14-3-3 proteins	0	Cytosol; nucleus	22780834
–	AT1G78380	–	–	Glutathione *S*-transferase TAU 19	26.9 misc.glutathione *S* transferases	0	Cytosol	15734904
–	AT1G78830	–	–	Curculin-like (mannose-binding) lectin family protein	26.16 misc.myrosinases-lectin-jacalin	1	S	22115780
AAM51555	AT1G79340	3.8e-139	–	Latex-abundant protein	29.5 protein.degradation	0	–	22924747
–	AT1G79550	–	–	Phosphoglycerate kinase	4.10 glycolysis.phosphoglycerate kinase	0	–	15734904
gi|147812715	AT2G01290	2.4e-78	–	Ribose 5-phosphate isomerase (PRI)	7.2.4 OPP.non-reductive PP.ribose 5-phosphate isomerase	0	Cytosol	19682288

(continued)

Table 6.4 (continued)

Original accession	*Arabidpsis* number	Blast e-value	Undistinguishable blast hits	Definition	MapManBin	TMHMM	TargetP	PMID number
Q9ZVF3	AT2G01520	3.4e-83	–	MLP328 (MLP-LIKE PROTEIN 328)	20.2.99 stress.abiotic.unspecified	0	–	19655309
P20346	AT2G02100	9.9e-22	AT2G02130, AT2G02120, AT1G61070, AT5G63660	Plant defensin-fusion protein	20.1 stress.biotic	1	S	22924747
ABC01903	AT2G02760	1.7e-81	AT1G14400	Ubiquiting-conjugating enzyme 2	29.5.11.3 protein.degradation. ubiquitin.E2	0	M	22924747
–	AT2G02990	–	–	Ribonuclease 1	27.1.19 RNA.processing.ribonucleases	1	S	15734904
gi|46207502	AT2G04160	8.5e-71	–	Subtilisin-like serine endopeptidase family protein	29.5.1 protein.degradation.subtilases	0	S	22780834
Os09g0249900	AT2G04700	2.5e-48	–	Ferredoxin-thioredoxin reductase subunit A (catalytic)	21.1 redox.thioredoxin	0	C	22106097
–	AT2G05710	–	–	Aconitase 3	8.2.3 TCA/org.transformation.other organic acid transformaitons.aconitase	0	Cytosol; mitochondria	15734904
gi|115477148, gi|18032028	AT2G13360	7.0e-179	–	Alanine-glyoxylate aminotransferase (AGT1)	1.2.3 PS.aminotransferases peroxisomal	0	Peroxisome	
AAN85431	AT2G16060	6.5e-57	–	Hemoglobin 1	21.3 redox.heme	0	–	22924747; 15367716
gi|260401128, gi|28616135; AAB51386, AAD22975	AT2G16600	5.1e-73	AT2G21130, AT4G38740, AT4G34870	Rotamase CYP 3 peptidyl-prolyl cis-trans isomerase	29.6 protein.folding	0	–	22780834; 22924747
Os03g0133400	AT2G17120	1.6e-51	–	LYM2 (LYSM DOMAIN GPI-ANCHORED PROTEIN 2 PRECURSOR)	35.1 not assigned.no ontology	1	S/plasma membrane	22106097
gi|1669291	AT2G18110	9.0e-28	AT1G30230	Elongation factor 1-beta	29.2.4 protein.synthesis.elongation	0	–	22780834
NP_001234138	AT2G19760	2.8e-56	AT4G29350, AT2G19770, AT4G29340	Profilin 1 (PRO1) (PFN1) (PRF1)/ allergen Ara t 8	31.1 cell.organisation	0	Cytosol	22924747
gi|110852417, gi|28615735	AT2G19900	1.4e-125	AT5G11670, AT1G79750, AT5G25880	NADP-malic enzyme 1 (NADP-ME1)	8.2.10 TCA/org.transformation.other organic acid transformaitons.malic	0	–	22780834
gi|63075439	AT2G20420	8.6e-94	–	Succinyl-CoA ligase beta subunit	8.1.6 TCA/org.transformation. TCA.succinyl-CoA ligase	0	M	22780834

(continued)

Table 6.4 (continued)

Original accession	*Arabidpsis* number	Blast e-value	Undistinguishable blast hits	Definition	MapManBin	TMHMM	TargetP	PMID number
Os02g0121300, gil119367479	AT2G21130	1.9e-66	AT4G38740, AT2G16600, AT4G34870	Peptidyl-prolyl cis-trans isomerase/cyclophilin (CYP2)/rotamase	31.3.1 cell.cycle.peptidylprolyl isomerase	0	–	22106097; 19682288
gil225427917; Os09g0535000; gil110853765	AT2G21170	5.6e-122	–	Triosephosphate isomerase-1 (TPI-1)	1.3.5 PS.calvin cycle.TPI	0	C	19682288; 22106097; 22780834; 15734904
ABB87126, BAA03741	AT2G21660	4.9e-36	AT4G39260	Glycine-rich RNA-binding protein (GRP7) (dual cytosol and nucleus)	27.4 RNA.RNA binding	0	Cytosol; nucleus	22924747; 15734904
Os02g0794700	AT2G24200	9.4e-207	–	Eucyl aminopeptidase (LAP1)	29.5 protein.degradation	0	–	22106097
Os08g0440800	AT2G24270	1.6e-234	–	NON-phosphorylating NADP glyceraldehyde 3-phosphate dehydrogenase (NP-GAPDH)	2 major CHO metabolism	0	Cytosol	22106097
Os04g0412200	AT2G27510	1.4e-36	AT1G10960, AT1G60950	Fd3—root plastid specific	7.3 OPP.electron transfer	0	C	22106097
–	AT2G27720	–	–	60S acidic ribosomal protein family	29.2.1.2.2.82 protein.synthesis.ribosomal protein.eukaryotic.60S subunit.P2	0	Cytosol	15734904
gil3790441; gil31672181	AT2G28000	2.1e-216	–	Cpn60-alpha-1	29.6 protein.folding	0	C	19682288; 22780834
Os12g0569500	AT2G28790	2.7e-28	–	Osmotin-like protein	20.2 stress.abiotic	0	S	22106097
gil63105013	AT2G29420	1.0e-42	AT3G09270, AT2G29460, AT2G29490, AT2G29480, AT1G10370, AT2G29450, AT2G29470	ATGSTU7 (GLUTATHIONE *S*-TRANSFERASE 25); glutathione transferase	26.9 misc.glutathione *S* transferases	0	Cytosol	22780834
gil45447426	AT2G30860	8.4e-80	–	Glutathione *S*-transferase class phi (AtGST9)	26.9 misc.glutathione *S* transferases	0	–	22780834
gil75114857	AT2G30950	2.9e-267	AT1G06430	FtsH2 (VAR2 and Pftf)	29.5.7 protein.degradation. metalloprotease	1	C	19682288
Os02g0236000	AT2G30970	1.3e-186	–	Putative aspartate aminotransferase Asp1	13.1.1.2.1 amino acid metabolism.synthesis.central amino acid metabolism.aspartate.aspartate aminotransferase	0	M	22106097

(continued)

Table 6.4 (continued)

Original accession	*Arabidpsis* number	Blast e-value	Undistinguishable blast hits	Definition	MapManBin	TMHMM	TargetP	PMID number
Os11g0256200	AT2G31725	6.4e-50	AT1G05730	Eukaryotic protein of unknown function (DUF842)	35.2 not assigned.unknown	0	–	22106097
Os07g0150200	AT2G32060	3.6e-47	AT1G15930	40S ribosomal protein S12 (RPS12C)	29.2.1.2.1.12 protein. synthesis.ribosomal protein.eukaryotic.40S subunit.S12	0	Cytosol	22106097
Os03g0452300	AT2G33800	3.1e-57	–	30S ribosomal protein S5	29.2.1.1.1.5 protein.synthesis. ribosomal protein.prokaryotic. chloroplast.30S subunit.S5	0	C	22106097
Os04g0591000	AT2G35410	4.7e-31	–	Putative RNA binding protein	27.4 RNA.RNA binding	0	C	22106097
–	AT2G36160	–	–	Ribosomal protein S11 family protein	29.2.1.2.1.14 protein.synthesis.ribosomal protein.eukaryotic.40S subunit.S14	0	–	15734904
gi|110858063, gi|110859418, gi|110871044; ABC01905	AT2G36460	2.7e-145	AT5G03690, AT3G52930	Aldolase	4.7 glycolysis.aldolase	0	Cytosol	22780834; 21782461; 22924747
gi|28619422, gi|1169534, gi|225455555, gi|533474, gi|209926760; P26330/AFP49323, AAB87127	AT2G36530	5.4e-211	–	Enolase (ENO2 also LOS2) dual targeted nucleus and cytosol	4.12 glycolysis.enolase	0	Cytosol; nucleus	22780834; 22924747; 15734904
–	AT2G37190	–	–	Ribosomal protein L11 family protein	29.2.1.2.2.12 protein.synthesis. ribosomal protein.eukaryotic. 60S subunit.L12	0	Cytosol	15734904
AAL39067	AT2G37220	9.2e-74	AT3G53460	RNA binding protein CP29 B'	27.4 RNA.RNA binding	0	C	22924747
NP_181358	AT2G38230	4.4e-145	–	PDX1.1 (Vitamine b6 synthesis)	18.20 Co-factor and vitamine metabolism.vitamin b6	0	Cytosol	22924747
Os12g0175500	AT2G38270	1.0e-81	–	CAX-interacting protein glutaredoxin-related protein	34.21 transport.calcium	0	C	22106097
–	AT2G38530	–	–	Lipid transfer protein 2	11.6 lipid metabolism.lipid transfer proteins etc.	0	S	15734904
P08454	AT2G38870	3.1e-11	AT5G43580	Potato inhibitor I-type family protein	20.1 stress.biotic	0	–	22924747

(continued)

Table 6.4 (continued)

Original accession	*Arabidpsis* number	Blast e-value	Undistinguishable blast hits	Definition	MapManBin	TMHMM	TargetP	PMID number
Os11g0707000; gi|115334981; O49074; gi|188379887, gi|31670830, gi|229045333, gi|31670123, gi|188265008	AT2G39730	3.5e-207	–	Rubisco activase	1.3.13 PS.calvin cyle.rubisco interacting	0	C	22106097; 19682288; 22115780; 22924747; 22780834; 15734904
gi|18405801	AT2G41770	0	AT3G57420	Protein of unknown function (DUF288)	35.2 not assigned.unknown	1	–	23238061
–	AT2G42530	–	–	Cold regulated 15b	20.2.2 stress.abiotic.cold	0	C	15734904
–	AT2G42540	–	–	Cold-regulated 15a	20.2.2 stress.abiotic.cold	0	Plastid stroma	15734904
–	AT2G44350	–	–	Citrate synthase	8.1.2 TCA/ org.transformation.TCA.CS	0	M	22115780
gi|225449424	AT2G44920	1.7e-72	–	Thylakoid lumen pentapeptide repeat	27.7 RNA.misc.pentatricopeptide (PPR) repeat-containing protein	0	C	19682288
–	AT2G47190	–	–	Myb domain protein 2	27.3.25 RNA.regulation of transcription.MYB domain transcription factor family	0	–	17686455
Os01g0303000	AT2G47400	8.1e-27	–	CP12 protein	1.3.14 PS.calvin cycle regulation CP12	0	C	22106097
gi|161788876	AT3G01280	1.6e-108	–	Voltage dependent ion channel VDAC1	34.20 transport.porins	0	Mitochondria	19682288
Q1SLS0	AT3G01410	1.2e-13	–	Ribonuclease H-like superfamily protein	27.1.19 RNA.processing. ribonucleases	0	–	19655309
Os01g0639900; gi|1354517; gi|21650244, gi|31669520; CAH60891, NP_001233847	AT3G01500	2.7e-115	AT5G14740	Beta carbonic anhydrase-1 (beta CA1)—this model lacks the cTP	8.3 TCA/org.transformation. carbonic anhydrases	0	Plastid stroma	22106097; 19682288; 19017644; 22780834; 22924747; 15734904
AAB55380	AT3G01580	804	–	Tetratricopeptide repeat (TPR)-like superfamily protein	35.1.5 not assigned.no ontology.pentatricopeptide (PPR) repeat-containing protein	0	M	22924747

(continued)

Table 6.4 (continued)

Original accession	*Arabidpsis* number	Blast e-value	Undistinguishable blast hits	Definition	MapManBin	TMHMM	TargetP	PMID number
–	AT3G02730	–	–	Thioredoxin F-type 1	21.1 redox.thioredoxin	0	C	15734904
gi|56536027; gi|120676; gi|120676; AAB07758	AT3G04120	3.7e-109	AT1G13440	Glyceraldehyde-3-phosphate dehydrogenase C-1 (GapC-1)	4.9 glycolysis.glyceraldehyde 3-phosphate dehydrogenase	0	Cytosol	22780834; 20397974; 18298409; 22924747; 15734904
Os07g0176900	AT3G04790	3.2e-94	–	Ribose 5-phosphate isomerase (PRI)	1.3.10 PS.calvin cyle.Rib5P Isomerase	0	C	22106097
–	AT3G06050	–	–	Peroxiredoxin IIF	21.5 redox.periredoxins	0	M	15734904
gi|108862740	AT3G07770	1.7e-269	–	Hsp90-6	29.6 protein.folding	0	M	23238061
gi|15231937	AT3G08580	–	–	Mitochondrial ADP	34.9 transport.metabolite transporters at the mitochondrial membrane	3	Mitochondria	23238061; 22115780
–	AT3G09840	–	–	Cell division cycle protein 48 (CDC48A))	31.2 cell.division	0	Cytosol; nucleus; plasma membrane	22115780; 15734904
gi|147945633; gi|20902	AT3G10920	1.1e-95	–	Manganese superoxide dismutase	21.6 redox.dismutases and catalases	0	M	19682288; 23238061
CAC48323; gi|19173077; gi|21553667	AT3G11630	1.3e-99	AT5G06290	2-Cys Peroxiredoxin A (Prx A; formerly named BAS1)	21.5 redox.periredoxins	0	C	22924747; 23238061; 19682288
Os03g0586800	AT3G11710	5.4e-227	–	Lys-tRNA synthetase	29.1.6 protein.aa activation.lysine-tRNA ligase	0	–	22106097
–	AT3G11940	–	–	40S ribosomal protein 5A (ATRPS5A)	29.2.1.2.1.5 protein.synthesis.ribosomal protein.eukaryotic.40S subunit.S5	0	Cytosol	22115780
gi|8099682	AT3G12490	1.4e-31	AT2G40880	Cysteine protease inhibitor	29.5.3 protein.degradation.cysteine protease	0	–	19682288
Os10g0542900; AAC24807	AT3G12500	6.3e-75	–	Basic endochitinase	20.1 stress.biotic	0	S	22106097; 22924747; 15734904
Os03g0196800	AT3G13120	1.3e-51	–	30S ribosomal protein S10	29.2.1.1.1.1.10 protein.synthesis.ribosomal protein.prokaryotic.chloroplast.30S subunit.S10	0	C	22106097
gi|157337766	AT3G13470	1.5e-254	AT5G56500	Cpn60-beta-1	29.6 protein.folding	0	C	19682288
–	AT3G13920	–	–	Eukaryotic translation initiation factor 4A1	29.2.3 protein.synthesis.initiation; 27.6 RNA.DEAD BOX helicase	0	–	15734904

(continued)

Table 6.4 (continued)

Original accession	*Arabidpsis* number	Blast e-value	Undistinguishable blast hits	Definition	MapManBin	TMHMM	TargetP	PMID number
gi|167961875; gi|121530; gi|359806771; gi|21651872	AT3G14420	1.4e-171	AT3G14415, AT4G18360	Glycolate oxidase-1 (GOX-1)—peroxisome and plastid	1.2.2 PS.photorespiration.glycolate oxydase	0	Peroxisome; plastid	22213812; 19682288; 23238061; 22780834
–	AT3G14590	–	–	Calcium-dependent lipid-binding (CaLB domain) family protein	30.3 signaling.calcium	0	M	20089767
–	AT3G14990	–	–	Class I glutamine amidotransferase-like superfamily protein	35.1 not assigned.no ontology	0	–	15734904
Os01g0200700	AT3G15353	2.1e-10	–	Metallothionein 3	15.2 metal handling.binding, chelation and storage	0	–	22106097
–	AT3G16420	–	–	PYK10-binding protein 1	26.16 misc.myrosinases-lectin-jacalin	0	Cytosol	15734904
ABA40424	AT3G16640	1.0e-49	AT3G05540	TCTP homolog—tumor protein homologue	35.2 not assigned.unknown	0	–	22924747; 15734904
–	AT3G17390	–	–	*S*-adenosylmethionine synthetase family protein	13.1.3.4.11 amino acid metabolism.synthesis.aspartate family.methionine.*S*-adenosylmethionine synthetase	0	–	15734904
–	AT3G20250	–	–	Pumilio 5	35.1.12 not assigned.no ontology.pumilio/Puf RNA-binding domain-containing protein	0	–	15734904
–	AT3G21510	–	–	Histidine-containing phosphotransmitter 1	30.10 signaling.phosphorelay	0	–	23443557
Q9LRL0	AT3G21980	5.2e-114	–	Domain of unknown function (DUF26)	30.2.99 signaling.receptor kinases.misc	1	S	19655309
O81148	AT3G22110	1.6e-124	–	20S proteasome alpha subunit C (PAC1)	29.5.11.20 protein.degradation.ubiquitin.proteasom	0	cytosol	19655309
–	AT3G23990	–	–	Heat shock protein 60	29.6 protein.folding	0	M	15734904
Os02g0813500	AT3G24170	4.7e-196	–	Gluthatione reductase—cytosol or peroxisome?	21.2.2 redox.ascorbate and glutathione.glutathione	0	–	22106097
Os09g0501850	AT3G25220	2.2e-56	AT5G48580	FKBP15-1 (FK506-binding protein 15 kD-1); FK506 binding/peptidyl-prolyl cis-trans isomerase	31.3.1 cell.cycle.peptidylprolyl isomerase	0	S	22106097

(continued)

Table 6.4 (continued)

Original accession	*Arabidpsis* number	Blast e-value	Undistinguishable blast hits	Definition	MapManBin	TMHMM	TargetP	PMID number
–	AT3G25770	–	–	Allene oxide cyclase 2	17.7.1.4 hormone metabolism.jasmonate. synthesis-degradation.allene oxidase cyclase	0	C	18297659
Os03g0219900	AT3G25920	3.8e-68	–	50S ribosomal protein L15	29.2.1.1.1.2.15 protein.synthesis. ribosomal protein.prokaryotic.chloroplast. 50S subunit.L15	0	C	22106097
–	AT3G27850	–	–	Ribosomal protein L12-C	29.2.1.1.1.2.12 protein.synthesis. ribosomal protein.prokaryotic.chloroplast. 50S subunit.L12	0	C	15734904
Os01g0279300	AT3G43810	1.5e-75	AT2G27030, AT2G41110, AT3G56800, AT5G21274, AT1G66410, AT5G37780,	Calmodulin-7 (CAM7)	30.3 signaling.calcium	0	–	22106097
gi|45420766; ACU27904	AT3G46520	6.8e-133	AT5G59370, AT2G37620, AT3G53750, AT5G09810, AT3G12110, AT1G49240, AT3G18780	Actin 12 (ACT12)	31.1 cell.organisation	0	–	22780834; 22924747; 15734904
gi|225434849	AT3G46780	9.0e-131	–	Unknown protein (pTAC16)	28.3 DNA.plastid nucleoid interacting	0	C	19682288
P04045	AT3G46970	3.5e-285	–	Heteroglycan phosphorylase-2 (PHS2) (cytosol)	2.2.2.2 major CHO metabolism. degradation.starch.starch phosphorylase	0	Cytosol	22924747
–	AT3G47070	–	–	LOCATED IN: thylakoid, chloroplast thylakoid membrane	1.1.30 PS.lightreaction.state transition	0	Thylakoid-peripheral-stromal-side	15734904
CAB89318	AT3G47370	4.9e-59	AT3G45030, AT5G62300	40S ribosomal protein S20 (RPS20B)	29.2.1.2.1.20 protein.synthesis.ribosomal protein.eukaryotic.40S subunit.S20	0	Cytosol	22924747; 22115780

(continued)

Table 6.4 (continued)

Original accession	*Arabidpsis* number	Blast e-value	Undistinguishable blast hits	Definition	MapManBin	TMHMM	TargetP	PMID number
gi|21650115	AT3G47470	9.1e-67	–	LHCI-4—LHCI-730	1.1.2.1 PS.lightreaction.photosystem I.LHC-I	0	C	22780834
gi|29373073; gi|21652369, gi|56536433	AT3G48000	4.4e-232	AT1G23800	Aldehyde dehydrogenase (ALDH2)	5.10 fermentation.aldehyde dehydrogenase	0	M	19682288; 22780834
–	AT3G48870	–	–	Clp ATPase	29.5.5 protein.degradation.serine protease	0	C	15734904
Os01g0501800; gi|108951295; gi|31670028; P23322	AT3G50820	4.9e-130	AT5G66570	psbO OEC33-like	1.1.1.2 PS.lightreaction.photosystem II.PSII polypeptide subunits	0	C	22106097; 19682288; 22780834; 22924747; 15734904
gi|4528263; P29449	AT3G51030	5.3e-46	–	Thioredoxin H-type 1	21.1 redox.thioredoxin	0	–	22780834; 22924747
gi|21652217	AT3G51420	1.2e-18	AT3G51430, AT3G51450	Strictosidine synthase family protein	16.4.1 secondary metabolism.N misc.alkaloid-like	0	S	22780834
–	AT3G51800	–	–	ATG2 (G2p-related protein)	29.5.7 protein.degradation. metalloprotease	0	Nucleus	22115780
Os09g0567300; gi|31671767, gi|63061661	AT3G52880	5.8e-175	AT5G03630	Monodehydroascorbate reductase	21.2.1 redox.ascorbate and glutathione.ascorbate	0	Peroxisome	22106097; 22780834; 18297659
–	AT3G52930	–	–	Aldolase superfamily protein	4.7 glycolysis.aldolase	0	Cytosol	15734904
Os02g0192700; gi|28617793	AT3G52960	4.0e-64	–	Peroxiredoxin IIE (PrxII E)	21.5 redox.periredoxins	0	C	22106097; 18165327; 22780834; 15734904
–	AT3G53870	–	–	40S ribosomal protein S3 (RPS3B)	29.2.1.2.1.3 protein.synthesis. ribosomal protein.eukaryotic.40S subunit.S3	0	Cytosol	22115780
Os03g0851200	AT3G54900	1.1e-40	–	Glutaredoxin (formerly CAX-interacting protein 1; CAXIP1)	21.5 redox.periredoxins	0	C	22106097
Os03g0283100	AT3G55040	3.4e-76	–	gluthatione Transferase Lamda -2 (GSTL2)	26.9 misc.glutathione S transferases	0	C	22106097
gi|188321391	AT3G55330	2.2e-72	–	OEC23-like-4 Tat ITP TL25.6	1.1.1.2 PS.lightreaction.photosystem II.PSII polypeptide subunits	0	C	22780834
–	AT3G55380	–	–	Ubiquitin-conjugating enzyme 14	29.5.11.3 protein.degradation. ubiquitin.E2	0	–	15734904

(continued)

Table 6.4 (continued)

Original accession	*Arabidpsis* number	Blast e-value	Undistinguishable blast hits	Definition	MapManBin	TMHMM	TargetP	PMID number
P48491; Os01g0147900; gi|157647066; ABB02628	AT3G55440	2.1e-122	–	Triosephosphate isomerase-2 (TPI-2)	4.8 glycolysis.TPI	0	–	19655309; 22115780; 22106097; 22780834; 22924747; 15734904
Q940F8; gi|31670972	AT3G55800	5.8e-191	–	Sedoheptulose-bisphosphatase (SBPase)	1.3.9 PS.calvin cyle.seduheptulose bisphosphatase	0	C	19655309; 22780834
–	AT3G56310	–	–	Alpha-galactosidase/hydrolase	3.8.2 minor CHO metabolism.galactose.alpha-galactosidases	1	S	22115780
gi|124361199	AT3G56690	7.4e-41	–	CIP111 (CAM INTERACTING PROTEIN 111); ATPase/calmodulin binding	30.3 signaling.calcium	0	C	19682288
Os05g0573700	AT3G58610	3.6e-237	–	Ketol-acid reductoisomerase	13.1.4.1.2 amino acid metabolism.synthesis.branched chain group.common.ketol-acid reductoisomerase	0	C	22106097
BAB20862	AT3G59760	3.5e-136	AT2G43750	Cysteine synthase	13.1.5.3.1 amino acid metabolism.synthesis.serine-glycine-cysteine group.cysteine.OASTL	0	Mitochondria	22924747
Os10g0566700	AT3G60210	2.1e-35	AT2G44650	Cpn10-2	29.6 protein.folding	0	C	22106097
Q9LZY8; Os06g0133800; gi|188283651; gi|225455509; Os04g0682300; gi|31669770	AT3G60750	0	AT2G45290	Transketolase-1 (TKL-1)	1.3.8 PS.calvin cyle.transketolase	0	C	19655309; 22115780; 22106097; 22780834; 19682288; 15734904
–	AT3G61440	–	–	Cysteine synthase (O-acetylserine (thiol)-lyase/O-acetylserine sulfhydrylase/AtcysC1)	13.1.5.3.1 amino acid metabolism.synthesis.serine-glycine-cysteine group.cysteine.OASTL	0	M	22115780
–	AT3G62030	–	–	Rotamase CYP 4	29.6 protein.folding	0	C	15734904
–	AT3G62980	–	–	F-box/RNI-like superfamily protein	17.2.2 hormone metabolism.auxin.signal transduction	0	–	22171938

(continued)

Table 6.4 (continued)

Original accession	*Arabidpsis* number	Blast e-value	Undistinguishable blast hits	Definition	MapManBin	TMHMM	TargetP	PMID number
gil110836843	AT4G00570	1.5e-119	–	Malate oxidoreductase	8.2.10 TCA/org.transformation.other organic acid transformaitons.malic	0	M	22780834
Os07g0638100	AT4G01870	1.4e-189	–	tolB protein-related	35.1 not assigned.no ontology	0	–	22106097
Os03g0314700	AT4G01940	1.3e-58	–	CnfU-Ivb -2Fe-2S assembly PSI and Fd	29.8.1 protein assembly and cofactor ligation.Fe-S assembly	0	C	22106097
–	AT4G02530	–	–	Chloroplast thylakoid lumen protein	35.2 not assigned.unknown	0	C	15734904
–	AT4G03280	–	–	Photosynthetic electron transfer C	1.1.3 PS.lightreaction.cytochrome b6/f	1	C	15734904
Os12g0188700	AT4G03520	2.1e-35	AT3G15360, AT1G03680	Thioredoxin m2	21.1 redox.thioredoxin	0	C	22106097
gil231610	AT4G04640	2.6e-133	–	CF1y—atpC	1.1.4 PS.lightreaction.ATP synthase	0	C	19682288
gil21650492	AT4G05180	1.1e-47	AT4G21280	psbQ OEC16-like Tat lTP	1.1.1.2 PS.lightreaction.photosystem II.PSII polypeptide subunits	0	C	22780834; 18297659
NP_001233851	AT4G08900	4.0e-151	AT4G08870	Arginase 2 (ARGAH1)	13.2.2.3 amino acid metabolism. degradation.glutamate family.arginine	0	M	22924747
gil19570344	AT4G09320	5.8e-72	–	NDPK1—very abundant and multiple localizations	23.4.10 nucleotide metabolism.phosphotransfer and pyrophosphatases.nucleoside diphosphate kinase	0	Cytosol; nucleus; peroxisome	19682288; 22115780; 15734904
Os08g0113100	AT4G10260	3.5e-129	–	pfkB family carbohydrate (Fructose) kinase	2.2.1.1 major CHO metabolism. degradation.sucrose. fructokinase	0	–	22106097
gil19184; gil21650938	AT4G10340	3.0e-107	–	LHCII-5—CP26	1.1.1.1 PS.lightreaction.photosystem II.LHC-II	0	C	19682288; 22780834
–	AT4G11150	–	–	V-ATPase subunit E (VATE)	34.1.1 transport.p- and v-ATPases.H^+-transporting two-sector ATPase	0	Vacuole-tonoplast	22115780
Os02g0664000	AT4G11600	5.6e-74	–	Glutathione peroxidase 6 (ATGPX6)	21.2.2 redox.ascorbate and glutathione.glutathione	0	Not plastid	22106097; 15734904
Os03g0663500; gil45448494, gil46206572	AT4G11650	2.3e-77	–	Osmotin (AtOSM34)	20.2 stress.abiotic	1	S	22106097; 22115780; 22780834

(continued)

Table 6.4 (continued)

Original accession	*Arabidpsis* number	Blast e-value	Undistinguishable blast hits	Definition	MapManBin	TMHMM	TargetP	PMID number
gi|223531971	AT4G12080	4.4e-52	AT4G00200	AT-hook motif nuclear-localized protein 1	27.3.67 RNA.regulation of transcription.putative transcription regulator	0	–	19682288
Os02g0125100	AT4G13430	–	–	Isopropylmalate isomerase (IPMI) LSU—also involved in Met-derived glucosinolate synthesis	13.1.4.4 amino acid metabolism.synthesis.branched chain group.leucine specific	0	C	22106097; 22115780
–	AT4G13850	–	–	Glycine-rich RNA-binding protein 2	27.4 RNA.RNA binding	0	M	15734904
gi|38035226	AT4G13930	1.7e-111	–	Glycine/serine hydroxymethyltransferase	25.1 C1-metabolism.glycine hydroxymethyltransferase	0	–	22780834
Os11g0455500; gi|31669668, gi|28616432	AT4G13940	–	AT3G23810	*S*-adenosyl-L-homocysteine hydrolase (HOG11);	13.2.3.4 amino acid metabolism. degradation.aspartate family.methionine	0	–	22106097; 22115780; 22780834; 15734904
–	AT4G14880	–	–	*O*-acetylserine (thiol) lyase (OAS-TL) isoform A1	13.1.5.3.1 amino acid metabolism.synthesis.serine-glycine-cysteine group.cysteine.OASTL	0	Cytosol	15734904
Os03g0659200	AT4G14890	5.2e-32	–	FdC2 2Fe-2S ferredoxin	26.30 misc.other Ferredoxins and Rieske domain	0	C	22106097
gi|464849	AT4G14960	1.9e-224	AT1G04820, AT1G50010	Tubulin alpha-6 chain (TUA6)	31.1 cell.organisation	0	Cytoskeleton	22780834; 15734904
Os05g0405000	AT4G15530	0	–	Pyruvate phosphate dikinase (PPDK)—splice form 1-plastid	4.4 glycolysis.PPFK	0	Plastid stroma	22106097
–	AT4G16500	–	–	Cystatin/monellin superfamily protein	29.5.3 protein.degradation.cysteine protease	1	S	15734904
Q8VYF1	AT4G17390	4.2e-101	AT4G16720	60S ribosomal protein L15 (RPL15B)	29.2.1.2.2.15 Protein.synthesis.ribosomal protein.eukaryotic.60S subunit.L15	0	Cytosol	19655309
–	AT4G20260	–	–	Plasma-membrane associated cation-binding protein 1	35.1 not assigned.no ontology	0	Plasma membrane	15734904
Os02g0595700	AT4G20360	5.2e-174	–	Elongation factor Tu (EF-Tu-1)	29.2.4 protein.synthesis.elongation	0	C	22106097; 15734904
Os05g0404200	AT4G21860	4.3e-60	–	Methionine sulfoxide reductase type B—2 (MSRB2)	29.11 protein.methione sulfoxide reductases	0	C	22106097

(continued)

Table 6.4 (continued)

Original accession	*Arabidpsis* number	Blast e-value	Undistinguishable blast hits	Definition	MapManBin	TMHMM	TargetP	PMID number
Os01g0662600	AT4G22220	5.1e-57	–	ISU1—scaffold Fe-S assembly	29.8 protein assembly and cofactor ligation	0	M	22106097
gi|62900641	AT4G22240	3.5e-97	AT4G04020	Fibrillin 1b (FBN1b)	26.31 misc.fibrillins	0	C/plastoglobules	19682288
Os07g0628700	AT4G23180	4.4e-138	–	Cysteine-rich RLK (RECEPTOR-like protein kinase) 10	30.2.17 signaling.receptor kinases.DUF 26	2	–	22106097
–	AT4G23670	–	–	Polyketide cyclase/dehydrase and lipid transport superfamily protein	20.2.99 stress.abiotic.unspecified	0	–	15734904
gi|55290624	AT4G24190	5.4e-117	–	HSP90 or GRP94—similar to Sheperd (Hsp90-7)	29.6 protein.folding	0	S/ER	22780834
Os12g0534200	AT4G25050	3.0e-27	AT1G54580, AT1G54630, AT5G27200, AT3G05020	Acyl carrier protein	11.1.12 lipid metabolism.FA synthesis and FA elongation.ACP protein	0	C	22106097; 15734904
Q9SEK0	AT4G26200	2.7e-170	–	1-amino-cyclopropane-1-carboxylate synthase 7	17.5.1.1 hormone metabolism.ethylene.synthesis-degradation.1-aminocyclopropane-1-carboxylate synthase	0	–	19655309
gi|21650493	AT4G26530	1.3e-104	–	Fructose-bisphosphate aldolase	4.7 glycolysis.aldolase	0	Cytosol	22780834
Os01g0191200	AT4G29260	1.7e-40	–	Subfamily IIIB acid phosphatase	26.13 misc.acid and other phosphatases	1	S	22106097
gi|147766607; gi|224092554	AT4G31990	2.8e-207	–	Aspartate aminotransferase	13.1.1.2.1 amino acid metabolism.synthesis.central amino acid metabolism.aspartate.aspartate aminotransferase	0	C	19682288; 22780834
gi|225430312	AT4G32260	1.2e-46	–	CFO-II—atpG	1.1.4 PS.lightreaction.ATP synthase	0	C	19682288
gi|121083; Os01g0711400	AT4G33010	0	AT2G26080	Glycine decarboxylase/glycine cleavage system P-protein (ATGLDP1)	13.2.5.2 amino acid metabolism.degradation.serine-glycine-cysteine group.glycine	0	M	22213812; 23238061; 22106097; 20089767
–	AT4G33030	–	–	UDP-sulfoquinovose synthase (SQD1)	11.10.3 lipid metabolism.glycolipid synthesis.UDP-sulfoquinovose synthase	0	C	22115780
gi|38035627, gi|42623464	AT4G34050	7.5e-95	–	Caffeoyl-CoA 3-*O*-methyltransferase	16.2.1.6 secondary metabolism. phenylpropanoids.lignin biosynthesis.CCoAOMT	0	–	22780834

(continued)

Table 6.4 (continued)

Original accession	*Arabidpsis* number	Blast e-value	Undistinguishable blast hits	Definition	MapManBin	TMHMM	TargetP	PMID number
–	AT4G34200	–	–	D-3-phosphoglycerate dehydrogenase	13.1.5.1.1 amino acid metabolism.synthesis.serine-glycine-cysteine group.serine.phosphoglycerate dehydrogenase	0	C	15734904
–	AT4G34870	–	–	Rotamase cyclophilin 5	29.6 protein.folding	0	Cytosol; nucleus; plasma membrane	15734904
gi|115705; CAA85470	AT4G35090	2.3e-242	AT1G20630	Catalase 2 (CAT2)	21.6 redox.dismutases and catalases	0	Peroxisome	22213812; 22924747
Os03g0851700	AT4G35450	6.8e-94	AT2G17390	Ankyrin repeat family protein (AKR2A)—chaperone envelope OEP	31.1 cell.organisation	0	Cytosol; nucleus	22106097
Os12g0567700	AT4G36130	3.3e-124	–	60S ribosomal protein L8 (RPL8C)	29.2.1.2.2.8 Protein.synthesis. ribosomal protein.eukaryotic. 60S subunit.L8	0	Cytosol	22106097
gi|45447792	AT4G36195	6.4e-82	AT4G36190, AT2G18080	Serine carboxypeptidase (S28)	29.5.5 protein.degradation.serine protease	1	Peroxisome	22780834
–	AT4G36540	–	–	BR enhanced expression 2	27.3.6 RNA.regulation of transcription.bHLH,Basic Helix-Loop-Helix family	0	–	20089767
gi|9794870; Q2MY50, P15477	AT4G37050	2.4e-114	AT2G26560, AT4G37070	PATATIN-like protein 4	33.1 development.storage proteins	0	–	19682288; 22924747
Os03g0265900	AT4G37300	9.0e-28	–	MEE59 (maternal effect embryo arrest 59)	35.2 not assigned.unknown	0	–	22106097
gi|1346156; gi|462187, gi|301631113	AT4G37930	4.5e-239	AT4G32520	Glycine/serine hydroxymethyltransferase; erine/threonine aldolase (SHM1)	25.1 C1-metabolism.glycine hydroxymethyltransferase	0	M	19682288; 23238061; 20089767
gi|188265008	AT4G37990	1.1e-61	AT4G37980, AT4G37970, AT4G39330	ELI3-2 (ELICITOR-ACTIVATED GENE 3)	16.2.1.10 secondary metabolism. phenylpropanoids.lignin biosynthesis.CAD	0	–	22780834
Os03g0128800	AT4G38225	1.2e-64	–	Unknown protein	35.2 not assigned.unknown	0	C	22106097
Os08g0129200	AT4G38680	5.6e-37	–	Glycine-rich protein (CSDP2/ AtGRP2)- dually targeted to cytosol and nucleus	20.2.2 stress.abiotic.cold	0	Cytosol; nucleus	22106097

(continued)

Table 6.4 (continued)

Original accession	*Arabidpsis* number	Blast e-value	Undistinguishable blast hits	Definition	MapManBin	TMHMM	TargetP	PMID number
Q9LLD7, Q944G9, Q3E716; Q1A7T7; Os11g0171300; gi|57932482; gi|55292638; ABY58016	AT4G38970	3.1e-160	AT2G21330	Fructose-bisphosphate aldolase-2 (SFBA-2)	1.3.6 PS.calvin cyle.aldolase	0	C/plastid stroma; plastoglobules	19655309; 18445036; 22106097; 22780834; 22780834; 22924747; 15734904
Os02g0469600	AT4G39090	3.3e-133	AT4G16190, AT2G21430	Cysteine-type peptidase (RD19 - RESPONSIVE TO DEHYDRATION 19);	29.5.3 protein.degradation.cysteine protease	0	S	22106097
gi|57572853	AT5G02500	–	AT3G09440, AT1G56410, AT5G02490	HSP70-1 (HSC70-1) (not plastid)—very abundant—dual localized nucleus and cytosol	29.6 protein.folding	0	Cytosol; nucleus	22780834; 22115780; 15734904
gi|28618466	AT5G02790	2.6e-85	AT5G02780	Gluthatione Transferase Lamda - 1B (GSTL1B)	26.9 misc.glutathione *S* transferases	0	Cytosol	22780834
gi|56531804	AT5G03300	3.8e-123	AT3G09820	adenosine kinase 2 (ADK2) (Adenosine to AMP)	23.3.2.1 nucleotide metabolism.salvage.nucleoside kinases.adenosine kinase	0	–	22780834; 22115780; 15734904
Os10g0545700	AT5G03455	4.0e-41	–	Rhodanese-like domain-containing protein (cdc25)	26.23 misc.rhodanese	0	M	22106097
Os07g0658400	AT5G04140	0	–	Ferredoxin-dependent glutamate synthase/glu1/Fd-GOGAT 1 (dual cTP and mTP)	12.2.1.1 N-metabolism.ammonia metabolism.glutamate synthase.ferredoxin dependent	0	C	22106097
–	AT5G04200	–	–	Metacaspase 9	29.5 protein.degradation	0	–	17110382
gi|71597361	AT5G05460	4.7e-54	AT3G11040	Glycosyl hydrolase family 85	35.1 not assigned.no ontology	0	C	22780834
gi|55404784, gi|31671693	AT5G06720	8.1e-82	AT5G06730, AT4G08780, AT3G49120, AT3G49110, AT3G32980, AT4G08770	Peroxidase 2	26.12 misc.peroxidases	1	S	22780834
gi|11596188	AT5G06740	7.9e-15	AT5G03140, AT5G55830, AT3G55550	Concanavalin A-like lectin protein kinase family protein	30.2.19 signaling.receptor kinases.legume-lectin	1	S	19682288
–	AT5G07440	–	–	Glutamate dehydrogenase 2	12.3.1 N-metabolism.N-degradation.glutamate dehydrogenase	0	M	15734904

(continued)

Table 6.4 (continued)

Original accession	*Arabidpsis* number	Blast e-value	Undistinguishable blast hits	Definition	MapManBin	TMHMM	TargetP	PMID number
Os04g0677500	AT5G08570	5.7e-232	AT5G63680	Pyruvate kinase	4.13 glycolysis.PK	0	Cytosol	22106097
gi|2116558; gi|114421, gi|34419031	AT5G08670	8.0e-210	AT5G08690, AT5G08680	H + -transporting ATP synthase beta chain	9.9 mitochondrial electron transport/ ATP synthesis.F1-ATPase	0	M	23238061; 22780834
gi|28616087	AT5G08690	1.5e-80	AT5G08670, AT5G08680	H + -transporting ATP synthase beta chain (mitochondrial)	9.9 mitochondrial electron transport/ ATP synthesis.F1-ATPase	0	M	22780834
gi|585272; gi|28617128	AT5G09590	1.3e-287	AT4G37910	heat shock protein mtHsc70-2 (Hsc70-5)	29.6 protein.folding	0	M	23238061; 22780834
gi|55291273	AT5G09640	4.3e-76	AT1G33540, AT3G12203, AT3G12220, AT1G73270, AT3G10450, AT1G73300, AT3G12240, AT3G12230, AT1G73290, AT5G36180, AT2G22920, AT2G23010, AT1G73280, AT1G73310, AT2G22980,	serine carboxypeptidase-like 19	29.5 protein.degradation	1	S	22780834
gi|3183079; gi|37509092	AT5G09660	4.6e-150	AT2G22780	peroxisomal NAD-malate dehydrogenase 2	6.3 gluconeogenesis.Malate DH	0	Peroxisome	22213812; 22780834; 15734904
Os03g0390000	AT5G10360	1.2e-96	–	40S ribosomal protein S6 (RPS6B)—EMB3010	29.2.1.2.1.6 protein.synthesis. ribosomal protein.eukaryotic. 40S subunit.S6	0	Cytosol	22106097
–	AT5G10860	–	–	Cystathionine beta-synthase (CBS) family protein	21.1 redox.thioredoxin	0	M	15734904
gi|255546447	AT5G11520	1.1e-187	AT5G19550	Aspartate aminotransferase (ASP3) (YLS4)	13.1.1.2.1 amino acid metabolism.synthesis.central amino acid metabolism.aspartate.aspartate aminotransferase	0	C	22780834
AAB54016	AT5G13420	5.0e-169	–	Transaldolase-1—very abundant	7.2.2 OPP.non-reductive PP.transaldolase	0	C	22924747

(continued)

Table 6.4 (continued)

Original accession	*Arabidpsis* number	Blast e-value	Undistinguishable blast hits	Definition	MapManBin	TMHMM	TargetP	PMID number
Os11g0525600	AT5G13980	0	–	Glycosyl hydrolase family 38 protein	26.3 misc.gluco-, galacto- and mannosidases	0	S	22106097
–	AT5G14040	–	–	Phosphate transporter (PHT3-1 or PIC1)	34.9 transport.metabolite transporters at the mitochondrial membrane	0	Mitochondria	22115780
–	AT5G14260	–	–	Rubisco methyltransferase family protein	27.3.69 RNA.regulation of transcription.SET-domain transcriptional regulator family	0	C	15734904
–	AT5G15490	–	–	UDP-glucose 6-dehydrogenase	10.1.4 cell wall.precursor synthesis.UGD	0	S	22115780
Os04g0439900	AT5G16620	4.7e-86	–	Tic40 (PDE120—PIGMENT DEFECTIVE EMBRYO)	29.3.3 protein.targeting.chloroplast	0	C	22106097
ABM69253	AT5G16710	2.2e-97	–	Dehydroascorbate reductase-2 (DHAR-2) plastid stroma	21.2.1 redox.ascorbate and glutathione.ascorbate	0	C	22924747
AAB71613	AT5G17310	1.6e-195	AT3G03250	UDP-glucose pyrophosphorylase	4.1 glycolysis.UGPase	0	S	22924747
Q6KCR2; Q9SRV5	AT5G17920	0	AT3G03780, AT5G20980	5-methyltetrahydropteroyl triglutamate–homocysteine *S*-methyltransferase (ATCIMS)—abundant	13.1.3.4 amino acid metabolism.synthesis.aspartate family.methionine	0	Cytosol	19655309; 18445036; 15734904
Os03g0219200	AT5G18100	1.1e-52	–	Copper/zinc superoxide dismutase (CSD3)	21.6 redox.dismutases and catalases	0	Peroxisome	22106097
AAT40505	AT5G19510	1.2e-64	AT5G12110	Elongation factor 1B alpha-subunit 2 (eEF1Balpha2)	29.2.4 protein.synthesis.elongation	0	–	22924747; 15734904
gi|157058854	AT5G20630	1.3e-72	–	Germin-like protein (GER3)	20.2.99 stress.abiotic.unspecified	0	S/extracellular	19682288; 18297659
gi|157342492	AT5G20720	3.1e-98	–	Cpn21 (also Cpn20)	29.6 protein.folding	0	C	19682288; 15734904
–	AT5G23120	–	–	Photosystem II stability/assembly factor, chloroplast (HCF136)	29.8 protein assembly and cofactor ligation	0	C	15734904
Os02g0634500	AT5G23140	1.4e-63	–	ClpP2	29.5.5 protein.degradation.serine protease	0	M	22106097
Os07g0577700	AT5G23250	1.8e-139	AT5G08300	Succinyl-CoA ligase, alpha subunit	8.1.6 TCA/org.transformation. TCA.succinyl-CoA ligase	0	M	22106097
Os07g0565000	AT5G23740	2.8e-65	AT3G48930, AT4G30800	RPS11-BETA (RIBOSOMAL PROTEIN S11-BETA); structural constituent of ribosome	29.2.1.2.1.11 protein.synthesis.ribosomal protein.eukaryotic.40S subunit.S11	0	Cytosol	22106097

(continued)

Table 6.4 (continued)

Original accession	*Arabidpsis* number	Blast e-value	Undistinguishable blast hits	Definition	MapManBin	TMHMM	TargetP	PMID number
gi|135444; Os03g0105600	AT5G23860	8.5e-190	AT5G12250, AT5G62690, AT5G62700, AT2G29550, AT1G75780, AT4G20890	Tubulin beta-8 chain (TUB8)	31.1 cell.organisation	0	–	22780834; 22106097; 22115780
NP_001233834	AT5G24490	2.5e-87	–	PSRP-1 associates with 30S (translation factor? Interact with RRF?)	29.2.1.1.1.1.530 protein.synthesis. ribosomal protein.prokaryotic.chloroplast. 30S subunit.S30A	0	C	22924747
–	AT5G25100	–	–	Endomembrane protein 70	35.1 not assigned.no ontology	10	S	22115780
Q56H06	AT5G26000	1.3e-207	–	Thioglucoside glucohydrolase 1 (TGG1) (myrosinase)	16.5.1 secondary metabolism.sulfur-containing.glucosinolates	0	S	19655309
gi|57873962	AT5G28010	5.5e-28	AT1G14930, AT1G70830, AT1G70850, AT1G14940, AT2G01520, AT1G70890, AT1G70840, AT4G23680, AT1G14950, AT3G26460, AT2G01530, AT4G23670, AT1G14960, AT4G14060, AT5G28000	Polyketide cyclase/dehydrase and lipid transport superfamily protein	20.2.99 stress.abiotic.unspecified	0	–	22780834
Os10g0417600	AT5G28840	2.2e-191	–	5-epimerase/NAD binding/ catalytic—GME =	21.2.1.1 redox.ascorbate and glutathione.ascorbate.GME	0	–	22106097
Os03g0315800	AT5G30510	6.8e-133	–	30S ribosomal protein S1	29.2.1.1.3.1.1 protein.synthesis.ribosomal protein.prokaryotic.unknown organellar.30S subunit.S1	0	C	22106097
–	AT5G35100	–	–	Cyclophilin-like peptidyl-prolyl cis-trans isomerase family protein	29.6 protein.folding	0	C	18297659

(continued)

Table 6.4 (continued)

Original accession	*Arabidpsis* number	Blast e-value	Undistinguishable blast hits	Definition	MapManBin	TMHMM	TargetP	PMID number
–	AT5G35630	–	–	Glutamine synthetase 2	12.2.2 N-metabolism.ammonia metabolism.glutamine synthase	0	C	15734904
gi|357460817	AT5G37510	0	–	NADH-DH 76 kDa subunit—Complex I	9.1.1 mitochondrial electron transport/ATP synthesis.NADH-DH.complex I	0	M	23238061
gi|110883423	AT5G37600	4.4e-145	AT5G16570, AT1G66200, AT3G17820	Glutamine synthetase/Glutamate–ammonia ligase (GS1)	12.2.2 N-metabolism.ammonia metabolism.glutamine synthase	0	–	22780834
P05346, Q43746, A1YN03; Q43746; gi|132097; Os12g0292400; gi|132106; gi|57932579, gi|56529771, gi|21650278; P26576	AT5G38410	3.7e-86	AT5G38420, AT5G38430, AT1G67090	Rubisco small subunit 3b (RBCS-3B)	1.3.2 PS.calvin cyle.rubisco small subunit	0	C	19655309; 18445036; 22213812; 22106097; 23238061; 22780834; 22924747; 15734904
ABU96710; gi|15637350	AT5G40370	3.1e-41	–	Glutaredoxin	21.4 redox.glutaredoxins	0	S	22924747; 19682288; 15734904
gi|224058097	AT5G42020	0	AT5G28540	Luminal binding protein 2 precursor (BiP-2) (AtBP2)	29.6 protein.folding	1	S/ER	22780834; 15734904
Os11g0456300, Os09g0539500	AT5G42190	7.9e-52	AT1G75950, AT4G34210, AT4G34470, AT2G03170	ASK2 (SKP1-LIKE 2); ubiquitin-protein ligase	29.5.11.4.3.1 protein.degradation.ubiquitin.E3.SCF.SKP	0	–	22106097
Os04g0670200	AT5G43060	1.9e-144	–	Cysteine proteinase	29.5.3 protein.degradation.cysteine protease	0	Cytosol; nucleus	22106097
Os10g0478200; gi|8789134	AT5G43330	1.8e-155	AT1G04410	Malate dehydrogenase	8.2.9 TCA/org.transformation.other organic acid transformaitons.cyt MDH	0	–	22106097; 22780834
–	AT5G44340	–	–	Tubulin beta-4 chain (TUB4)	31.1 cell.organisation	0	–	22115780; 15734904
Os04g0208200	AT5G45890	8.1e-66	AT5G50260, AT3G48340	Senescence-associated gene 12	29.5.3 protein.degradation.cysteine protease	1	S	22106097
Os03g0563300	AT5G45930	1.5e-146	–	Mg-protoporphyrin IX chelatase—CHLI-2	19.10 tetrapyrrole synthesis.magnesium chelatase	0	C	22106097
gi|15238002	AT5G47070	3.9e-208	–	Protein kinase superfamily protein	29.4.1.57 protein.postranslational modification.kinase.receptor like cytoplasmatic kinase VII	0	C	23238061

(continued)

Table 6.4 (continued)

Original accession	*Arabidpsis* number	Blast e-value	Undistinguishable blast hits	Definition	MapManBin	TMHMM	TargetP	PMID number
Q2A9K1	AT5G47090	2.3e-113	–	Coiled coil-type protein (InterPro:IPR018613); Has 30201 Blast hits to 17322 proteins in 780 species: Archae—12; Bacteria—1396; Metazoa—17338; Fungi—3422; Plants—5037; Vir	35.2 not assigned.unknown	0	–	19655309
Os12g0244100; gi|124245039, gi|1143427; gi|110861913	AT5G49910	8.6e-284	AT4G24280	cpHSP70-2 (Dnak homologue)	29.6 protein.folding	0	C	22106097; 19682288; 21964330; 22780834
Os12g0564600	AT5G49930	1.9e-275	–	Zinc knuckle (CCHC-type) family protein	35.1 not assigned.no ontology	0	–	22106097
Os11g0267400	AT5G50920	0	–	ClpC1 (also named HSP93-V)—highly similar to ClpC2	29.5.5 protein.degradation.serine protease	0	C	22106097
gi|46402892; gi|195548074; O65327; gi|70732151	AT5G51100	4.7e-86	AT4G25100	Iron superoxide dismutase 2 (FSD2)	21.6 redox.dismutases and catalases	0	C	19682288; 22780834; 19655309; 23238061
–	AT5G52640	–	–	Heat shock protein 90	20.2.1 stress.abiotic.heat	0	–	15734904
Os10g0502000	AT5G53490	5.6e-67	–	Peptapetide repeat TL17.4 (PPR)	27.7 RNA.misc.pentatricopeptide (PPR) repeat-containing protein	0	C	22106097
gi|34431011	AT5G54160	1.0e-90	–	Quercetin 3-*O*-methyltransferase 1/caffeic acid/5-hydroxyferulic acid *O*-methyltransferase (OMT1) - dual localized in cytosol and nucleus	16.2.1.9 secondary metabolism. phenylpropanoids.lignin biosynthesis.COMT	0	–	22780834
–	AT5G55220	–	–	Trigger factor type chaperone family protein	29.6 protein.folding	0	C	15734904
gi|37509117	AT5G55480	8.6e-78	AT4G26690	Glycerophosphoryl diester phosphodiesterase family protein	11.9.3.3 lipid metabolism. lipid degradation. lysophospholipases. glycerophosphodiester phosphodiesterase	1	S/plasma membrane	22780834
Os07g0585000	AT5G55530	3.5e-72	–	Calcium-dependent lipid-binding (CaLB domain) family protein	20.2.2 stress.abiotic.cold	0	C	22106097

(continued)

Table 6.4 (continued)

Original accession	*Arabidpsis* number	Blast e-value	Undistinguishable blast hits	Definition	MapManBin	TMHMM	TargetP	PMID number
–	AT5G56010	–	–	Heat shock protein 81-3	20.2.1 stress.abiotic.heat	0	S	15734904
gi|225462013	AT5G56030	0,005	AT5G56010, AT5G56000	Hsp90.2	20.2.1 stress.abiotic.heat	0	S	22780834
gi|110845815; Os02g0102900	AT5G56500	8.2e-121	AT1G55490, AT3G13470	Cpn60-beta-3—model.2 has higher score than model.1	29.6 protein.folding	0	C	22780834; 22106097
gi|225469736	AT5G58050	2.5e-220	AT5G58170	SHV3-like 4	11.9.3.3 lipid metabolism. lipid degradation. lysophospholipases. glycerophosphodiester phosphodiesterase	1	S	19682288
gi|38051404; AAA33837	AT5G58390	8.5e-71	AT5G58400, AT5G05340	Peroxidase, putative	26.12 misc.peroxidases	0	S	22780834; 22924747
–	AT5G59880	–	–	Actin depolymerizing factor 3	31.1 cell.organisation	0	–	15734904
Os09g0442300	AT5G60360	1.1e-139	–	Cysteine proteinase (AALP) - similar to barley aleurain	29.5.3 protein.degradation.cysteine protease	1	Vacuole	22106097
–	AT5G60390	–	–	GTP binding Elongation factor Tu family protein	29.2.4 protein.synthesis.elongation	0	Cytosol; nucleus	15734904
gi|28617228	AT5G60600	2.0e-103	–	4-hydroxy-3-methylbutyl diphosphate synthase (HDS; also named GcpE	16.1.1.6 secondary metabolism.isoprenoids.non-mevalonate pathway.HDS	0	C	22780834
gi|57569897	AT5G61410	1.3e-81	–	Ribulose-5-phosphate-3-epimerase (RPE)	1.3.11 PS.calvin cyle.RPE	0	C	22780834
–	AT5G61790	–	–	Calnexin 1 (CNX1)	30.3 signaling.calcium	1	S/ER	22115780
Q84W15	At5g62750	2.0e-16	–	Unknown protein; Has 24942 Blast hits to 11726 proteins in 897 species: Archae—48; Bacteria—2086; Metazoa—8361; Fungi—2243; Plants—1040; Viruses—210; Other Eukaryotes—10954 (*source* NCBI BLink).	35.2 not assigned.unknown	0	–	19655309
Os05g0578000	AT5G62810	4.7e-16	–	PEX14 (PEROXISOME DEFECTIVE 2)	35.1 not assigned.no ontology	1	Peroxisome	22106097
Q852S5	AT5G63310	3.3e-69	–	NDPK2-stromal	23.4.10 nucleotide metabolism.phosphotransfer and pyrophosphatases.nucleoside diphosphate kinase	0	C	22924747

(continued)

Table 6.4 (continued)

Original accession	*Arabidpsis* number	Blast e-value	Undistinguishable blast hits	Definition	MapManBin	TMHMM	TargetP	PMID number
gi|225456404	AT5G63570	9.6e-198	AT3G48730	1-aminomutase 1 (GSA 1)	19.3 tetrapyrrole synthesis.GSA	0	C	19682288
gi|37509367	AT5G64040	2.1e-51	–	psaN—TAT LTP	1.1.2.2 PS.lightreaction.photosystem I.PSI polypeptide subunits	0	C	22780834
Os03g0385400	AT5G64080	7.9e-13	–	Bifunctional inhibitor/lipid-transfer protein/seed storage 2S albumin superfamily protein	26.21 misc.protease inhibitor/seed storage/lipid transfer protein (LTP) family protein	2	S	22106097
–	AT5G65210	–	–	bZIP transcription factor family protein	27.3.35 RNA.regulation of transcription.bZIP transcription factor family	0	–	20716698
gi|5802794; gi|46213374; ACV69975	AT5G65430	1.2e-126	AT5G10450	14-3-3 protein GF14 kappa (grf8)	30.7 signaling.14-3-3 proteins	0	Nucleus	19682288; 22780834; 22924747
AAZ85313	AT5G66190	6.0e-141	AT1G20020	FNR-1	1.1.7 PS.lightreaction.ferredoxin reductase	0	C	22924747
gi|195647178	AT5G66760	2.0e-293	–	Succinate dehydrogenase (ubiquinone) flavoprotein subunit	8.1.7 TCA/org.transformation. TCA.succinate dehydrogenase	0	M	23238061; 22115780
gi|4271228/ gi|122166198; gi|21217723, gi|114329641; ABB90028	ATCG00120	7.5e-230	–	CF1a—atpA	1.1.4 PS.lightreaction.ATP synthase	0	Thylakoid-peripheral-stromal-side	19682288; 22780834; 22924747; 15734904
gi|28630971	ATCG00270	3.4e-115	–	psbD D2	1.1.1.2 PS.lightreaction.photosystem II.PSII polypeptide subunits	6	Thylakoid-integral	22780834; 15734904
gi|122166187	ATCG00280	1.0e-234	–	psbC CP43	1.1.1.2 PS.lightreaction.photosystem II.PSII polypeptide subunits	5	Thylakoid-integral	22780834
gi|114329655	ATCG00340	–	–	PsaA/PsaB protein	1.1.2.2 PS.lightreaction.photosystem I.PSI polypeptide subunits	11	Thylakoid-integral	22780834; 22115780
gi|114329656	ATCG00350	0	–	PsaA/PsaB protein	1.1.2.2 PS.lightreaction.photosystem I.PSI polypeptide subunits	9	Thylakoid-integral	22780834

(continued)

Table 6.4 (continued)

Original accession	*Arabidpsis* number	Blast e-value	Undistinguishable blast hits	Definition	MapManBin	TMHMM	TargetP	PMID number
gi|6634488	ATCG00480	8.4e-238	–	CF1b—atpB	1.1.4 PS.lightreaction.ATP synthase	0	Thylakoid-integral	22780834; 15734904
O03042, Q6Y9Y8; Q33557; gi|113374116; Os11g0532750, Os12g0207600; gi|21634141; gi|118572704; gi|4049496; YP_635647	ATCG00490	1.3e-257	–	Rubisco large subunit (RBCL)	1.3.1 PS.calvin cyle.rubisco large subunit	0	Thylakoid-integral	19655309; 18445036; 22213812; 22115780; 22106097; 23238061; 19682288; 22780834; 22924747; 15734904
gi|114329673	ATCG00580	1.7e-40	–	psbE cytb559a	1.1.1.2 PS.lightreaction.photosystem II.PSII polypeptide subunits	1	Thylakoid-integral	22780834
gi|113952647	ATCG00680	5.5e-273	–	psbB CP47	1.1.1.2 PS.lightreaction.photosystem II.PSII polypeptide subunits	5	Thylakoid-integral	19682288; 15734904
gi|114329685	ATCG00720	3.1e-112	–	petB—Cytochrome b6	1.1.3 PS.lightreaction.cytochrome b6/f	4	Thylakoid-integral	22780834
gi|7525086	ATCG01060	2.6e-44	–	psaC—subunit VII—stromal side	1.1.2.2 PS.lightreaction.photosystem I.PSI polypeptide subunits	0	Thylakoid-peripheral-stromal-side	22780834
gi|224020956; gi|114408, gi|1381685	ATMG01190	1.2e-240	AT2G07698	atp1 ATPase subunit 1	9.9 mitochondrial electron transport/ ATP synthesis.F1-ATPase	0	Mitochondria	23238061; 22780834; 20089767
AAM22686	–	–	–	Pathogenesis related protein 10				22924747
Os01g0910900	–	–	–	Uncharacterized protein				22106097
Os02g0121100	–	–	–	Predicted protein				22106097

The blast gives sometimes several hits with the same (or very close) e-value leading to ambigous identification. Functional categories, number of predicted trans-membrane domains, and cellular location have been reported using MapmanBin (http://mapman.gabipd.org/ and http://ppdb.tc.cornell.edu/), TMHMM (http://www.cbs.dtu.dk/services/TMHMM/) and TargetP (http://www.cbs.dtu.dk/services/TargetP/) respectively

For protein location, *bold* indicates that the prediction has been verified experimentally, *S* for secreted, *C* for plastid, and *M* for mitochondria. A total of 373 nitrosylated proteins have been identified showing a wide range of functions and cellular location functions. In nearly 90 % of all plant proteomic studies, BS was used to identify nitrosothiols and *Arabidopsis thaliana* was the model plant in 68 % studies (Table 6.3). Unfortunately, nitrosylation sites were rarely identified and a single study so far quantified nitrosylation peptides (Fares et al. 2011) and two studies identified nitrosylation sites without chemical derivation (Lindermayr et al. 2006; Elviri et al. 2010). The development of approaches permitting direct reduction of nitrosothiols as it is the case with phosphine switch could reduce time and losses induced by this multiple step experiment and facilitate the identification and quantitation of nitrosothiols if isotopic tags are linked to these reagents.

Acknowledgments This work was supported by a research grant to A. F. from Région Languedoc-Roussillon. We thank Drs. Claude Nespoulous and Wojtek Szponarski for critical reading of the manuscript.

References

Abat JK, Mattoo AK, Deswal R (2008) *S*-nitrosylated proteins of a medicinal CAM plant *Kalanchoe pinnata*- ribulose-1 5-bisphosphate carboxylase/oxygenase activity targeted for inhibition. FEBS J 275:2862–2872

Abat JK, Deswal R (2009) Differential modulation of *S*-nitrosoproteome of *Brassica juncea* by low temperature: change in *S*-nitrosylation of Rubisco is responsible for the inactivation of its carboxylase activity. Proteomics 9:4368–4380

Arnold WP, Mittal CK, Katsuki S, Murad F (1977) Nitric oxide activates guanylate cyclase and increases guanosine 3′:5′-cyclic monophosphate levels in various tissue preparations. Proc Natl Acad Sci USA 74:3203–3207

Astier J, Lindermayr C (2012) Nitric oxide-dependent posttranslational modification in plants: an update. Int J Mol Sci 13:15193–15208

Bechtold E, Reisz JA, Klomsiri C et al (2010) Water-soluble triarylphosphines as biomarkers for protein *S*-nitrosation. ACS Chem Biol 5:405–414

Belenghi B, Romero-Puertas MC, Vercammen D et al (2007) Metacaspase activity of *Arabidopsis thaliana* is regulated by *S*-nitrosylation of a critical cysteine residue. J Biol Chem 282:1352–1358

Benhar M, Thompson JW, Moseley MA, Stamler JS (2010) Identification of *S*-nitrosylated targets of thioredoxin using a quantitative proteomics approach. Biochem 49:6963–6969

Camejo D, Romero-Puertas Mdel C, Rodríguez-Serrano M et al (2013) Salinity-induced changes in *S*-nitrosylation of pea mitochondrial proteins. J Proteomics 79:87–99

Camerini S, Polci ML, Restuccia U et al (2007) A novel approach to identify proteins modified by nitric oxide: the HIS-TAG switch method. J Proteome Res 6:3224–3231

Chen YY, Chu HM, Pan KT et al (2008) Cysteine *S*-nitrosylation protects protein-tyrosine phosphatase 1B against oxidation-induced permanent inactivation. J Biol Chem 283: 35265–35272

Chouchani ET, Hurd TR, Nadtochiy SM et al (2010) Identification of *S*-nitrosated mitochondrial proteins by *S*-nitrosothiol difference in gel electrophoresis (SNO-DIGE): implications for the regulation of mitochondrial function by reversible *S*-nitrosation. Biochem J 430:49–59

Ckless K, Reynaert NL, Taatjes DJ et al (2004) In situ detection and visualization of *S*-nitrosylated proteins following chemical derivatization: identification of Ran GTPase as a target for *S*-nitrosylation. Nitric Oxide 11:216–227

Delledonne M, Xia Y, Dixon RA, Lamb C (1998) Nitric oxide functions as a signal in plant disease resistance. Nature 394:585–588

Doulias PT, Greene JL, Greco TM et al (2010) Structural profiling of endogenous *S*-nitrosocysteine residues reveals unique features that accommodate diverse mechanisms for protein *S*-nitrosylation. Proc Natl Acad Sci USA 107:16958–16963

Durner J, Wendehenne D, Klessig DF (1998) Defense gene induction in tobacco by nitric oxide, cyclic GMP, and cyclic ADP-ribose. Proc Natl Acad Sci USA 95:10328–10333

Elviri L, Speroni F, Careri M et al (2010) Identification of in vivo nitrosylated phytochelatins in *Arabidopsis thaliana* cells by liquid chromatography-direct electrospray-linear ion trap-mass spectrometry. J Chromatogr A 1217:4120–4126

Faccenda A, Bonham CA, Vacratsis PO et al (2010) Gold nanoparticle enrichment method for identifying *S*-nitrosylation and*S*-glutathionylation sites in proteins. J Am Chem Soc 132:11392–11394

Fares A, Rossignol M, Peltier JB (2011) Proteomics investigation of endogenous *S*-nitrosylation in *Arabidopsis*. Biochem Biophys Res Commun 416:331–336

Feng J, Wang C, Chen Q et al (2013) *S*-nitrosylation of phosphotransfer proteins represses cytokinin signaling. Nat Commun 4:1529

Forrester MT, Thompson JW, Foster MW et al (2009) Proteomic analysis of *S*-nitrosylation and denitrosylation by resin-assisted capture. Nat Biotechnol 27:557–559

Foster MW, Forrester MT, Stamler JS (2009) A protein microarray-based analysis of *S*-nitrosylation. Proc Natl Acad Sci USA 106:18948–18953

Furchgott RF, Zawadzki JV (1980) The obligatory role of endothelial cells in the relaxation of arterial smooth muscle by acetylcholine. Nature 288:373–376

Gow AJ, Chen Q, Hess DT et al (2002) Basal and stimulated protein *S*-nitrosylation in multiple cell types and tissues. J Biol Chem 277:9637–9640

Han P, Zhou X, Huang B et al (2008) On-gel fluorescent visualization and the site identification of *S*-nitrosylated proteins. Anal Biochem 377:150–155

Hao G, Derakhshan B, Shi L et al (2006) SNOSID a proteomic method for identification of cysteine *S*-nitrosylation sites in complex protein mixtures. Proc Natl Acad Sci USA 103:1012–1017

Hess DT, Matsumoto A, Nudelman R, Stamler JS (2001) *S*-nitrosylation: spectrum and specificity. Nat Cell Biol 3:E46–E49

Holtgrefe S, Gohlke J, Starmann J et al (2008) Regulation of plant cytosolic glyceraldehyde 3-phosphate dehydrogenase isoforms by thiol modifications. Physiol Plant 133:211–228

Ignarro LJ, Buga GM, Wood KS et al (1987) Endothelium-derived relaxing factor produced and released from artery and vein is nitric oxide. Proc Natl Acad Sci USA 84:9265–9269

Jaffrey SR, Snyder SH (2001) The biotin switch method for the detection of *S*-nitrosylated proteins. Sci STKE 86:Pl1

Kato H, Takemoto D, Kawakita K (2012) Proteomic analysis of *S*-nitrosylated proteins in potato plant. Physiol Plant. doi: 10.1111/j.1399-3054.2012.01684.x

Lee TY, Chen YJ, Lu TC et al (2011) SNOSite: exploiting maximal dependence decomposition to identify cysteine *S*-nitrosylation with substrate site specificity. PLoS One 6:e21849

Li YX, Shao YH, Jing L, Deng NY (2011) An efficient support vector machine approach for identifying protein *S*-nitrosylation sites. Protein Pept Lett 18:573–587

Lin A, Wang Y, Tang J et al (2012) Nitric oxide and protein *S*-nitrosylation are integral to hydrogen peroxide-induced leaf cell death in rice. Plant Physiol 158:451–464

Lindermayr C, Saalbach G, Durner J (2005) Proteomic identification of *S*-nitrosylated proteins in *Arabidopsis*. Plant Physiol 137:921–930

Lindermayr C, Saalbach G, Bahnweg G, Durner J (2006) Differential inhibition of *Arabidopsis* methionine adenosyltransferases by protein *S*-nitrosylation. J Biol Chem 281:4285–4291

Lindermayr C, Sell S, Müller B et al (2010) Redox regulation of the NPR1-TGA1 system of Arabidopsis thaliana by nitric oxide. Plant Cell 22:2894–2907

Liu M, Hou J, Huang L et al (2010) Site-specific proteomics approach for study protein *S*-nitrosylation. Anal Chem 82:7160–7168

Liu Z, Cao J, Ma Q et al (2011) GPS-YNO2: computational prediction of tyrosine nitration sites in proteins. Mol Biosyst 7:1197–1204

Mur LA, Mandon J, Persijn S et al (2013) Nitric oxide in plants: an assessment of the current state of knowledge. AoB Plants 5:pls052

Murray CI, Uhrigshardt H, O'Meally RN et al (2012) Identification and quantification of *S*-nitrosylation by cysteine reactive tandem mass tag switch assay. Mol Cell Proteomics 11:M111013441

Ortega-Galisteo AP, Rodríguez-Serrano M, Pazmiño DM et al (2012) *S*-nitrosylated proteins in pea (*Pisum sativum* L) leaf peroxisomes: changes under abiotic stress. J Exp Bot 63:2089–2103

Paige JS, Xu G, Stancevic B, Jaffrey SR (2008) Nitrosothiol reactivity profiling identifies *S*-nitrosylated proteins with unexpected stability. Chem Biol 15:1307–1316

Palmer RM, Ferrige AG, Moncada S (1987) Nitric oxide release accounts for the biological activity of endothelium-derived relaxing factor. Nature 327:524–526

Palmieri MC, Lindermayr C, Bauwe H et al (2010) Regulation of plant glycine decarboxylase by *S*-nitrosylation and glutathionylation. Plant Physiol 152:1514–1528

Perazzolli M, Dominici P, Romero-Puertas MC et al (2004) *Arabidopsis* nonsymbiotic hemoglobin AHb1 modulates nitric oxide bioactivity. Plant Cell 16:2785–2794

Romero-Puertas MC, Laxa M, Mattè A et al (2007) *S*-nitrosylation of peroxiredoxin II E promotes peroxynitrite-mediated tyrosine nitration. Plant Cell 19:4120–30

Romero-Puertas MC, Campostrini N, Mattè A et al (2008) Proteomic analysis of *S*-nitrosylated proteins in Arabidopsis thaliana undergoing hypersensitive response. Proteomics 8:1459–1469

Santhanam L, Gucek M, Brown TR et al (2008) Selective fluorescent labeling of *S*-nitrosothiols (*S*-FLOS): a novel method for studying *S*-nitrosation. Nitric Oxide 19:295–302

Serpa V, Vernal J, Lamattina L et al (2007) Inhibition of AtMYB2 DNA-binding by nitric oxide involves cysteine *S*-nitrosylation. Biochem Biophys Res Commun 361:1048–1053

Sun J, Morgan M, Shen RF et al (2007) Preconditioning results in *S*-nitrosylation of proteins involved in regulation of mitochondrial energetics and calcium transport. Circ Res 101: 1155–1163

Tada Y, Spoel SH, Pajerowska-Mukhtar K et al (2008) Plant immunity requires conformational changes of NPR1 via *S*-nitrosylation and thioredoxins. Science 321:952–956

Tanou G, Job C, Rajjou L et al (2009) Proteomics reveals the overlapping roles of hydrogen peroxide and nitric oxide in the acclimation of citrus plants to salinity. Plant J 60:795–804

Tello D, Tarín C, Ahicart P et al (2009) A fluorescence switch technique increases the sensitivity of proteomic detection and identification of *S*-nitrosylated proteins. Proteomics 9:5359–5370

Terrile MC, París R, Calderón-Villalobos LI et al (2012) Nitric oxide influences auxin signaling through *S*-nitrosylation of the Arabidopsis TRANSPORT INHIBITOR RESPONSE 1 auxin receptor. Plant J 70:492–500

van der Linde K, Gutsche N, Leffers HM et al (2011) Regulation of plant cytosolic aldolase functions by redox-modifications. Plant Physiol Biochem 49:946–957

Wang YQ, Feechan A, Yun BW et al (2009) *S*-nitrosylation of AtSABP3 antagonizes the expression of plant immunity. J Biol Chem 284:2131–2137

Wawer I, Bucholc M, Astier J et al (2010) Regulation of *Nicotiana tabacum* osmotic stress-activated protein kinase and its cellular partner GAPDH by nitric oxide in response to salinity. Biochem J 429:73–83

Wimalasekera R, Villar C, Begum T, Scherer GF (2011) Copper Amine Oxidase1 (CuAO1) of *Arabidopsis thaliana* contributes to abscisic acid- and polyamine-induced nitric oxide biosynthesis and abscisic acid signal transduction. Mol Plant 4:663–678

Wu C, Parrott AM, Liu T et al (2011) Distinction of thioredoxin transnitrosylation and denitrosylation target proteins by the ICAT quantitative approach. J Proteomics 74:2498–2509

Xue Y, Liu Z, Gao X et al (2010) GPS-SNO: computational prediction of protein *S*-nitrosylation sites with a modified GPS algorithm. PLoS One 5:e11290

Yun BW, Feechan A, Yin M et al (2011) *S*-nitrosylation of NADPH oxidase regulates cell death in plant immunity. Nature 478:264–268

Zhang J, Li S, Zhang D et al (2010) Reductive ligation mediated one-step disulfide formation of *S*-nitrosothiols. Org Lett 12:4208–4211

Chapter 7
Nitric Oxide: Detection Methods and Possible Roles During Jasmonate-Regulated Stress Response

Palmiro Poltronieri, Marco Taurino, Stefania Bonsegna, Stefania De Domenico and Angelo Santino

Abstract Nitric oxide (NO) has been involved in modulation of signaling pathways under biotic and abiotic stress responses. NO synthesis, detection methods and mechanism of action are briefly introduced. Studies on the comparison of jasmonates, 12-oxophytodienoic acid (OPDA), and ABA content and of gene expression variation in chickpea roots from a drought tolerant and a responsive variety have extended studies on drought and salt stress on other chickpea varieties, confirming the opinion of involvement of upregulation of specific LOX, AOS, and hydroperoxyde lyase (HPL) isoforms. In this context, various levels of regulation of NO on jasmonate (JA) signaling and JA biosynthesis pathway are reviewed and discussed. Finally, an additional level of regulation by epigenetics and microRNAs, with the involvement of abscisic acid and NO-responsive elements in promoters of transcription factor genes, is briefly introduced.

Keywords ABA-responsive elements · Jasmonate signaling · MicroRNAs · Nitric oxide responsive promoters · Oxylipins

7.1 Introduction

Nitrogen monoxide, or nitric oxide (NO), is a free radical and a gasotransmitter involved in cell-to-cell communication. NO, carbon monoxide (CO), and hydrogen sulfide (H_2S), are volatile gasotransmitters permeable to cell membranes, being implicated in the communication between bacteria and roots, in the regulation of root growth and in plant-pathogen interactions.

P. Poltronieri (✉) · M. Taurino · S. Bonsegna · S. De Domenico · A. Santino
CNR-ISPA, Institute of Sciences of Food Productions, via Monteroni 73100 Lecce, Italy
e-mail: palmiro.poltronieri@ispa.cnr.it

M. N. Khan et al. (eds.), *Nitric Oxide in Plants: Metabolism and Role in Stress Physiology*, DOI: 10.1007/978-3-319-06710-0_7,

NO action is dependent on its concentration and spatial generation patterns. NO shares great similarity and common properties of O_2, such as the strong binding to heme and to iron-sulfur groups in enzymes. NO, locally produced in a very low amount, from picomolar to nanomolar range, has a relatively short half-life, about 1–10 s, in the in vivo environment, since it is sequestered by reactive oxygen species, proteins and peptides (glutathione), and haem- and iron-sulfur prosthetic groups. NO has been found to play important roles in plant physiology (Durner and Klessig 1999), such as root elongation (Fernándes-Marcos et al. 2012; Mur et al. 2013), interactions with beneficial microorganisms and rhizobia and response to biotic and abiotic stresses.

NO is known to be involved in plant defence (Leitner et al. 2009) and in biotic stress response (Bellin et al. 2013) either as a secondary messenger or as a modulator of hormone signaling pathways. Pathogen Associated Molecular Pattern (PAMP) triggered immunity (PTI), bacterial compound-sensing, receptor-mediated mechanisms that enables plants to protect from nonpathogenic microbes are also involved, with an attenuated plant response, in the sensing of endophytes, bacteria, and rhizobia. Plants perceive pathogen-derived effector molecules via disease resistance (R) proteins, involving effector-triggered immunity (ETI). ETI is a more rapid and stronger type of response than PTI, often resulting in the hypersensitive response (HR). NO was shown to potentiate plant defence (Delledonne et al. 2001), as in the case of soybean defence responses (Durner et al. 1998), either by synergizing not only with reactive oxygen species (ROS) or by salicylic acid (SA) signaling pathway (Delledonne et al. 1998).

7.2 Biological Activities of Nitric Oxide

NO can diffuse across biological membranes, thus acting at short distances. NO, due to its short half-life, about 1–10 s, exerts its action through a very rapid activity, after that it is sequestered by heme-containing proteins, sulfhydryl groups in proteins, and by glutathione. NO-related signaling is assigned to various NO derivatives, collectively referred to as reactive nitrogen species (RNS). RNS comprise not only the NO radical (NO.) and its nitroxyl (NO^-) and nitrosonium (NO^+) ions, but also peroxynitrite ($ONOO^-$), *S*-nitrosothiols (SNO), higher oxides of nitrogen, and dinitrosyl–iron complexes; that mediate NO-dependent modifications.

Specific modifications driven by NO are generally reversible, such as the covalent modification of cysteine (*S*-nitrosylation) and tyrosine (tyrosine 3-nitration) in enzymes and proteins. *S*-nitrosylation refers to the addition of an NO moiety to a reactive thiol in cysteines (Cys) to form an SNO group (Astier et al. 2012). This redox modification is a central route for NO bioactivity, based on Cys residues and their ability to serve as a molecular switch, enabling protein function to be responsive to cellular redox status. Many Cys targets subjected to *S*-nitrosylation are embedded within a consensus motif, which is susceptible to redox-based posttranslational modifications. This situation is similar to other distinct post-translational modifications, such as phosphorylation. Hydrophobic regions can also drive some of these modifications because the reaction between NO and oxygen is promoted in such environments, producing species that support Cys modification.

NO accumulates in response to attempted pathogen ingress. SA stalls plant growth and stimulates an accompanying immune response. Most SA-inducible genes are controlled by the transcriptional activator NPR1. NPR1 is a protein target for NO-mediated cysteine nitrosylation. NPR1 proteins are normally present as cytosolic oligomers linked by intramolecular disulfide bonds. Upon SA treatment, NPR1 oligomers are monomerized due to a change in the intracellular redox status. NPR1 monomers are translocated to the nucleus where they activate gene expression. Recently, NPR1 was shown to bind directly to SA through a metal (possibly copper) via two cysteine residues.

S-nitrosylation under physiological conditions (basal NO) is assumed to inhibit methionine adenosyltransferase 1 (MAT1) (Lindermayr et al. 2005), to modify NPR1 and other targets identified with a proteomic approach (Lindermayr et al. 2006). Induction of defence-related NO production leads to *S*-nitrosylation of glyceraldehyde 3-phosphate dehydrogenase (GAPDH), salicylic acid binding protein 3 (SABP3) and peroxiredoxin II E (PrxIIE). GAPDH activity is blocked upon *S*-nitrosylation and the protein is thought to translocate to the nucleus. *S*-nitrosylation of SABP3 causes the loss of SA-binding capacity as well as inhibition of carbonic anhydrase_activity. PrxIIE has been found to be *S*-nitrosylated during the hypersensitive response, resulting in the inhibition of its hydroperoxide-reducing peroxidase activity (Romero-Puertas et al. 2007). *S*-nitrosylation of PrxIIE leads to increased $ONOO^-$ levels and protein nitration.

7.3 Methods of NO Detection

7.3.1 Detection of NO and NO Measurement in Cell Culture and in Planta

Several technologies are needed to validate *in planta* measurements of spatio-temporal patterns of NO production. It is advisable to employ several alternative approaches in parallel, in order to be confident with the NO measurements.

No definite pathway of plant NO signaling has been identified yet, but rather NO appears to act as a modifier of other signaling pathways. Since plants are exposed to NO from a number of external sources, it would be useful to use NO scavenging systems, such as nsHbs and other NO sinks. For instance, a tight control in intracellular compartment of ROS, glutathione concentration, and other NO sinks could confirm the data obtained.

Useful NO detection methods, applied to cells, tissue sections or *in planta* are: Quantum Cascade Lasers (QCL)-based detection (Mur et al. 2011), MS, Laser Photoacoustic Spectroscopy; Electron Paramagnetic Resonance (EPR) Spin Trap, NO fluorescent probes and NO electrodes methods are deeply described in several reviews (Vandelle and Delledonne 2008; Vitecek et al. 2008). NO electrodes have been often used in plant tissue cultures. It is difficult to obtain and maintain a stable baseline, avoiding fluctuations over time. Highly pure reagents without traces of nitrites need to be used to perform the set up of the electrode and a curve of NO concentrations. A Faraday cage, an electromagnetic shield or a screen room are necessary to isolate the system from the environment.

NO fluorescent probes for confocal microscopy allowed detection and imaging of nitric oxide, such as diaminofluoresceins (DAF), 3-Amino-4-(N-methylamino)-2,7-difluofluorescein (DAF-FM) (Kojima et al. 2001), rhodamine B (2-amino-3′,6′-bis(diethylamino)-2,3-dihydrospiro [iso-indole-1,9′-xanthene]-3-one)-based dyes and Europium (III) chelate (Vandelle and Delledonne 2008). The fluorescence in the presence of DAF dyes, if suppressed by co-application of NO scavengers for example, 2-(4-carboxyphenyl)-4,4,5,5-tetramethylimidazoline-l-oxyl-3-oxide (cPTIO) may confirm the nature of the signal (Gupta and Igamberdiev 2013).

7.3.2 Methods of Detection of Nitrosylated Proteins

The method of choice for detecting proteins containing nitrosothiols was the Biotin Switch method (Jaffrey and Snyder 2001). In a three-step procedure, nitrosylated cysteines are converted to biotinylated cysteines (Forrester et al. 2009). Biotinylated cysteines can then be detected by immunoblotting or can be purified by avidin-affinity chromatography.

Redox-sensitive cysteine residues in proteins may serve as important components of oxidative signaling or sensors of oxidative stress. Cysteine sulfenic acid formation in proteins results from the oxidative modification of susceptible cysteine residues by mild oxidizing agents such as hydrogen peroxide, alkyl hydroperoxides and $ONOO^-$. These modifications are of considerable biological interest as important players in redox catalysis and redox regulation.

NBD chloride is a trapping agent capable of discriminating between thiol and sulfenic acid moieties on proteins (Ellis and Poole 1997), but the utility of this approach was restricted to purified proteins with only one or a small number of cysteines.

Reagents have been developed to tag sulfenic acids in proteins using a dimedone-like reagent (DCP-Bio1, KeraFast Inc), containing a cleavable biotin tag linked to a reporter group. These probes are compatible with custom methods in the biology laboratory, such as Western blotting, ELISA and affinity isolation. DCP-N3 was developed to be used in detecting the formation of cysteine sulfenic acid in the redox regulation of proteins (Klomsiri et al. 2010). The azide group can be used for selective conjugation to phosphine- or alkynyl- containing reagents such as biotin or common fluorophores for desired analytical techniques, and can be used to label protein sulfenic acids in cellular proteins, either by in situ labeling of intact cells or by labeling at the time of lysis.

The Alkyne β-ketoester probe (Alk-β-KE) can be utilized as a robust chemical probe for the labeling and analysis of sulfenic acid (-SOH) modified proteins. This type of probe is ideal for in vitro and in vivo applications and mass spectrometric (MS) analyses. Alk-β-KE is non-dimedone-based probe that is customizable through the addition of biotin or other tags. The tags can be subsequently cleaved using NH_2OH making this probe ideal for quantitative analysis via mass spectrometry (MS).

Biotin-1,3-cyclopentanedione (BP1) is a non-dimedone-based probe that contains a biotin tag making it compatible with several techniques and forms of analyses. At physiological pH its reactivity is comparable with dimedone and has increasing activity at lower pH. Non-hydrolysable amide linkage prevents the possibility of cleavage. In addition, the ongoing synthesis and evaluation of new organic compounds as HNO donors such as 1-nitrosocyclohexyl Acetate and the development of new organic-based methods of detection will support NO studies.

7.3.3 NO Donors and NO Scavengers

Pharmaceutical NO donors have been used sometimes with no consideration of the kinetics of NO production. NO action is dependent on its concentration and spatial generation patterns. However, currently not a single of available technology is able to provide accurate *in planta* measurements of spatial-temporal patterns of NO production. It is also the case that pharmaceutical NO donors are used in plant studies. NO donors include NONOates (spermidine- or diethylamine-NONOate), diethylamine nitric oxide, *S*-nitroso-N-acetylpenicillamine (SNAP), *S*-nitrosoglutathione (GSNO), and sodium nitroprusside (SNP). Certain endogenous compounds such as SNO, certain organic nitrates, nitrosylated metal complexes, dinitrosyl-iron complexes (DNIC), or nitrite anions (NO_2^-), under hypoxic conditions, can act as NO donors or elicit NO-like reactions in vivo.

SNP is a highly stable electrophile that can be stabilized by coordination with metals. A problem with SNP is the activity of the "spent" donor. SNP and the "spent donor" produce potassium ferricyanide (Fe (III) CN) and potassium ferrocyanide (Fe (II) CN) with generation of cyanide (CN_2).

GSNO is often used as NO donor in nitrosylation. It undergoes spontaneous homolytic cleavage of the Cys-based S–NO bond to release NO (Ederli et al. 2009). The results obtained using NO donors should be validated by the addition of NO scavengers, such as 2-(4-carboxy-2-phenyl)-4,4,5,5-tetramethylimidazoline-1-oxyl-3-oxide (cPTIO).

7.3.4 Reporter Genes

Nowadays a widespread use of mutant lines and studies on plant functional genomics make use of knockout (i.e., through RNA silencing) or overexpressing plantlets and tissue cultures. It may be helpful to transform plant cells with suitable vectors for expression of reporter genes. For instance, a gene coding for the enzyme sGC, such as a mammalian sGC or the *Arabidopsis* H-NOX family flavin monooxygenase, could detect the NO-dependent synthesis of cGMP,)measuring the increase in intracellular Ca^{2+} with a fluorescent probe. An inducible suppression of Hbs, in particular in root tissue, would increase the detection of NO. A suppression of Hb could aid to further elevate NO concentration (Mur et al. 2011).

7.4 Potentiation of Nitric Oxide and Jasmonates Signaling in Abiotic Stress Responses

Abiotic stress is one of the primary factors of crop loss worldwide, causing average yield losses of more than 50 % in major crops. Tolerance and susceptibility to abiotic stresses are very complex. Plants can resist abiotic stresses through different distinct mechanisms; however, traits that are associated with resistance mechanisms are multigenic, often converging on genes shared by different stresses.

Wounding is a stress involving the Jasmonate (JA) biosynthesis pathway and JA signaling (Koo et al. 2009). In *Arabidopsis*, the wounding induced the production of JA, and at early stages the production of NO (Huang et al. 2004), while exogenous NO induced three genes of jasmonic acid (JA) biosynthesis allene oxide synthase (AOS), lipoxygenase (LOX2) and 12-oxophytodienoic acid (OPDA) Reductase (OPR3) (Huang et al. 2004). However, NO resulted in accumulation of salicylic acid (SA), that could limit the synthesis of JA, since in transgenic NahG plants (impaired in SA accumulation and/or signaling), NO did increase JA production.

Dehydration-related stresses such as drought and salinity have ionic as well as osmotic attributes that elicit signal transduction cascades resulting in activation of effector genes to adapt the metabolism of the plant to the stress. In the model arising from research in *Arabidopsis* or rice, the first step of signaling is the perception of the stress through G-protein coupled receptors (GPCR), inositol polyphosphates, or receptor-like kinases (RLKs).

Abiotic below-ground stresses are early signaling affecting root growth regulation, resource acquisition, and root-shoot communication (Schachman and Goodger 2008). Abiotic stresses elicit early signals that need to be transduced at distance to affect protection mechanisms, such as growth regulation, resource acquisition synthesis of osmoprotectants, water potential, stomatal closure, among others. There are several signaling compounds (RNAs, lipids, PGPs, and peptide factors) involved in root-shoot communication (Seki et al. 2007; Schachman and Goodger 2008).

7.4.1 NO Regulation of JA Signaling, Epigenetics, and Role of microRNAs

In *Arabidopsis* plants NO was found to induce key enzymes of JA biosynthesis such as allene oxide synthase (AOS) and LOX (Huang et al. 2004). Allene oxide cyclase (AOC) has been found *S*-nitrosylated by NO in a cysteine proximal to the catalytic site, during the hypersensitive response (HR) (Delker et al. 2006). AOC oligomerization has been found necessary in the synthesis of JA (Stenzel et al. 2012) with the requirement of specific isoforms to form heterodimers, and transcriptional regulation through RNAs, and RNA binding complexes.

Nitrosylation of cysteines in enzymes of the SA/JA synthesis have been found to be important in regulating and controlling JA production and JA signaling. In plants, NO-mediated nitrosylation activates transcription factors such as MYB, involved in JA-dependent signaling. SABP3, modulating the SA response and integrating the JA signaling, was nitrosylated by NO during the hypersensitive response (HR) (Wang et al. 2009). Furthermore, nitrosylation of cysteines in enzymes of the SA/JA signaling have been found to be important in regulating and controlling JA production. It is thus plausible to hypothesize that NO provides an *S*-nitrosylation control of the R2R3-MYB class of transcription factors (Serpa et al. 2007), inhibiting DNA binding of MYB TFs. Thus, the NO-specific transcriptional output contributes to the modulation of JA signaling pathway in abiotic stress responses.

NO induction of JA biosynthesis genes did not result in elevated levels of JA in *Arabidopsis* (Huang et al. 2004). JA-responsive genes such as defensin (PDF1.2) were not induced in that system, so that expressed genes may not be paralleled by activation of MYB transcription factors. The intracellular production/release and containment of JA intermediates is conducted in specific and often strictly localized reactions, to allow for spatially and temporally regulated signaling events.

The transcription factor TCP4 regulates several genes of the lipoxygenase pathway, in *Arabidopsis* (Schommer et al. 2008). The miR-159/miR-319 family of signaling molecules moving through the phloem (Buhtz et al. 2010) to the roots targets transcription factors of the MYB and TCP family of transcription factors. It is proposed that an early activation by TCP4 of JA biosynthesis pathway may be followed by a negative feedback determined by miR-319 binding to TCP4. This coordinated activity may orchestrate timely and localized differential gene expression of LOX, OPR, and AOS in roots responding to different stresses. "NO-responsive" promoters were identified bioinformatically, and showed that salicylate- and jasmonate-responsive cis-elements were prominent (Palmieri et al. 2008).

The results of oxylipin profiling in chickpea root (De Domenico et al. 2012) indicated that JA-Ile and OPDA may act coordinately for the full activation of root response to drought stress in the tolerant ICC 4958 variety and preceded ABA synthesis. The pattern of ABA accumulation was similar in drought-sensitive and tolerant cultivars, though ABA content in drought stressed roots was constitutively higher (about 20 %) in the tolerant cultivar ICC 4958. ABA concentration showed a sharp increase within 24 h, after which ABA content remained constant in the tolerant variety, whereas it decreased in the susceptible one. After 72 h from stress onset, ABA levels were about 37 % higher in ICC 4958 than in ICC 1882. The integration between JA and Abscisic Acid (ABA) mediated signals along multiple abiotic stresses remains mainly unknown. An interesting crossroad between the ABA and JA signaling are the *NAC* family of transcription factors (TF), formed by the subfamilies ATAF, NAM, and CUC TFs (Santino et al. 2013). *ATAF2* is relevant in response to wounding, salinity stress and JA treatment, while *ATAF1* responds to dehydration, wounding, and ABA treatment.

Long distance signaling is fundamental in plants for the regulation of several processes including leaf development, flowering, and pathogen defence. Small RNAs, among them several microRNAs, have been detected in the phloem sap of plant species. The small RNA population found in phloem sap includes miRNA and small interfering RNA (siRNA). As a prototype of mobile signals, miR399 is a phloem-mobile long distance signal (Franco-Zorrilla et al. 2007) responding to phosphate deficiency, moving from leaves to roots via phloem, and targeting PHO2/UBC24, an E2 ubiquitin ligase, thus freeing MYB/PHR1 in the roots.

Several findings have established a fundamental role of miRNAs in plant stress response to abiotic stresses and nutrient deprivation (Khraiwesh et al. 2012). The expression profiles of several miRNAs involved in plant growth and development are significantly altered during stress. These findings imply a control of stress-responsive miRNA on the attenuation of plant growth and development under stress that is strictly related to phytohormones perception and signaling.

A direct link between miRNAs and stress response has emerged with the identification of the target genes of each microRNA: miR-159/miR-319 (MYB33, MYB101, TCPs), miR-166 (HD-ZIP TFs), mir-172 (AP2 transcription factors), miR395 (ATP sulfurylase); miR-396 (GRF TFs), miRNA398 (SOD), and miR399 (PHO2/MYB complexes), miR-393 (the auxin-dependent Transport Inhibitor Response 1, TIR1). TIR1 F-box protein, and auxin receptor is *S*-nitrosylated by NO that thus influences auxin signaling (Terrile et al. 2012).

Both ABA and NO are involved in many elements of plant physiology including stomatal closure, root formation and seed dormancy. ABA and NO signaling pathways often involve ROS and have interactions with other hormones and signaling molecules (Hancock et al. 2011). In particular, ABA signaling promotes the expression of the drought regulated miR-159, miR-393 and miR-398. ABA signaling acts through the ABA-Responsive element (ABRE), present in the promoter of miR-169n, targeting the nuclear factor Y subunit (NF-YA) that is downregulated in wheat leaves by drought.

7.4.2 Roots in the Sensing of Drought and Salt Stresses: A Role of Nitric Oxide and Jasmonates

Root is the first plant organ sensing and responding to environmental and soil conditions. Plant growth and development are largely dependent on the plant root system, due to its crucial role in water and mineral uptake. Below-ground sensing of nutrient soil status affects root growth regulation, resource acquisition and root-shoot communication under abiotic stress (Seki et al. 2007; Schachman and Goodger 2008).

Jasmonates (JAs) directly induce *nod* gene expression in rhizobia, and indirectly promote bacterial Nod factor production by inducing (iso) flavonoid biosynthesis genes (Zhang et al. 2007). As a feedback, Nod factor induces Ca^{2+} spiking in root hairs and inhibition of JA synthesis (Oldroyd 2007). Regulation of the redistribution of nutrients is one of the roles of jasmonates in Arbuscular Mycorrhiza (AM) roots. In plants such as *M. truncatula* and barley, developing a mutualistic symbiosis, ultimately leads to a promoted growth, JAs might help to regulate the nutrient exchange between both partners. JA, in its methylated form (Me-JA), is involved in the regulation of growth of lateral roots (Hsu et al. 2013).

Nitric oxide plays a central role in determining lateral root development in tomato (Correa-Aragunde et al. 2004). Kolbert et al. (2010) showed the involvement of nitrate reductase (NR) in osmotic stress-induced NO generation in *Arabidopsis thaliana* roots (Kolbert et al. 2010). On the interaction between NO signaling with genes and protein products in the biosynthesis pathway of oxylipins, supporting data and possible mechanisms will be discussed.

Plants are being continually exposed to NO from the soil. NO production is a feature of the oxido-reductive steps ranging from NH_4^+ to NO_3^- that form the nitrogen cycle. Various factors also influence NO production in soil such as temperature, oxygen availability, humidity, soil pH, and nitrogen status. These influence the activities of nitrifying and denitrifying bacteria, which under different conditions can produce NO at differing rates. Bacteria assimilate nitrate, which is central in nitrogen metabolism, and reduce it to nitrite (NO_2^-) through a two-electron reduction reaction. The accumulation of cellular nitrite can be harmful because nitrite can be reduced to nitric oxide with a potential cytotoxic effect. Nitrite is removed from the cell by channels and transporters, or reduced to ammonium or N_2 through the action of assimilatory enzymes. Nitrate reductase and NOS-oxy in bacteria and rhizobia may contribute to the production of NO and signaling between bacteria and roots, and may have an important role in drought and salt sensing in nodules.

Endosymbiotic bacteria hosted by AM provide beneficial properties such as protection from pests and functions for the growth of plants and trees, such as plant growth promotion, plant elicitation, nutrient acquisition, and competition for pathogens, priming, and preconditioning of induced systemic resistance (Jung et al. 2012). Root bacteria can produce NO, CO, and H_2S, three gasotransmittors that may contribute to signaling a stress alert in roots and nodules. In mammals, hydrogen sulfide (H_2S), is generated by cystathionine γ-lyase, acting as a physiologic vasorelaxant (Mustafa et al. 2009), similarly to nitric oxide. Mechanisms involving H_2S signaling are still elusive. H_2S was shown to modify physiologically cysteines in a large number of proteins by *S*-sulfhydration. About 10–25 % of many liver proteins, including actin, tubulin, and glyceraldehyde-3-phosphate dehydrogenase (GAPDH), are sulfhydrated under physiological conditions. Sulfhydration augments GAPDH activity and enhances actin polymerization. Sulfhydration thus appears to be a physiologic posttranslational modification for proteins that can compete with NO-dependent nitrosylation. CO and NO compete with antagonistic effects for binding to heme and iron-sulfur groups in proteins. CO is involved in lateral root formation. BnHO1, a heme oxygenase-1 gene from *Brassica napus*, is required for salinity and osmotic stress-induced lateral root formation (Cao et al. 2011). Then, a Heme oxygenase was shown to be involved in nitric oxide- and auxin-induced lateral root formation in rice (Chen et al. 2012).

In legumes, leghemoglobins accumulate in symbiosomes. Non-symbiotic Hb genes are expressed in specific plant tissues, and overexpressed in organs of stressed tissues. These proteins may function as additional O_2 transporters and in buffering of NO that may be released at later times.

The experiments on salt response in rhizobia inoculated roots of salt tolerant chickpea INRAT-93, the salt sensitive Amdoun control, and ICC4958 and the ICC6098 weakly tolerant variety, were performed (Molina et al. 2011). Differential expression of JA biosynthesis genes in nodules and in root apexes was shown. Salt stress sensed in nodules by locally producing lipoxygenase (LOX) isoforms expression levels was higher than in the roots, implicate a nodule-localized involvement of NO production with effects on up-regulation of JA biosynthesis pathway. In intact soybean nodules, presence of nitric oxide-leghemoglobin complexes was shown together with the production of radicals (Mathieu et al. 1998).

Thus, it is conceivable to suppose a larger involvement of bacteria in stress signaling, with the production of NO, and NO-induced amplification of JA synthesis through specific promoter activation and *S*-nitrosylation of enzymes and transcription factors.

7.5 Conclusion

Thus, it can be postulated that TFs, playing a key role in hormone signaling and stress response, are either positively or negatively regulated by NO under the control of other stress regulated hormones in a feedback signaling network. In this context, a vivid interplay of phytohormone signaling at local and distal tissues have been suggested that regulate transcription factors belonging to different families, which are fine-tuned by the levels of specific classes of small RNAs. Increasing our knowledge about the molecular, physiological, and metabolic aspects of plant response to multiple stresses will be vital to develop new varieties able to better cope with future global climate changes.

References

Astier J, Kulik A, Koen E et al (2012) Protein *S*-nitrosylation: what's going on in plants? Free Radic Biol Med 53:1101–1110

Bellin D, Asai S, Delledonne M, Yoshioka H (2013) Nitric oxide as a mediator for defense responses. Mol Plant Microbe Interact 26:271–277

Buhtz A, Pieritz J, Springer F, Kehr J (2010) Phloem small RNAs, nutrient stress responses, and systemic mobility. BMC Plant Biol 10:64

Cao Z, Geng B, Xu S, Xuan W et al (2011) BnHO1, a haem oxygenase-1 gene from *Brassica napus*, is required for salinity and osmotic stress-induced lateral root formation. J Exp Bot 62:4675–4689

Chen YH, Chao YY, Hsu YY et al (2012) Heme oxygenase is involved in nitric oxide- and auxin-induced lateral root formation in rice. Plant Cell Rep 31:1085–1091

Correa-Aragunde N, Graziano M, Lamattina L (2004) Nitric oxide plays a central role in determining lateral root development in tomato. Planta 218:900–905

De Domenico S, Bonsegna S, Horres R et al (2012) Transcriptomic analysis of oxylipin biosynthesis genes and chemical profiling reveal an early induction of jasmonates in chickpea roots under drought stress. Plant Physiol Biochem 61:115–122

Delker C, Stenzel I, Hause B, Miersch O, Feussner I, Wasternack C (2006) Jasmonate biosynthesis in *Arabidopsis thaliana*—enzymes, products, regulation. Plant Biol 8:297–306

Delledonne M, Xia Y, Dixon RA, Lamb C (1998) Nitric oxide signal functions in plant disease resistance. Nature 394:585–588

Delledonne M, Zeier J, Marocco A, Lamb C (2001) Signal interactions between nitric oxide and reactive oxygen intermediates in the plant hypersensitive disease-resistance response. Proc Natl Acad Sci USA 98:13454–13459

Durner J, Klessig DF (1999) Nitric oxide as a signal in plants. Curr Opin Plant Biol 2:369–374

Durner J, Wendehenne D, Klessig DF (1998) Defense gene induction in tobacco by nitric oxide, cyclic GMP, and cyclic ADP-ribose. Proc Natl Acad Sci USA 95:10328–10333

Ederli L, Reale L, Madeo L et al (2009) NO release by nitric oxide donors in vitro and *in planta*. Plant Physiol Biochem 47:42–48

Ellis HR, Poole LB (1997) Novel application of 7-chloro-4-nitrobenzo-2-oxa-1,3-diazole to identify cysteine sulfenic acid in the AhpC component of alkyl hydroperoxide reductase. Biochemistry 36:15013–15018

Fernándes-Marcos M, Sanz L, Lorenzo O (2012) Nitric oxide an emerging regulator of cell elongation during primary root growth. Plant Signal Behav 7:196–200

Forrester MT, Foster MW, Benhar M, Stamler JS (2009) Detection of protein *S*-nitrosylation with the biotin-switch technique. Free Radic Biol Med 46:119–126

Franco-Zorrilla JM, Valli A, Todesco M et al (2007) Target mimicry provides a new mechanism for regulation of microRNA activity. Nat Genet 39:1033–1037

Gupta KJ, Igamberdiev AU (2013) Recommendations of using at least two different methods for measuring NO. Frontiers Plant Sci 4:58

Hancock JT, Neill SJ, Wilson ID (2011) Nitric oxide and ABA in the control of plant function. Plant Sci 181:555–559

Hsu YY, Chao YY, Kao CH (2013) Methyl jasmonate-induced lateral root formation in rice: the role of heme oxygenase and calcium. J Plant Physiol 170:63–69

Huang X, Stettmaier K, Michel C et al (2004) Nitric oxide is induced by wounding and influences jasmonic acid signaling in *Arabidopsis thaliana*. Planta 218:938–946

Jaffrey SR, Snyder SH (2001) The biotin switch method for the detection of *S*-nitrosylated proteins. Sci STKE 86:11

Jung SC, Martienz-Medina A, Lopez-Raez JA, Pozo MJ (2012) Mycorrhiza-induced resistance and priming of plant defenses. J Chem Ecol 38:651–664

Khraiwesh B, Zhu J-K, Zhu J (2012) Role of miRNAs and siRNAs in biotic and abiotic stress responses of plant. Biochim Biophys Acta 1819:137–148

Klomsiri C, Nelson KJ, Bechtold E et al (2010) Use of dimedone-based chemical probes for sulfenic acid detection: evaluation of conditions affecting probe incorporation into redox-sensitive proteins. Methods Enzymol 473:77–94

Kojima H, Hirotani M, Nakatsubo N et al (2001) Bioimaging of nitric oxide with fluorescent indicators based on the rhodamine chromophore. Anna Chem 73:1967–197373

Kolbert Z, Ortega L, Erdei L (2010) Involvement of nitrate reductase (NR) in osmotic stress-induced NO generation of *Arabidopsis thaliana* L. roots. J Plant Physiol 167:77–80

Koo AJ, Gao X, Jones AD, Howe GA (2009) A rapid wound signal activates the systemic synthesis of bioactive jasmonates in Arabidopsis. Plant J 59:974–986

Leitner M, Vandelle E, Gaupels F et al (2009) NO signals in the haze. Nitric oxide signalling in plant defence. Curr Opin Plant Biol 12:451–458

Lindermayr C, Saalbach G, Durner J (2005) Proteomic identification of *S*-nitrosylated proteins in Arabidopsis. Plant Physiol 137:921–930

Lindermayr C, Saalbach G, Bahnweg G, Durner J (2006) Differential inhibition of Arabidopsis methionine adenosyltransferases by protein *S*-nitrosylation. J Biol Chem 281:4285–4291

Mathieu C, Moreau S, Frendo P et al (1998) Direct detection of radicals in intact soybean nodules: presence of nitric oxide-leghemoglobin complexes. Free Rad Biol Med 24:1242–1249

Molina C, Zaman-Allah M, Khan F et al (2011) The salt-responsive transcriptome of chickpea roots and nodules via deepSuperSAGE. BMC Plant Biol 11:31

Mur LAJ, Mandon J, Cristescu SM et al (2011) Methods of nitric oxide detection in plants: a commentary. Plant Sci 181:509–519

Mur LAJ, Mandon J, Persijn S et al (2013) Nitric oxide in plants: an assessment of the current state of knowledge. AoB PLANTS 5: pls052; doi:10.1093/aobpla/pls052

Mustafa AK, Gadalla MM, Sen N et al (2009) H2S signals through protein *S*-Sulfhydration. Sci Signal 2(92): ra72

Oldroyd GE (2007) Nodules and hormones. Science 315:52–53

Palmieri MC, Sell S, Huang X et al (2008) Nitric oxide-responsive genes and promoters in *Arabidopsis thaliana:* a bioinformatics approach. J Exp Bot 59:177–186

Romero-Puertas MC, Laxa M, Mattè A et al (2007) *S*-nitrosylation of peroxiredoxin II E promotes peroxynitrite mediated tyrosine nitration. Plant Cell 19:4120–4130

Santino A, Taurino M, De Domenico S et al (2013) Jasmonate signalling in plant defense response to multiple (a)biotic stresses. Plant Cell Rep 32:1085–1098

Schachman DP, Goodger JQD (2008) Chemical root to shoot signalling under drought. Trends Plant Sci 13:281–287

Schommer C, Palatnik JF, Aggarwal P et al (2008) Control of Jasmonate biosynthesis and senescence by miR319 targets. PLoS Biol 6:e230

Seki M, Umezawa T, Urano K, Shinozaki K (2007) Regulatory metabolic networks in drought stress responses. Curr Opin Plant Biol 10:296–302

Serpa V, Vernal J, Lamattina L et al (2007) Inhibition of AtMYB2 DNA-binding by nitric oxide involves cysteine *S*-nitrosylation. Biochem Biophys Res Commun 361:1048–1053

Stenzel I, Otto M, Delker C et al (2012) *ALLENE OXIDE CYCLASE* (*AOC*) gene family members of *Arabidopsis thaliana*: tissue- and organ-specific promoter activities and *in vivo* heteromerization. J Exp Bot 63:6125–6138

Terrile MC, París R, Calderón-Villalobos LI et al (2012) Nitric oxide influences auxin signaling through *S*-nitrosylation of the Arabidopsis transport inhibitor response 1 auxin receptor. Plant J 70:492–500

Vandelle E, Delledonne M (2008) Methods for nitric oxide detection during plant-pathogen interactions. Methods Enzymol 437:575–594

Vitecek J, Reinohl V, Jones RL (2008) Measuring NO production by plant tissues and suspension cultured cells. Mol Plant 1:270–284

Wang YQ, Feechan A, Yun BW et al (2009) *S*-nitrosylation of AtSABP3 antagonizes the expression of plant immunity. J Biol Chem 284:2131–2137

Zhang J, Subramanian S, Zhang Y, Yu O (2007) Flavone synthases from *Medicago truncatula* are fl avanone-2-hydroxylases and are important for nodulation. Plant Physiol 144:741–751

Chapter 8
S-Nitrosoglutathione Reductase: Key Regulator of Plant Development and Stress Response

Mounira Chaki and Christian Lindermayr

Abstract It is now recognized that an evolutionarily conserved, glutathione-dependent formaldehyde dehydrogenase (FALDH; EC 1.2.1.1), a type III alcohol dehydrogenase, has activity as an *S*-nitrosoglutathione reductase (GSNOR). This enzyme reduces *S*-nitrosoglutathione (GSNO) to glutathione disulfide and ammonia in a NADH-dependent reaction. In plants, GSNOR has been found in both monocotyledonous and dicotyledonous species where it is involved in development, fertility, and in the adaptive response to biotic and abiotic stresses. These discoveries greatly extend our knowledge in the metabolism of nitric oxide and nitric oxide-derived molecules where GSNO is an important component. An overview of the function of GSNOR in plant development and stress response is given in this chapter.

Keywords Formaldehyde dehydrogenase · Nitric oxide · Reactive nitrogen species · *S*-nitrosoglutathione · *S*-nitrosoglutathione reductase

8.1 Introduction

Nitric oxide (NO), a hydrophobic diffusible gaseous molecule, participates in a wide range of physiological processes during plant growth and development such as seed germination, flowering, primary and lateral root growth, fruit ripening, pollen tube growth, and senescence (Bethke et al. 2004; Corpas et al. 2004; Shapiro 2005). Moreover, NO is an important signaling molecule in biotic and

M. Chaki · C. Lindermayr (✉)
German Research Center for Environmental Health, Institute of Biochemical Plant Pathology, Helmholtz Zentrum München, Neuherberg, Germany
e-mail: lindermayr@helmholtz-muenchen.de

M. N. Khan et al. (eds.), *Nitric Oxide in Plants: Metabolism and Role in Stress Physiology*, DOI: 10.1007/978-3-319-06710-0_8,

abiotic stress responses including pathogens, salinity, wounding, and extreme temperature (Airaki et al. 2012; Chaki et al. 2011b; Delledonne et al. 1998; Durner et al. 1998; Valderrama et al. 2007).

NO can exert its biological function through different ways, it can interact with different biomolecules such as nucleic acids, lipids and proteins and affect their function/activity. Interaction of NO with proteins is the most studied and probably also the most important aspect of NO interaction with biomolecules. NO can react with transition metals of metalloproteins to form metal-nitrosyl complexes. In more detail, NO binds to iron, zinc, or copper centers of metalloproteins through coordination bonding (Ford 2010). The analysis of NO binding to plant metal-containing protein has been done mainly with plant hemoglobins (Gupta et al. 2011). Protein tyrosine nitration is a posttranslational modification mediated by reactive nitrogen species (RNS) that is linked to nitro-oxidative damages in plant cells. It is the result of addition of a nitro ($-NO_2$) group to one of the two equivalent ortho carbons in the aromatic ring of tyrosine residues (Gow et al. 2004). There are several compounds that are known to mediate protein tyrosine nitration in vivo involving either peroxynitrite ($ONOO^-$) or nitrite/H_2O_2/heme peroxidase or transition metals (Radi 2004). Protein *S*-nitrosylation is another NO-dependent posttranslational modification. Here sulfur groups of cysteine residues are modified by NO. In this way the function of a wide spectrum of proteins can be modified (Lindermayr and Durner 2009; Lindermayr et al. 2005; Stamler et al. 2001).

S-nitrosoglutathione reductase (GSNOR) also known as glutathione-dependent formaldehyde dehydrogenase (FALDH) belongs to the alcohol dehydrogenase family class III. It has been proposed that the major role of GSNOR/FALDH is to control *S*-nitrosoglutathione (GSNO) and *S*-nitrosothiols (SNO) levels rather than to detoxify formaldehyde in living cells. In the presence of excess of glutathione (GSH), a major intracellular low-molecular-mass antioxidant, GSNOR catalyzes the NADH-dependent reduction of GSNO to glutathione disulfide and ammonia (Lamotte et al. 2005; Liu et al. 2001).

8.2 Reactive Nitrogen Species

Recently, the term reactive nitrogen species (RNS) was introduced in the biological literature to designate NO and other NO-related molecules, such as dinitrogen trioxide (N_2O_3), $ONOO^-$, SNOs, and GSNO, among others, which have important roles in multiple physiological processes of animal and plant cells. However, in higher plants, the information available on the metabolism of RNS is very restricted compared with animal systems. Therefore, the free radical molecule NO and RNS have been found to be of notable relevance in plants under physiological and stress conditions (Delledonne et al. 1998; Durner et al. 1998). These molecules can mediate many physiological processes such as germination, plant

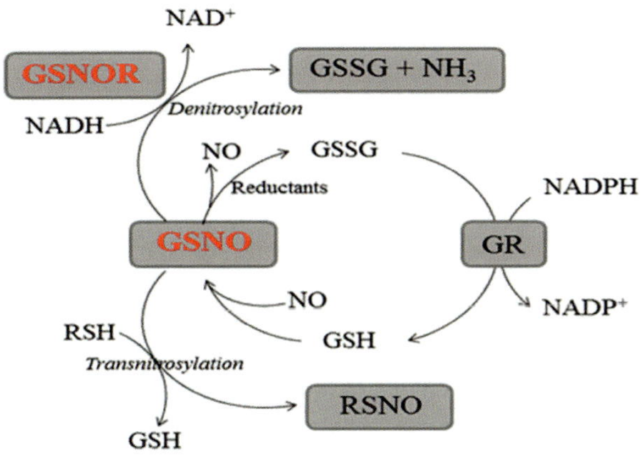

Fig. 8.1 Schematic presentation of the metabolism of *S*-nitrosoglutathione (GSNO) in plants. Nitric oxide (NO) under aerobic conditions *S*-nitrosylates glutathione (GSH) to yield GSNO, which in presence of reductant agents, such as ascorbic acid, GSH, or Cu^+ can be decomposed to produce NO and oxidized glutathione (GSSG). GSNO also can be converted by *S*-nitrosoglutathione reductase (GSNOR) by a process of NADH-dependent denitrosylation into GSSG and ammonia (NH_3). Another pathway for GSNO decomposition is a process of transnitrosylation reactions where GSNO can transfer the NO to cysteine residues of proteins leads to the reversible formation of an *S*-nitrosothiol. GSH levels restored by enzymatic reduction of GSSG leading to consumption of NADPH

development, and senescence, directly or through posttranslational modification (Begara-Morales et al. 2013; Besson-Bard et al. 2008). GSNO, the most abundant low-molecular weight SNO in plant cells, is formed by *S*-nitrosylation reaction of NO with GSH in the presence of oxygen which could have a great physiological importance for plants since GSNO is thought to function as a mobile reservoir of NO bioactivity (Diaz et al. 2003; Durner and Klessig 1999). GSNO can mediate the signaling pathway throughout specific posttranslational modification of redox-sensitive proteins by a reaction of trans-nitrosylation where GSNO can transfer NO to cysteine residues of proteins. However, little information is available on the metabolism of SNOs in plant cells and still less is known about its modulation under physiological and environmental stress conditions (Corpas et al. 2008; Feechan et al. 2005). There are few reports that have studied the presence and distribution of GSNO in plant species under normal and stress conditions by immunohistochemical analysis using an antibody against GSNO (Barroso et al. 2006; Chaki et al. 2009, 2011b). Nevertheless, the stability of GSNO depends upon the presence of other compounds in the medium such as ascorbate, glutathione or Cu^+ that can decompose GSNO to produce NO and oxidized glutathione (Gorren et al. 1996; Smith and Dasgupta 2000) (Fig. 8.1). The knowledge of GSNO metabolism and its localization in the cells/tissues is crucial to understand NO function.

8.3 GSNO Reductase Controls GSNO Turnover

As mentioned previously, GSNO is formed by the reaction of NO with GSH. In this way NO is more stable (Fernandez et al. 2003; Gaston et al. 1993; Singh et al. 1996). GSNO reductase is considered as a key determinant in controlling SNO cellular homeostasis and in the metabolism of RNS (Liu et al. 2001). In animal cells, it has been shown that the GSNOR activity controls intracellular levels of both GSNO and *S*-nitrosylated proteins and enhances cellular resistance to nitrosative stress (Liu et al. 2001, 2004). In plants, GSNOR activity has been found in *Arabidopsis* (Achkor et al. 2003; Lee et al. 2008; Sakamoto et al. 2002), tobacco (Diaz et al. 2003), pea (Barroso et al. 2006) and sunflower (Chaki et al. 2009). In rice (*Oryza sativa*), over-expression of OsGSNOR in *noe1* (*nitric oxide excess1)* plants reduced SNO levels, alleviated leaf cell death, which is consistent with a key role for this enzyme in SNO homeostasis (Lin et al. 2012). More recent studies in *Arabidopsis* transgenic lines with modified levels of the GSNOR provide evidence that this enzyme controls the level of intracellular SNO and NO under physiological conditions (Frungillo et al. 2013).

8.4 GSNO Reductase in Animals

In animal cells, GSNO is being intensively studied since this molecule is considered as a natural reservoir of NO (Padgett and Whorton 1995; Steffen et al. 2001; Zhang and Hogg 2004) and one of the most relevant compounds to perform *S*-nitrosylation reactions under physiological conditions (Steffen et al. 2001). Several evidences support the idea that GSNO is a biomolecule with physiological and clinical implications more than simply a source of NO, it has been found to be biologically active by itself as a vasodilator in animals, preventing platelet adhesion and aggregation (Bauer and Fung 1991; Ignarro et al. 1981). As mentioned above, GSNOR activity is highly specific for GSNO. It has been shown to control intracellular levels of both GSNO and SNO proteins (Liu et al. 2001). In mice, deletion of GSNOR impacted vascular function and its silencing leads to increased damage in the lymphatic and liver tissues after being challenged with bacterial endotoxin (Liu et al. 2004). Recently, it has showed that GSNOR is a key regulator of cardiovascular function and vascular tone, regulating a dynamic nitrosylation/denitrosylation cycle of proteins (Beigi et al. 2012; Lima et al. 2009). Moreover, Beigi et al. (2012) have been demonstrated that mice deficient in GSNOR exist in a persistent state of systemic vasodilatation. Thus, GSNOR governs two essential cardiovascular responses, systemic vasodilatation, and β-agonist–induced inotropic responses, indicating that GSNO plays a key role in classic physiology customarily ascribed to NO/cGMP (Furchgott and Zawadzki 1980; Ignarro et al. 1999; Murad 2006; Palmer et al. 1988).

As described above, FALDH has activity as a GSNOR is located not only in cytoplasm but also in the nucleus of rat cells (Iborra et al. 1992), where it could regulate the levels of GSNO spatiotemporally and play a role in protecting the genetic material from NO-induced damage. Besides, in mammal cells the FALDH-encoding gene is expressed in every tissue, but the level of expression varies considerably in different cell types with higher levels in liver and kidney (Adinolfi et al. 1984; Duley et al. 1985). However, in mice knockout mutants, the deletion of FALDH-encoding gene increases the cell susceptibility to nitrosative stress and initiates accumulation of *S*-nitrosylated proteins (Liu et al. 2001).

8.5 GSNO Reductase in Plants

In higher plants, GSNOR has been described as an important enzyme in NO signaling, plant development and response to adverse environmental stresses. However, the molecular mechanisms of how GSNOR regulates processes in plants are still vague. Until now, little is known about GSNOR either from the perspective of its formaldehyde-detoxifying activity (Dixon et al. 1998; Giese et al. 1994; Martinez et al. 1996) or from its function in GSNO reduction (Feechan et al. 2005; Rusterucci et al. 2007; Sakamoto et al. 2002). Plants, deficient in GSNOR activity, not only contain higher levels of nitroso species but also a higher level of other NO species. Interestingly, these plants, demonstrated specific traits such as acclimation to heat, failure to grow on nutrient plates, and increased reproductive shoots and reduced fertility (Lee et al. 2008). Moreover, it was noted that systemic acquired resistance was impaired in plants overexpressing GSNOR and enhanced in the antisense plants (Rusterucci et al. 2007). In recent years, the presence of GSNOR activity has been reported in different plant species including *Arabidopsis thaliana* (Achkor et al. 2003; Espunya et al. 2006; Feechan et al. 2005; Sakamoto et al. 2002); tobacco (Diaz et al. 2003), pea (Barroso et al. 2006; Corpas et al. 2008), sunflower (Chaki et al. 2009), pepper (Airaki et al. 2012) and tomato (Kubienova et al. 2013).

Arabidopsis GSNOR is a cytosolic protein which is encoded by a single copy gene (At5g43940) previously named *alcohol dehydrogenase2* due to its activity versus primary alcohols (Martinez et al. 1996). The gene seems to be expressed through the plant, reduced by jasmonic acid and wounding, and activated by salicylic acid (Diaz et al. 2003). Furthermore, first study of biochemical and structural characterization of a tomato (*Solanum lycopersicum*) GSNOR was reported by Kubienová et al. (2013). The GSNOR cDNA (1140 bp) of tomato encodes for a protein of 379 amino acids with a predicted molecular mass of 42.5 kDa. SlGSNOR shows 90 % sequence identity with GSNORs from *Arabidopsis* and *Zea mays*, and is highly homologous to GSNOR sequences of animal or yeast.

8.6 Functions of GSNO Reductase During Plant Development

GSNOR activity appears to be necessary for normal development and fertility under optimal growth conditions of plants (Lee et al. 2008). Thus, in *Arabidopsis*, the analysis of GSNOR activity, protein, and gene expression showed that this protein is differentially expressed, being highest in roots and leaves from the first developmental stages (Espunya et al. 2006). Furthermore, both GSNOR over-expressing and knock-down transgenic *Arabidopsis* plants had a short-root phenotype, which was correlated with a lower intracellular *S*-glutathione level and a change in its spatial distribution in the roots, suggesting that GSNOR and consequently GSNO might be involved in the regulation of the organ redox state (Espunya et al. 2006). In addition, *AtGSNOR1*is implicated in the control of shoot branching, seed yield, hypocotyl growth, flowering time, and root development (Kwon et al. 2012). Moreover, Holzmeister et al. (2011) have shown that *AtGSNOR* knockout *Arabidopsis* plants are showing delayed seed germination and reduced plant growth and have increased numbers of highly branched shoots compared to wild-type plants. Furthermore, these knockout plants have many more flowers, which are smaller and develop to smaller siliques containing smaller seeds. Also, shoot leaf morphology is altered and trichome density is reduced. The authors suggested that this pleiotropic phenotype of GSNOR knockout plants demonstrates the regulatory function of GSNOR during plant growth and development.

The overexpressing (GSNOR^{OE}) and the antisense (GSNOR^{AS}) *Arabidopsis* cell suspension lines showed an imbalance of GSNOR activity. Under optimal growth conditions (At 5th day cell cultures were in the middle of the linear growth phase), GSNOR expression was highly increased in the GSNOR^{OE} line, and slightly decreased in the GSNOR^{AS} line compared to wild-type. Under nutritional stress conditions (Reduction in the dry weight at the 10th day after subculture), GSNOR activity was reduced in all genotypes. However, small increases of *GSNOR* transcript levels were measured. The authors proposed that nutritional stress might negatively modulate GSNOR activity by posttranscriptional mechanisms (Frungillo et al. 2013). Moreover, Frungillo and colleagues proposed the importance of adequate GSNOR activity for mitochondrial bioenergetics, as the activities of complex I and external NADH dehydrogenase were shown to be responsive to changes in GSNOR levels under optimal growth conditions as well as under nutritional stress. In tomatoes, GSNOR expression in the early stage of development was higher in cotyledons than in roots. However, at later stages, the expression was higher in roots and stem compared to leaves and the shoot apex (Kubienová et al. 2013). In pepper plants, the roots showed the highest content of NO and the lowest GSNOR activity, whereas stems and leaves showed a totally opposite behavior (Airaki et al. 2011).

8.7 GSNO Reductase during Stress Response

Higher plants are frequently exposed to unfavorable conditions such as pathogens, extreme temperature, ozone, heavy metals, wounding, ultraviolet radiation, light intensity, and salinity. These conditions affect many physiological aspects including germination, development, and reproduction (Potters et al. 2007). Many researchers have described the implication of GSNOR in the mechanism of response against different biotic and abiotic stress conditions (Airaki et al. 2012; Barroso et al. 2006; Chaki et al. 2011a; Chen et al. 2009; Corpas et al. 2008; Diaz et al. 2003; Feechan et al. 2005; Holzmeister et al. 2011; Lee et al. 2008; Leterrier et al. 2012; Rusterucci et al. 2007; Wunsche et al. 2011). A summary of the multiple functions of GSNOR in plants is shown in Table 8.1 and Fig. 8.2.

8.7.1 Biotic Stress

An increasing number of publications suggest a critical role for GSNOR in plant pathogenesis (Feechan et al. 2005; Rusterucci et al. 2007). The loss of *AtGSNOR1* function increased SNO levels in *Arabidopsis* plants, which seems to negatively affect plant defense responses (Feechan et al. 2005). However, *Arabidopsis* plants with reduced GSNOR expression (antisense line) showed enhanced basal resistance against *Peronospora parasitica*. This was correlated with higher levels of intracellular SNOs and constitutive activation of PR-1 (Rusterucci et al. 2007). Furthermore, systemic acquired resistance was impaired in plants overexpressing GSNOR and enhanced in the antisense plants. In contrast, the activity of *AtGSNOR* was not affected after virulent and avirulent infections with *Pseudomonas syringae* DC3000 strains, suggesting that GSNOR is required for R gene-mediated as well as basal resistance. These contrary observations are just a result of the different infection methods and conditions (Holzmeister et al. 2011). An inverse correlation between GSNO content and GSNOR activity was detected in sunflower (*Helianthus annuus* L.) resistant to *Plasmopara halstedii*. In the control plants, GSNO was present in the epidermal and cortex cells of sunflower hypocotyls and a low amount of GSNOR was localized in the cortex cells. After infection GSNO was located exclusively in a layer of cells closer to epidermal cells and the GSNOR was induced in both cortex and epidermal cells (Chaki et al. 2009).

8.7.2 Abiotic Stress

The relevance of GSNOR has been also described under different abiotic stress conditions. Diaz et al. (2003) and Espunya et al. (2012) noted that after wounding, expression and enzymatic activity of *AtGSNOR* was downregulated, and SNO

Table 8.1 Summary of GSNOR functions under biotic and abiotic stress conditions

Stress	GSNO/SNOs level	GSNOR	References
Biotic stress			
P. parasítica	SNOs (+)	Transcript (–)	Rusterucci et al. (2007)
P. halstedii	GSNO and SNOs (–)	Activity and protein (+)	Chaki et al. (2009)
P. syringae vir.	ND	(=)	Holzmeister et al. (2011)
P. syringae avir.	ND	(=)	Holzmeister et al. (2011)
Abiotic stress			
Wounding			
A. thaliana	SNOs (+)	Transcript and activity (–)	Diaz et al. (2003), Rusterucci et al. (2007)
P. sativum	SNOs (+)	Activity (+)	Corpas et al. (2008)
H. annuus	SNOs (+)	Transcript, protein, and activity (–)	Chaki et al. (2011b)
N. attenuata	ND	(=)	Wunsche et al. (2011)
High temperature			
P. sativum	SNOs (+)	Activity (+)	Corpas et al. (2008)
A. thaliana (WT)	Nitroso species (=)	Protein (=)	Lee et al. (2008)
H. annuus	GSNO and SNOs (+)	Transcript, protein, and activity (–)	Chaki et al. (2011a)
Low temperature			
P. sativum	SNOs (+)	Activity (+)	Corpas et al. (2008)
H. annuus	SNOs (=)	Activity (=)	Chaki et al. (2011b)
C. annuum	SNOs (+)	Activity (+)	Airaki et al. (2012)
Salt			
P.sativum-mitochondria	ND	Activity (+)	Camejo et al. (2013)
Cadmium			
P. sativum	GSNO (–)	Transcript and activity (–)	Barroso et al. (2006)
Arsenic			
A. thaliana	GSNO (–)	Activity (+)	Leterrier et al. (2012)

GSNOR *S*-nitrosoglutathione reductase, *SNO* total *S*-nitrosothiol level, *GSNO* *S*-nitrosoglutathione content, *ND* not determined, (=) no change, (+) increased, (–) decreased

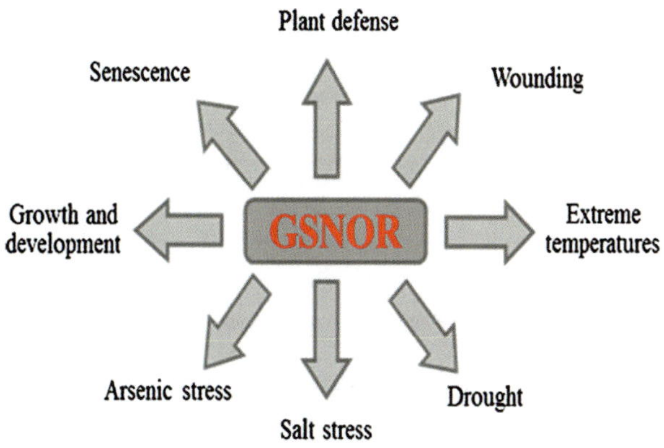

Fig. 8.2 Physiological functions/processes GSNOR is involved in

content was increased (Diaz et al. 2003; Espunya et al. 2012). Similar results have been reported in sunflower hypocotyls where after mechanical wounding a reduction of GSNOR has been observed at different levels—transcript, protein, and enzymatic activity. This resulted in accumulation of SNOs, specifically GSNO, and the authors proposed that GSNOR and SNOs were the new key elements in the wound signaling pathway (Chaki et al. 2011b). However, in pea leaves, wounding induces GSNOR activity, accompanied by an accumulation of NO and SNOs contents (Corpas et al. 2008). In contrast, Wünsche et al. (2011) have been shown that wounding does not affect the activity of GSNOR in *Nicotiana attenuata*.

High temperature is considered as one of the major abiotic stresses that negatively affects both vegetative and reproductive growth. High temperature triggers the reduction of GSNOR activity and accumulation of GSNO and SNOs in sunflower hypocotyls, leading to an increase in protein nitration, which is considered as a marker of nitrosative stress (Chaki et al. 2011a). However, in pea seedlings exposed to the same stress, GSNOR activity and SNO content were increased (Corpas et al. 2008). In addition, in the *Arabidopsis* mutant HOT5 (sensitive to hot temperatures), which encodes a GSNOR, it has been revealed that this enzyme is required for thermotolerance (Lee et al. 2008). However, in wild-type plants subjected to the same stress, GSNOR protein and nitroso species were similar in control and heat-stressed leaves (Lee et al. 2008).

Low temperature is another harmful abiotic stress that affects plants (Janska et al. 2010; Sharma et al. 2005). In *Pisum sativum* plants exposed to 8 °C for 48 h, GSNOR activity as well as SNOs content were enhanced with consequent increase in tyrosine-nitrated proteins (Corpas et al. 2008). Similar trends were observed in pepper plants exposed to low temperature for 24 h (Airaki et al. 2012). However, this stress had no significant influence on the GSNOR activity and SNO content in sunflower hypocotyls (Chaki et al. 2011b). Nonetheless, pea mitochondrial GSNOR activity was also enhanced in response to short and long-term NaCl treatment, where a higher number of nitrated proteins were also detected (Camejo et al. 2013).

Cadmium is a toxic trace pollutant (He et al. 2005). Different metabolic routes such as respiration and photosynthesis are affected by this metal (Sandalio et al. 2001). In leaves of *P. sativum*, grown with 50 μM cadmium, GSNOR activity and gene expression were reduced (Barroso et al. 2006). Under arsenic stress, plants suffer alterations at different levels including gene expression, transport and metabolism (Abercrombie et al. 2008; Verbruggen et al. 2009; Zhao et al. 2009). GSNOR activity and NO content were elevated and GSNO content was reduced in *Arabidopsis* seedlings subjected to this stress (Leterrier et al. 2012).

8.8 Conclusions

GSNOR is present in nearly all organisms and highly conserved during evolution. GSNOR regulates the cellular level of GSNO which is a natural reservoir of NO, and in this way also the cellular SNO content. Therefore, the study of the mechanisms which regulate the activity of GSNOR during plant development and stress response is a crucial aspect to understand the complex metabolism of NO in higher plants. So, further research is needed for better understanding of the correlation between NO and GSNOR.

Acknowledgments This work was supported by Marie Curie Intra-European Fellowships (FP7-PEOPLE-2011-IEF) under grant agreement no. 300176.

References

Abercrombie JM, Halfhill MD, Ranjan P et al (2008) Transcriptional responses of Arabidopsis thaliana plants to As (V) stress. BMC Plant Biol 8:87

Achkor H, Diaz M, Fernandez MR et al (2003) Enhanced formaldehyde detoxification by overexpression of glutathione-dependent formaldehyde dehydrogenase from Arabidopsis. Plant Physiol 132:2248–2255

Adinolfi A, Adinolfi M, Hopkinson DA (1984) Immunological and biochemical characterization of the human alcohol dehydrogenase chi-ADH isozyme. Ann Human Gen 48:1–10

Airaki M, Sanchez-Moreno L, Leterrier M et al (2011) Detection and quantification of *S*-nitrosoglutathione (GSNO) in pepper (Capsicum annuum L.) plant organs by LC-ES/MS. Plant Cell Physiol 52:2006–2015

Airaki M, Leterrier M, Mateos RM et al (2012) Metabolism of reactive oxygen species and reactive nitrogen species in pepper (Capsicum annuum L.) plants under low temperature stress. Plant Cell Environ 35:281–295

Barroso JB, Corpas FJ, Carreras A et al (2006) Localization of *S*-nitrosoglutathione and expression of *S*-nitrosoglutathione reductase in pea plants under cadmium stress. J Exp Bot 57:1785–1793

Bauer JA, Fung HL (1991) Differential hemodynamic effects and tolerance properties of nitroglycerin and an *S*-nitrosothiol in experimental heart failure. JPET 256:249–254

Begara-Morales JC, Chaki M, Sanchez-Calvo B et al (2013) Protein tyrosine nitration in pea roots during development and senescence. J Exp Bot 64:1121–1134

Beigi F, Gonzalez DR, Minhas KM et al (2012) Dynamic denitrosylation via *S*-nitrosoglutathione reductase regulates cardiovascular function. PNAS 109:4314–4319

Besson-Bard A, Pugin A, Wendehenne D (2008) New insights into nitric oxide signaling in plants. Annu Rev Plant Biol 59:21–39

Bethke PC, Badger MR, Jones RL (2004) Apoplastic synthesis of nitric oxide by plant tissues. Plant Cell 16:332–341

Camejo D, Romero-Puertas Mdel C, Rodriguez-Serrano M et al (2013) Salinity-induced changes in *S*-nitrosylation of pea mitochondrial proteins. J Proteomics 79:87–99

Chaki M, Fernandez-Ocana AM, Valderrama R et al (2009) Involvement of reactive nitrogen and oxygen species (RNS and ROS) in sunflower-mildew interaction. Plant Cell Physiol 50:265–279

Chaki M, Valderrama R, Fernandez-Ocana AM et al (2011a) High temperature triggers the metabolism of *S*-nitrosothiols in sunflower mediating a process of nitrosative stress which provokes the inhibition of ferredoxin-NADP reductase by tyrosine nitration. Plant Cell Environ 34:1803–1818

Chaki M, Valderrama R, Fernandez-Ocana AM et al (2011b) Mechanical wounding induces a nitrosative stress by down-regulation of GSNO reductase and an increase in *S*-nitrosothiols in sunflower (Helianthus annuus) seedlings. J Exp Bot 62:1803–1813

Chen R, Sun S, Wang C et al (2009) The Arabidopsis PARAQUAT RESISTANT2 gene encodes an *S*-nitrosoglutathione reductase that is a key regulator of cell death. Cell Res 19:1377–1387

Corpas FJ, Barroso JB, Carreras A et al (2004) Cellular and subcellular localization of endogenous nitric oxide in young and senescent pea plants. Plant Physiol 136:2722–2733

Corpas FJ, Chaki M, Fernandez-Ocana A et al (2008) Metabolism of reactive nitrogen species in pea plants under abiotic stress conditions. Plant Cell Physiol 49:1711–1722

Delledonne M, Xia Y, Dixon RA, Lamb C (1998) Nitric oxide functions as a signal in plant disease resistance. Nature 394:585–588

Diaz M, Achkor H, Titarenko E, Martinez MC (2003) The gene encoding glutathione-dependent formaldehyde dehydrogenase/GSNO reductase is responsive to wounding, jasmonic acid and salicylic acid. FEBS Lett 543:136–139

Dixon DP, Cummins L, Cole DJ, Edwards R (1998) Glutathione-mediated detoxification systems in plants. Curr Opin Plant Biol 1:258–266

Duley JA, Harris O, Holmes RS (1985) Analysis of human alcohol- and aldehyde-metabolizing isozymes by electrophoresis and isoelectric focusing. Alcohol Clin Exp Res 9:263–271

Durner J, Klessig DF (1999) Nitric oxide as a signal in plants. Curr Opin Plant Biol 2:369–374

Durner J, Wendehenne D, Klessig DF (1998) Defense gene induction in tobacco by nitric oxide, cyclic GMP, and cyclic ADP-ribose. PNAS 95:10328–10333

Espunya MC, Diaz M, Moreno-Romero J, Martinez MC (2006) Modification of intracellular levels of glutathione-dependent formaldehyde dehydrogenase alters glutathione homeostasis and root development. Plant Cell Environ 29:1002–1011

Espunya MC, De Michele R, Gomez-Cadenas A, Martinez MC (2012) *S*-Nitrosoglutathione is a component of wound- and salicylic acid-induced systemic responses in Arabidopsis thaliana. J Exp Bot 63:3219–3227

Feechan A, Kwon E, Yun BW et al (2005) A central role for *S*-nitrosothiols in plant disease resistance. PNAS 102:8054–8059

Fernandez MR, Biosca JA, Pares X (2003) *S*-nitrosoglutathione reductase activity of human and yeast glutathione-dependent formaldehyde dehydrogenase and its nuclear and cytoplasmic localisation. CMLS 60:1013–1018

Ford PC (2010) Reactions of NO and nitrite with heme models and proteins. Inorg Chem 49:6226–6239

Frungillo L, de Oliveira JF, Saviani EE et al (2013) Modulation of mitochondrial activity by *S*-nitrosoglutathione reductase in Arabidopsis thaliana transgenic cell lines. Biochim Biophys Acta 1827:239–247

Furchgott RF, Zawadzki JV (1980) The obligatory role of endothelial cells in the relaxation of arterial smooth muscle by acetylcholine. Nature 288:373–376

Gaston B, Reilly J, Drazen JM et al (1993) Endogenous nitrogen oxides and bronchodilator *S*-nitrosothiols in human airways. PNAS 90:10957–10961

Giese M, Bauer-Doranth U, Langebartels C, Sandermann H Jr (1994) Detoxification of formaldehyde by the spider plant (Chlorophytum comosum L.) and by soybean (Glycine max L.) cell-suspension cultures. Plant Physiol 104:1301–1309

Gorren AC, Schrammel A, Schmidt K, Mayer B (1996) Decomposition of *S*-nitrosoglutathione in the presence of copper ions and glutathione. Arch Biochem Biophys 330:219–228

Gow AJ, Farkouh CR, Munson DA et al (2004) Biological significance of nitric oxide-mediated protein modifications. Am J Physiol Cell Physiol Lung Cell Mol Physiol 287:L262–L268

Gupta KJ, Hebelstrup KH, Mur LA, Igamberdiev AU (2011) Plant hemoglobins: important players at the crossroads between oxygen and nitric oxide. FEBS Lett 585:3843–3849

He ZL, Yang XE, Stoffella PJ (2005) Trace elements in agroecosystems and impacts on the environment. J Trace Elem Med Biol 19:125–140

Holzmeister C, Frohlich A, Sarioglu H et al (2011) Proteomic analysis of defense response of wildtype Arabidopsis thaliana and plants with impaired NO- homeostasis. Proteomics 11:1664–1683

Iborra FJ, Renau-Piqueras J, Portoles M et al (1992) Immunocytochemical and biochemical demonstration of formaldhyde dehydrogenase (class III alcohol dehydrogenase) in the nucleus. J Histochem Cytochem 40:1865–1878

Ignarro LJ, Lippton H, Edwards JC et al (1981) Mechanism of vascular smooth muscle relaxation by organic nitrates, nitrites, nitroprusside and nitric oxide: evidence for the involvement of *S*-nitrosothiols as active intermediates. JPET 218:739–749

Ignarro LJ, Cirino G, Casini A, Napoli C (1999) Nitric oxide as a signaling molecule in the vascular system: an overview. J Cardiovasc pharmacol 34:879–886

Janska A, Marsik P, Zelenkova S, Ovesna J (2010) Cold stress and acclimation—what is important for metabolic adjustment? Plant Biol 12:395–405

Kubienova L, Kopecny D, Tylichova M et al (2013) Structural and functional characterization of a plant *S*-nitrosoglutathione reductase from Solanum lycopersicum. Biochimie 95:889–902

Kwon E, Feechan A, Yun BW et al (2012) AtGSNOR1 function is required for multiple developmental programs in Arabidopsis. Planta 236:887–900

Lamotte O, Courtois C, Barnavon L et al (2005) Nitric oxide in plants: the biosynthesis and cell signalling properties of a fascinating molecule. Planta 221:1–4

Lee U, Wie C, Fernandez BO et al (2008) Modulation of nitrosative stress by *S*-nitrosoglutathione reductase is critical for thermotolerance and plant growth in Arabidopsis. Plant Cell 20:786–802

Leterrier M, Airaki M, Palma JM et al (2012) Arsenic triggers the nitric oxide (NO) and *S*-nitrosoglutathione (GSNO) metabolism in Arabidopsis. Environ Pollut 166:136–143

Lima B, Lam GK, Xie L et al (2009) Endogenous *S*-nitrosothiols protect against myocardial injury. PNAS 106:6297–6302

Lin A, Wang Y, Tang J et al (2012) Nitric oxide and protein *S*-nitrosylation are integral to hydrogen peroxide-induced leaf cell death in rice. Plant Physiol 158:451–464

Lindermayr C, Durner J (2009) *S*-Nitrosylation in plants: pattern and function. J Proteomics 73:1–9

Lindermayr C, Saalbach G, Durner J (2005) Proteomic identification of *S*-nitrosylated proteins in Arabidopsis. Plant Physiol 137:921–930

Liu L, Hausladen A, Zeng M et al (2001) A metabolic enzyme for *S*-nitrosothiol conserved from bacteria to humans. Nature 410:490–494

Liu L, Yan Y, Zeng M et al (2004) Essential roles of *S*-nitrosothiols in vascular homeostasis and endotoxic shock. Cell 116:617–628

Martinez MC, Achkor H, Persson B et al (1996) Arabidopsis formaldehyde dehydrogenase. Molecular properties of plant class III alcohol dehydrogenase provide further insights into the origins, structure and function of plant class p and liver class I alcohol dehydrogenases. Eur J Biochem FEBS 241:849–857

Murad F (2006) Shattuck Lecture. Nitric oxide and cyclic GMP in cell signaling and drug development. N Eng J Med 355:2003–2011

Padgett CM, Whorton AR (1995) *S*-nitrosoglutathione reversibly inhibits GAPDH by *S*-nitrosylation. Am J Physiol 269:C739–C749

Palmer RM, Ashton DS, Moncada S (1988) Vascular endothelial cells synthesize nitric oxide from L-arginine. Nature 333:664–666

Potters G, Pasternak TP, Guisez Y et al (2007) Stress-induced morphogenic responses: growing out of trouble? Trends Plant Sci 12:98–105

Radi R (2004) Nitric oxide, oxidants, and protein tyrosine nitration. PNAS 101:4003–4008

Rusterucci C, Espunya MC, Diaz M et al (2007) *S*-nitrosoglutathione reductase affords protection against pathogens in Arabidopsis, both locally and systemically. Plant Physiol 143:1282–1292

Sakamoto A, Ueda M, Morikawa H (2002) Arabidopsis glutathione-dependent formaldehyde dehydrogenase is an *S*-nitrosoglutathione reductase. FEBS Lett 515:20–24

Sandalio LM, Dalurzo HC, Gomez M et al (2001) Cadmium-induced changes in the growth and oxidative metabolism of pea plants. J Exp Bot 52:2115–2126

Shapiro AD (2005) Nitric oxide signaling in plants. Vit Horm 72:339–398

Sharma P, Sharma N, Deswal R (2005) The molecular biology of the low-temperature response in plants. BioEssays News Rev Mol Cell Dev Biol 27:1048–1059

Singh SP, Wishnok JS, Keshive M et al (1996) The chemistry of the *S*-nitrosoglutathione/glutathione system. PNAS 93:14428–14433

Smith JN, Dasgupta TP (2000) Kinetics and mechanism of the decomposition of *S*-nitrosoglutathione by l-ascorbic acid and copper ions in aqueous solution to produce nitric oxide. Nitric Oxide Biol Chem 4:57–66

Stamler JS, Lamas S, Fang FC (2001) Nitrosylation. The prototypic redox-based signaling mechanism. Cell 106:675–683

Steffen M, Sarkela TM, Gybina AA et al (2001) Metabolism of *S*-nitrosoglutathione in intact mitochondria. Biochem J 356:395–402

Valderrama R, Corpas FJ, Carreras A et al (2007) Nitrosative stress in plants. FEBS Lett 581:453–461

Verbruggen N, Hermans C, Schat H (2009) Mechanisms to cope with arsenic or cadmium excess in plants. Vitam Horm 12:364–372

Wünsche H, Baldwin IT, Wu J (2011) *S*-Nitrosoglutathione reductase (GSNOR) mediates the biosynthesis of jasmonic acid and ethylene induced by feeding of the insect herbivore Manduca sexta and is important for jasmonate-elicited responses in Nicotiana attenuata. J Exp Bot 62:4605–4616

Zhang Y, Hogg N (2004) The mechanism of transmembrane *S*-nitrosothiol transport. PNAS 101:7891–7896

Zhao FJ, Ma JF, Meharg AA, McGrath SP (2009) Arsenic uptake and metabolism in plants. New Phytol 181:777–794

Chapter 9
Nitro-Fatty Acids: Synthesis, Properties, and Role in Biological System

Homero Rubbo and Andrés Trostchansky

Abstract Recent developments have implicated nitric oxide-derived reactive species that react with unsaturated fatty acids to yield novel electrophilic derivatives (NO_2-FA, nitroalkenes). Of relevance, we show for a first time the presence of NO_2-FA in plants, specifically in olives and extra virgin olive oil (EVOO) and discuss the additional formation of NO_2-FA from EVOO under acidic gastric conditions. Since NO_2-FA may induce salutary anti-inflammatory gene expression and metabolic responses, it is speculated that the dietary consumption and endogenous generation of electrophilic anti-inflammatory lipids can contribute to the cardiovascular benefits associated with Mediterranean diet. Based on this information as well as recent in vivo data, NO_2-FA will also be addressed in the context of their promising pharmacological utility. Herein, NO_2-FA will be discussed in the context of their biochemical activities and cell signaling actions.

Keywords Antioxidants • Nitration • Nitric oxide • Nitro-fatty acids • Olive oil

9.1 Introduction

Nitration is the addition of NO_2 group to specific intracellular biomolecules, which is a most biologically relevant redox mechanism in animals. Nonetheless, we do not know much about peroxynitrite-mediated nitration in plants. According to

H. Rubbo (✉)
Facultad de Medicina, Departamento de Bioquímica,
Center for Free Radical and Biomedical Research, Universidad de la República,
Avda. Gral. Flores 2125, 11800 Montevideo, Uruguay
e-mail: hrubbo@fmed.edu.uy

A. Trostchansky
Facultad de Medicina, Departamento de Bioquímica,
Center for Free Radical and Biomedical Research, Universidad de la República,
Montevideo, Uruguay

M. N. Khan et al. (eds.), *Nitric Oxide in Plants: Metabolism and Role in Stress Physiology*, DOI: 10.1007/978-3-319-06710-0_9,

Radi (2013) the reaction of $^{\bullet}NO$ with superoxide anion radical ($O_2^{\bullet -}$) forms peroxynitrite which is a strong oxidizing and nitrating compound. Nitration by peroxynitrite causes the addition of a nitro ($-NO_2$) group in key intracellular targets, including proteins, lipids, or DNA. In fact, a recognized oxidative protein modification left by peroxynitrite is the formation of 3-nitrotyrosine, a posttranslational modification that modifies protein function (Radi 2004). Although originally described in biology to occur via peroxynitrite-dependent reactions, protein tyrosine nitration can also be due to other $^{\bullet}NO$-mediated processes, most notably in heme peroxidase-catalyzed reactions (Radi 2013). In higher plants, addition of nitrating agents such as peroxynitrite produced a rise of immunoreactive proteins (Morot-Gaudry-Talarmain et al. 2002; Saito et al. 2006). Nitrated proteins have been detected in leaves of olive plants exposed to nitrooxidative stress conditions (Valderrama et al. 2007). Proteomics combined with mass spectrometry techniques identified few nitrated proteins in higher plants (for a review see; Corpas and Barroso 2013). As an example, in *Arabidopsis* leaves, up to eight different proteins undergoing tyrosine nitration and mainly involved in photosynthesis were identified (Cecconi et al. 2009). Since the presence of nitrotyrosine has been considered as a footprint of protein nitration in mammals, same approach has been applied to the plants (Corpas et al. 2007). In fact, nitrotyrosine formation increases under abiotic and biotic stresses (Corpas et al. 2008) in olives leaves, *Arabidopsis* roots, citrus, sunflowers, and peppers among others (Valderrama et al. 2007; Corpas et al. 2009; Tanou et al. 2012; Chaki et al. 2009a, b; Cellini et al. 2011; Airaki et al. 2012; Leterrier et al. 2011; Signorelli et al. 2013). Nitration can cause a loss of function as observed for different plant enzyme activities including ascorbate peroxidase and catalase (Clark et al. 2000), *S*-adenosyl homocysteine hydrolase (Chaki et al. 2009a); and *O*-acetylserine(thiol)lyase A1 (Alvarez et al. 2011).

9.2 Fatty Acid Nitration

In addition to proteins, $^{\bullet}NO$-derived species react with unsaturated fatty acids yielding nitro-fatty acids (nitroalkenes, NO_2-FA) detected in human plasma, cell membranes, and tissues (Rubbo et al. 1994; Baker et al. 2005; Lima et al. 2002). These novel species represent redox anti-inflammatory signaling mediators that are able to modulate a variety of cell signaling pathways by interaction with specific cellular targets (Freeman et al. 2008; Trostchansky et al. 2013; Trostchansky and Rubbo 2008). Reaction mechanisms for fatty acid nitration include oxidation of polyunsaturated fatty acids (PUFAs) by $^{\bullet}NO$ secondary products, e.g., nitrogen dioxide ($^{\bullet}NO_2$), nitrite (NO_2^-) and peroxynitrite ($ONOO^-$). Nitrogen dioxide can be formed from both $^{\bullet}NO$ autooxidation (Radi et al. 2000) and acidic NO_2^-, present in physiological fluids at high concentrations (Pannala et al. 2003; Lundberg and Weitzberg 2012), when exposed to low pH in the gastric compartment as well as in phagocytic lysosomes; indeed, the human stomach is a source of $^{\bullet}NO$ and bioactive nitrogen oxides from precursors present in food and saliva (Lundberg and Weitzberg 2012). The currently accepted mechanisms for $^{\bullet}NO_2$-mediated oxidation and nitration of PUFAs involve hydrogen atom abstraction and addition reactions leading to the generation of isomerized, oxidized, and/or nitro-allylic, nitroalkene, dinitro, or nitro-hydroxy lipid derivatives (Trostchansky and Rubbo 2008). The allylic or bis-allylic hydrogen abstraction generates a carbon-centered lipid radical and nitrous acid (HONO), which rapidly decomposes turning the hydrogen abstraction reaction irreversible. Unlike the hydrogen abstraction pathway, the addition of $^{\bullet}NO_2$ to an unsaturated carbon center is a reversible reaction. Nonetheless, when the concentration of $^{\bullet}NO_2$ is fairly high, the addition mechanism is more likely to account for $^{\bullet}NO_2$-dependent fatty acid nitration/oxidation (Trostchansky and Rubbo 2008). An additional lipid nitration mechanism involves peroxynitrite. Peroxynitrite anion ($ONOO^-$) and its conjugate acid (ONOOH, pKa = 6.8 at 37 °C) are strong one-and two electron oxidizing species that react with a wide variety of biological targets, including protein tyrosine residues, thiols, and PUFAs (Baker et al. 2005; Rubbo et al. 1994) being tyrosine residues unable to compete with PUFAs nitration when present at similar levels (Bonacci et al. 2012). The main route for peroxynitrite-dependent fatty acid nitration is the generation of $^{\bullet}NO_2$ following ONOOH homolysis (Baker et al. 2005; Rubbo et al. 1994). For linolenic acid (18:3), major products include nitroso-peroxo-linolenate, hydroxyl-nitroso-peroxo-linolenate, and hydro-peroxo-nitroso-peroxo-linolenate (Rubbo et al. 1994). Since both peroxynitrite and $^{\bullet}NO_2$ readily diffuse through the membrane bilayers, reactions leading to $^{\bullet}NO_2$ generation may take place in the aqueous environment in proximity to the membrane or inside the lipid bilayer. Nitrated fatty acids formation has been reported in vivo under physiological and pathophysiological conditions as free, esterified, and nucleophilic-adducted species (Cui et al. 2006; Ferreira et al. 2009; Rudolph et al. 2010; Nadtochiy et al. 2009) at concentrations ranging from the micromolar to the picomolar range (Ferreira et al. 2009; Nadtochiy et al. 2009; Schopfer et al. 2009; Tsikas et al. 2009).

9.3 Electrophilic and Therapeutical Properties of NO_2-FA

Whether NO_2-FA are generated in situ or esterified following nitration, A_2-type phospholipases are capable of releasing NO_2-FA during inflammatory conditions or in response to other stimuli, being able to reach specific cellular targets and exert signaling actions through the formation of reversible covalent adducts with cytosolic and/or plasma proteins and low molecular weight thiols (Batthyany et al. 2006). Nitroalkenes stability is greatly altered in aqueous environments where they spontaneously release $^{\bullet}NO$, activating sGC and exerting vasorelaxation (Blanco et al. 2011; Schopfer et al. 2005a; Baker et al. 2005; Trostchansky et al. 2007). Despite the well-ascertained role of NO_2-FA as $^{\bullet}NO$ reservoirs in vivo, the most relevant cell signaling activities of nitroalkenes are linked to their receptor-mediated reactions and strong electrophilic nature. The highly electronegative nitro functional group facilitates reaction of the carbon β adjacent to the nitro group with nucleophilic cellular targets via Michael addition reversible reactions (Baker et al. 2007; Batthyany et al. 2006). Nitrated fatty acids are able to covalently alkylate susceptible thiols of multiple transcriptional regulatory proteins, affecting downstream gene expression and the metabolic and inflammatory responses under their regulation. Most of the nitroalkenes released into the cytosol are expected to be initially found as glutathione (GSH) adducts, given the high intracellular concentration of this major antioxidant. Nitro-oleic (NO_2-OA-GSH) and nitro-linoleic acids (NO_2-LA-GSH) adducts have been detected in healthy human red blood cells (Baker et al. 2007; Batthyany et al. 2006). Nitroalkenes display the largest second-order rate constants for the bimolecular reaction with cysteine and GSH ($k = 183$ and $355\ M^{-1}s^{-1}$ for nitro-oleic acid (NO_2-OA) and nitro-linoleic acid (NO_2-LA), respectively, at pH 7.4 and 37 °C), when compared to other lipid electrophiles (Baker et al. 2007).

When the strong electrophilic reactivity of NO_2-FA was described, posttranslational modifications of proteins emerged as a plausible mechanism for the observed anti-inflammatory effects on inflammatory cells. Particularly, nitroalkylation of nuclear transcription factors seems to be the main mechanism for NO_2-FA to modulate inflammatory cells responses. In this regard, the transcription factor peroxisome proliferator-activated receptor γ (PPARγ) contains a critical cysteine residue in the ligand-binding domain subject to nitroalkylation by NO_2-OA and NO_2-LA (Li et al. 2008). In the vasculature, PPARγ is expressed in monocytes/macrophages, smooth muscle cells, and endothelium and plays key roles in the regulation of energy balance and adipogenesis. Nitroalkenes are potent agonists for PPARγ, of significance since its activation has been associated with anti-inflammatory actions, resulting in macrophage CD36 expression, adipocyte differentiation, and glucose uptake at potency comparable to thiazolidineadiones (Schopfer et al. 2005b; Li et al. 2008; Villacorta et al. 2009; Schopfer et al. 2010). The transcription factor NF-κB is also subject to negative regulation by several naturally occurring electrophiles through alkylation of highly conserved cysteine residues in the DNA-binding domains p50 and p65 (Cui et al. 2006). NF-κB plays a crucial role in the induction of inflammatory cytokines and enzymes, chemokines, cell adhesion molecules, acute phase proteins, and growth factors. Adduction of the p65 subunit by NO_2-FA inhibits NF-κB-mediated pro-inflammatory responses (Cui et al. 2006). Also, NO_2-FA inhibit endotoxin-mediated STAT proinflammatory signaling through the induction of mitogen-activated protein kinase-1, a MAPK phosphatase known to contribute to anti-inflammatory signaling through alteration in the translation of mRNA to proteins (Ichikawa et al. 2008).

Effects of NO_2-FA on inflammatory stress also involve the modulation of phase II antioxidant enzymes, i.e, hemoxygenase-1 (HO-1). Induction of HO-1 represents a cytoprotective pathway triggered by a variety of stress-related signals and electrophilic species. Nitroalkenes induce HO-1 in endothelial cells (Wright et al. 2006), RAW264.7 (Cui et al. 2006) and J774.1 macrophages (Ferreira et al. 2009), linked to the MAPK/ERK signaling pathway (Ichikawa et al. 2008) whose expression is subject to complex gene regulation via antioxidant response elements (ARE) and binding sites for PPAR α or γ and NF-κB (Wright et al. 2006). E2-related nuclear factor 2 (Nrf2), a transcription factor is in an inactive form at the cytosol due to the activity of Keap1. When activated, Nrf2 migrates to the nucleus and binds as a heterodimer to the ARE in DNA, activating the expression of phase 2 enzymes. Potential activators for Nrf2 include lipid electrophiles which react with cysteine-rich protein Keap1 thiols, dissociating Nrf2 from ubiquitin E3 ligase complex and facilitating nuclear accumulation and downstream effects on gene transcription (Kansanen et al. 2011; Jyrkkanen et al. 2008; Kansanen et al. 2009).

A recently described protective mechanism for the nitrated derivative of arachidonic acid (NO_2-AA) is the regulation of superoxide radical ($O_2^{-\bullet}$) production via the NADPH oxidase (NOX) isoforms. In fact, NO_2-AA inhibits the phagocytic NOX-2-mediated O_2^- production in activated macrophages (Gonzalez-Perilli et al. 2013) by preventing the migration of the cytosolic subunits to the membrane, thus inhibiting the correct assembly of the active enzyme.

There are several reports using NO_2-FA as pharmacological modulators of inflammatory-related diseases in animal models (Ichikawa et al. 2008; Rudolph et al. 2010; Zhang et al. 2010; Cole et al. 2009; Villacorta et al. 2013). Nitro-fatty acid subcutaneous administration to angiotensin-II-treated mice significantly lowered the increase in blood pressure as well as the contractile responses through NO_2-FA binding to the AT1 Receptor, modulating intracellular signaling cascades (Zhang et al. 2010). Nitroalkenes were also tested in C57/BL6 mice subjected to coronary artery ligation followed by 30 min reperfusion (I/R), reducing the infarct size as well as preserving the left ventricular function when administered exogenously during the ischemic episodes (Rudolph et al. 2010). Moreover, acute administration of NO_2-FA is effective to reduce vascular inflammation in vivo (Villacorta et al. 2013). The mechanism involves a direct role of NO_2-FA in the disruption of the toll-like receptor 4 signaling complex in lipid rafts, leading to resolution of pro-inflammatory activation of NF-κB in the vasculature (Villacorta et al. 2013).

9.4 Formation of NO_2-FA in Extra Virgin Olive Oil

Extra virgin olive oil (EVOO) is the main source of lipids in the Mediterranean diet, which contain a substantial amount of diverse PUFA (ω-3, -6, -9), and characterized by a high intake of fruit, nuts, vegetables, cereals that are rich in the inorganic anions nitrite (NO_2^-) and nitrate (NO_3^-) (Lundberg and Weitzberg 2008; Lundberg et al. 2008; Nadtochiy and Redman 2011). These species undergo further reactions in the blood and tissues via both reductive and oxidizing metabolism, nitrosating, and nitrating conditions promoted by digestion and inflammation (Lundberg and Weitzberg 2008). In the case of NO_3^-, the commensal bacteria of the enterosalivary system reduce dietary NO_3^- to physiologically significant levels of NO_2^-, $^\bullet NO$, and secondary species (Jansson et al. 2008).

Major components of EVOO are triacylglycerides (98–99 %) predominantly esterified with monounsaturated oleic acid (OA). Fresh olives contain both unsaturated fatty acids and various oxides of nitrogen, raising the possibilities that electrophilic NO_2-FA are: (a) endogenously present in olives (b) extractable into the EVOO fraction and (c) generated after consumption of olive lipids by the acidic conditions of digestion. Olive oil should be a source of NO_2-FA generation during digestion. In fact, vegetables are the dominant source of dietary NO_3^- in humans and contribute 60–80 % of the total NO_3^- intake, with other sources being the drinking water and including animal-based products to which NO_3^- and NO_2^- are added as preservatives (Nadtochiy and Redman 2011; Lundberg et al. 2009; Lundberg and Weitzberg 2009). In plants, NO_3^- plays a crucial role in nutrition and function and is the main growth-limiting factor. The human stomach can be viewed as a bioreactor where a variety of bioactive nitrogen oxides are formed from precursors present in food and saliva, catalyzed by hydrochloric acid secreted from parietal cells in the gastric mucosa. Nitrite at the acidic stomach environment is protonated to nitrous acid that spontaneously decomposes to $^{\bullet}NO$ and other nitrogen oxides with nitrating properties (Nadtochiy and Redman 2011; Lundberg et al. 2009; Lundberg and Weitzberg 2009; Rocha et al. 2012). Thus, the Mediterranean diet rich in NO_2^- and unsaturated fatty acids (e.g., linoleic, oleic, and conjugated linoleic), and supplemented with acidic vinegar may favor intragastric generation of NO_2-FA as proposed previously from EVOO upon exposure to NO_2^- in mild acidic conditions. Overall, although nitration reactions are viewed as harmful (e.g., detection of nitrotyrosine in tissues), the increase of NO_2-FA formation may have tissue protective effects.

We have preliminary data showing the small amounts of NO_2-FA in olives and EVOO, being the main products nitrated derivatives of conjugated linoleic acid (NO_2-cLA) as well as NO_2-OA, detected as their esterified and conjugated to protein forms (Fazzari et al 2014). Conjugated linoleic acid (cLA) consists of a series of positional- and regioisomers derived from linoleic acid having conjugated dienes in the *cis* and/or *trans* configurations (Chin et al. 1992; Reynolds and Roche 2010; Cheng et al. 2004). There are plant and mammalian sources of cLA in the human diet; plant-derived oils have cLA levels of up to ~ 1.0 mg cLA/g fat (Chin et al. 1992), with levels of cLA in olive oils of up to 0.2 mg CLA/g fat (Chin et al. 1992), predominantly as the *cis*9-, *trans*11-, and *trans*10-, *cis*12-isomers (Chin et al. 1992). Conjugated linoleic acid displays both immunomodulatory and anti-inflammatory effects in a wide range of inflammatory diseases, including atherosclerosis and diabetes (Cheng et al. 2004; Reynolds and Roche 2010). The proposed actions for cLA include a reduction of pro-inflammatory cytokine levels via inhibition of NF-κB-dependent gene expression and activation expression of PPAR-regulated genes (Cheng et al. 2004; Reynolds and Roche 2010). Thus, formation of NO_2-cLA should increase its protective actions.

9.5 Potential Pitfalls

Despite the mounting evidence for endogenous formation of NO_2-FA, several key aspects remain unclear. Disagreements of the reported tissue concentrations can be ascribed to a lack of appropriate standards since the identification and quantification of NO_2-FA in biological samples has been performed using nitroalkenes standards. However, other structural possibilities (i.e., nitroalkanes) may be present in biological samples. The hydrophobic nature of NO_2-FA also poses an additional methodological complexity that should be considered during the experimental design, taking into account that various cellular pools of nitrated lipids are expected to be found in vivo: free, esterified to complex lipids in hydrophobic compartments and protein-adducted NO_2-FA. Thus, handling of biological samples should include protocols for de-esterification from proteins and complex lipids. Finally, controls for artifactual nitration due to acidic extraction have to be included during sample processing preventing overestimated concentrations reports.

Acknowledgements This work was supported by grants from ICGEB (Italy) and CSIC-Uruguay to HR and ANII (FCE-6353, Uruguay) and Fundación Mapfre (Spain) to AT.

References

Airaki M, Leterrier M, Mateos RM et al (2012) Metabolism of reactive oxygen species and reactive nitrogen species in pepper (*Capsicum annuum* L.) plants under low temperature stress. Plant Cell Environ 35:281–295

Alvarez C, Lozano-Juste J, Romero LC et al (2011) Inhibition of Arabidopsis *O*-acetylserine(thiol)lyase A1 by tyrosine nitration. J Biol Chem 286:578–586

Baker LM, Baker PR, Golin-Bisello F et al (2007) Nitro-fatty acid reaction with glutathione and cysteine. Kinetic analysis of thiol alkylation by a Michael addition reaction. J Biol Chem 282:31085–31093

Baker PR, Lin Y, Schopfer FJ et al (2005) Fatty acid transduction of nitric oxide signaling: multiple nitrated unsaturated fatty acid derivatives exist in human blood and urine and serve as endogenous peroxisome proliferator-activated receptor ligands. J Biol Chem 280: 42464–42475

Batthyany C, Schopfer FJ, Baker PR et al (2006) Reversible post-translational modification of proteins by nitrated fatty acids in vivo. J Biol Chem 281:20450–20463

Blanco F, Ferreira AM, Lopez GV et al (2011) 6-Methylnitroarachidonate: a novel esterified nitroalkene that potently inhibits platelet aggregation and exerts cGMP-mediated vascular relaxation. Free Radic Biol Med 50:411–418

Bonacci G, Baker PR, Salvatore SR et al (2012) Conjugated linoleic acid is a preferential substrate for fatty acid nitration. J Biol Chem 287:44071–44082

Cecconi D, Orzetti S, Vandelle E et al (2009) Protein nitration during defense response in Arabidopsis thaliana. Electrophoresis 30:2460–2468

Cellini A, Corpas FJ, Barroso JB, Masia A (2011) Nitric oxide content is associated with tolerance to bicarbonate-induced chlorosis in micropropagated Prunus explants. J Plant Physiol 168:1543–1549

Chaki M, Fernandez-Ocana AM, Valderrama R et al (2009a) Involvement of reactive nitrogen and oxygen species (RNS and ROS) in sunflower-mildew interaction. Plant Cell Physiol 50:265–279
Chaki M, ValderramaR Fernandez-Ocana AM et al (2009b) Protein targets of tyrosine nitration in sunflower (Helianthus annuus L.) hypocotyls. J Exp Bot 60:4221–4234
Cheng WL, Lii CK, Chen HW et al (2004) Contribution of conjugated linoleic acid to the suppression of inflammatory responses through the regulation of the NF-kappaB pathway. J Agric Food Chem 52:71–78
Chin SF, Liu W, Storkson JM et al (1992) Dietary Sources of conjugated dienoic isomers of linoleic acid, a newly recognized class of anticarcinogens. J Food Comp Anal 5:185–197
Clark D, Durner J, Navarre DA, Klessig DF (2000) Nitric oxide inhibition of tobacco catalase and ascorbate peroxidase. Mol Plant Microbe Interact 13:1380–1384
Cole MP, Rudolph TK, Khoo NK et al (2009) Nitro-fatty acid inhibition of neointima formation after endoluminal vessel injury. Circ Res 105:965–972
Corpas FJ, Barroso NK (2013) Nitro-oxidative stress vs oxidative or nitrosative stress in higher plants. New Phytol 199:633–635
Corpas FJ, del Rio LA, Barroso JB (2007) Need of biomarkers of nitrosative stress in plants. Trends Plant Sci 12:436–438
Corpas FJ, Chaki M, Fernandez-Ocana A (2008) Metabolism of reactive nitrogen species in pea plants under abiotic stress conditions. Plant Cell Physiol 49:1711–1722
Corpas FJ, Chaki M, Leterrier M, Barroso JB (2009) Protein tyrosine nitration: a new challenge in plants. Plant Signal Behav 4:920–923
Cui T, Schopfer FJ, Zhang J et al (2006) Nitrated fatty acids: Endogenous anti-inflammatory signaling mediators. J Biol Chem 281:35686–35698
Fazzari M, Trostchansky A, Schopfer FJ et al (2014) Olives and olive oil are sources of electrophilic fatty acid nitroalkenes. PLoS ONE 9:e84884
Ferreira AM, Ferrari MI, Trostchansky A et al (2009) Macrophage activation induces formation of the anti-inflammatory lipid cholesteryl-nitrolinoleate. Biochem J 417:223–234
Freeman BA, Baker PR, Schopfer FJ et al (2008) Nitro-fatty acid formation and signaling. J Biol Chem 283:15515–15519
Gonzalez-Perilli L, Alvarez MN, Prolo C et al (2013) Nitroarachidonic acid prevents NADPH oxidase assembly and superoxide radical production in activated macrophages. Free Radic Biol Med 58:126–133
Ichikawa T, Zhang J, Chen K et al (2008) Nitroalkenes suppress lipopolysaccharide-induced signal transducer and activator of transcription signaling in macrophages: a critical role of mitogen-activated protein kinase phosphatase 1. Endocrinol 149:4086–4094
Jansson EA, Huang L, Malkey R et al (2008) A mammalian functional nitrate reductase that regulates nitrite and nitric oxide homeostasis. Nat Chem Biol 4:411–417
Jyrkkanen HK, Kansanen E, Inkala M et al (2008) Nrf2 regulates antioxidant gene expression evoked by oxidized phospholipids in endothelial cells and murine arteries in vivo. Circ Res 103:e1–e9
Kansanen E, Bonacci G, Schopfer FJ et al (2011) Electrophilic nitro-fatty acids activate nrf2 by a keap1 cysteine 151-independent mechanism. J Biol Chem 286:14019–14027
Kansanen E, Jyrkkanen HK, Volger OL et al (2009) Nrf2-dependent and -independent responses to nitro-fatty acids in human endothelial cells: identification of heat shock response as the major pathway activated by nitro-oleic acid. J Biol Chem 284:33233–33241
Leterrier M, Chaki M, Airaki M et al (2011) Function of *S*-nitrosoglutathione reductase (GSNOR) in plant development and under biotic/abiotic stress. Plant Signal Behav 6:789–793
Li Y, Zhang J, Schopfer FJ et al (2008) Molecular recognition of nitrated fatty acids by PPAR gamma. Nat Struct Mol Biol 15:865–867
Lima ES, Di Mascio P, Rubbo H, Abdalla DS (2002) Characterization of linoleic acid nitration in human blood plasma by mass spectrometry. Biochem 41:10717–10722
Lundberg JO, Weitzberg E (2008) Nitrite reduction to nitric oxide in the vasculature. Am J Physiol Heart Circ Physiol 295:H477–H478

Lundberg JO, Weitzberg E (2009) NO generation from inorganic nitrate and nitrite: Role in physiology, nutrition and therapeutics. Arch Pharm Res 32:1119–1126

Lundberg JO, Weitzberg E (2012) Biology of nitrogen oxides in the gastrointestinal tract. Gut 62:616–629

Lundberg JO, Weitzberg E, Gladwin MT (2008) The nitrate-nitrite-nitric oxide pathway in physiology and therapeutics. Nat Rev Drug Discov 7:156–167

Lundberg JO, Gladwin MT, Ahluwalia A et al (2009) Nitrate and nitrite in biology, nutrition and therapeutics. Nat Chem Biol 5:865–869

Morot-Gaudry-Talarmain Y, Rockel P, Moureaux T et al (2002) Nitrite accumulation and nitric oxide emission in relation to cellular signaling in nitrite reductase antisense tobacco. Planta 215:708–715

Nadtochiy SM, Redman EK (2011) Mediterranean diet and cardioprotection: the role of nitrite, polyunsaturated fatty acids, and polyphenols. Nutrition 27:733–744

Nadtochiy SM, Baker PR, Freeman BA, Brookes PS (2009) Mitochondrial nitroalkene formation and mild uncoupling in ischaemic preconditioning: implications for cardioprotection. Cardiovasc Res 82:333–340

Pannala AS, Mani AR, Spencer JP et al (2003) The effect of dietary nitrate on salivary, plasma, and urinary nitrate metabolism in humans. Free Radic Biol Med 34:576–584

Radi R (2004) Nitric oxide, oxidants, and protein tyrosine nitration. Proc Natl Acad Sci USA 101:4003–4008

Radi R (2013) Peroxynitrite, a stealthy biological oxidant. J Biol Chem 288:26464–26472

Radi R, Denicola A, Alvarez B et al (2000) The biological chemistry of peroxynitrite. In: Ignarro LJ (ed) Nitric oxide biology and pathobiology. Academic press, San Diego

Reynolds CM, Roche HM (2010) Conjugated linoleic acid and inflammatory cell signalling. Prostaglandins Leukot Essent Fatty Acids 82:199–204

Rocha BS, Gago B, Barbosa RM et al (2012) Intragastric nitration by dietary nitrite: implications for modulation of protein and lipid signaling. Free Radic Biol Med 52:693–698

Rubbo H, Radi R, Trujillo M et al (1994) Nitric oxide regulation of superoxide and peroxynitrite-dependent lipid peroxidation. Formation of novel nitrogen-containing oxidized lipid derivatives. J Biol Chem 269:26066–26075

Rudolph V, Rudolph TK, Schopfer FJ et al (2010) Endogenous generation and protective effects of nitro-fatty acids in a murine model of focal cardiac ischaemia and reperfusion. Cardiovasc Res 85:155–166

Saito S, Yamamoto-Katou A, Yoshioka H et al (2006) Peroxynitrite generation and tyrosine nitration in defense responses in tobacco BY-2 cells. Plant Cell Physiol 47:689–697

Schopfer FJ, Batthyany C, Baker PR et al (2009) Detection and quantification of protein adduction by electrophilic fatty acids: mitochondrial generation of fatty acid nitroalkene derivatives. Free Radic Biol Med 46:1250–1259

Schopfer FJ, Cole MP, Groeger AL et al (2010) Covalent peroxisome proliferator-activated receptor gamma adduction by nitro-fatty acids: selective ligand activity and anti-diabetic signaling actions. J Biol Chem 285:12321–12333

Schopfer FJ, Baker PR, Giles G et al (2005a) Fatty acid transduction of nitric oxide signaling. Nitrolinoleic acid is a hydrophobically stabilized nitric oxide donor. J Biol Chem 280:19289–19297

Schopfer FJ, Lin Y, Baker PR et al (2005b) Nitrolinoleic acid: an endogenous peroxisome proliferator-activated receptor gamma ligand. Proc Natl Acad Sci USA 102:2340–2345

Signorelli S, Corpas FJ, Borsani O et al (2013) Water stress induces a differential and spatially distributed nitro-oxidative stress response in roots and leaves of Lotus japonicus. Plant Sci 201–202:137–146

Tanou G, Filippou P, Belghazi M et al (2012) Oxidative and nitrosative-based signaling and associated post-translational modifications orchestrate the acclimation of citrus plants to salinity stress. Plant J 72:585–599

Trostchansky A, Bonilla L, González-Perilli L et al (2013) Nitro-fatty acids: formation, redox signaling, and therapeutic potential. Antioxid Redox Signal. 19:1257–1265

Trostchansky A, Rubbo H (2008) Nitrated fatty acids: mechanisms of formation, chemical characterization, and biological properties. Free Radic Biol Med 44:1887–1896

Trostchansky A, Souza JM, Ferreira A et al (2007) Synthesis, isomer characterization, and anti-inflammatory properties of nitroarachidonate. Biochem 46:4645–4653

Tsikas D, Zoerner A, Mitschke A et al (2009) Specific GC-MS/MS stable-isotope dilution methodology for free 9- and 10-nitro-oleic acid in human plasma challenges previous LC-MS/MS reports. J Chromatogr B Analyt Technol Biomed Life Sci 877:2895–2908

Valderrama R, Corpas FJ, Carreras A et al (2007) Nitrosative stress in plants. FEBS Lett 581:453–461

Villacorta L, Schopfer FJ, Zhang J et al (2009) PPARgamma and its ligands: therapeutic implications in cardiovascular disease. Clin Sci (Lond) 116:205–218

Villacorta L, Chang L, Salvatore SR et al (2013) Electrophilic nitro-fatty acids inhibit vascular inflammation by disrupting LPS-dependent TLR4 signalling in lipid rafts. Cardiovasc Res 98:116–124

Wright MM, Schopfer FJ, Baker PR et al (2006) Fatty acid transduction of nitric oxide signaling: Nitrolinoleic acid potently activates endothelial heme oxygenase 1 expression. Proc Natl Acad Sci USA 103:4299–4304

Zhang J, Villacorta L, Chang L et al (2010) Nitro-oleic acid inhibits angiotensin II-induced hypertension. Circ Res 107:540–548

Part II
Nitric Oxide: Properties, Mode of Action and Functional Role in Stress Physiology

Chapter 10
Nitric Oxide and Reactive Nitrogen Species

Magdalena Arasimowicz-Jelonek, Jolanta Floryszak-Wieczorek, Dariusz Abramowski and Karolina Izbiańska

Abstract Nitric oxide (NO) plays a key role in plant metabolism, signaling, defense, and development. However, a fundamental question arises how the NO message is converted into a physiological response. NO-related signaling may be attributed to various NO derivatives, collectively named reactive nitrogen species (RNS). An increasing body of evidence suggests that nitroxyl (HNO) as a one-electron reduced and protonated state of NO and nitrosonium (NO^+), a one-electron oxidized form of NO, may be important factors. Thus, the impact of NO in plant biology should be re-evaluated in the light of chemical properties of these compounds as they are different from that of NO. Nitroxyl, unlike NO, can interact directly with thiols, prioritize ferric rather than ferrous heme proteins and it is resistant to scavenging by superoxides. Experimental data revealed that NO^+ can facilitate *S*-nitrosylation, while it also appears to be a key intermediate in the trans-nitrosylation reaction. In contrast to NO, it may also take part in the degradation of SNOs (*S*-nitrosothiols). In turn, peroxynitrite ($ONOO^-$) is a potent oxidant and nitrating agent, generated by the reaction of nitric oxide and superoxide in one of the most rapid reactions known in biology. Overproduction of $ONOO^-$ contributes to oxidative and nitrosative stress, however, in the physiological state or under low metabolic stress, peroxynitrite triggered NO-dependent signals behave as a potent modulator of the redox regulation in various cell transduction pathways.

Keywords Nitric oxide · Nitrosonium cation · Nitroxyl anion · Peroxynitrite

---

M. Arasimowicz-Jelonek · K. Izbiańska
Faculty of Biology, Department of Plant Ecophysiology, Adam Mickiewicz University, Poznan, Poland

J. Floryszak-Wieczorek (✉) · D. Abramowski
Department of Plant Physiology, Poznan University of Life Sciences, Poznan, Poland
e-mail: florysza@jay.up.poznan.pl

M. N. Khan et al. (eds.), *Nitric Oxide in Plants: Metabolism and Role in Stress Physiology*, DOI: 10.1007/978-3-319-06710-0_10,

10.1 Introduction

The discovery of Furchgott, Ignarro, and Murad in the late 1990s, which resulted in the Nobel Prize in Physiology and Medicine, made nitric oxide (NO) one of the most ubiquitous endogenous molecules implicated in signaling biochemistry. Since 1998, NO as a bioactive signaling molecule have been extensively studied when it was first reported as a regulating agent during plant defense (Delledonne et al. 1998; Durner et al. 1998). Although contribution of NO during various stages of plant life cycle under both physiological and pathophysiological conditions is well established, but the information regarding the synthesis of endogenous NO and NO-derived compounds is still limited. It is generally accepted that interactions between NO and biomolecules at precise molecular sites constitute the specific language of NO action facilitating transfer of NO message into a cellular response. Moreover, recent research has shown that NO mainly mediates biological functions through chemical reactions between spatially controlled accumulations of different reactive nitrogen species (RNS). Thus, the metabolic status of NO in the cellular milieu is not only governed by the systems involved in NO generation, but it is also exposed to negative regulation.

The term RNS refers to various nitrogenous products, including NO, nitroxyl (HNO/NO^-), nitrosonium cation (NO^+), higher oxides of nitrogen, peroxynitrite ($ONOO^-$), *S*-nitrosothiols (RSNOs), and dinitrosyl iron complexes (Table 10.1). Each of these compounds shows distinctive physicochemical properties determining their reactivity in the biological milieu. Without doubt NO is the main RNS product of the cells, while simultaneously it is the primary source for the other RNS. In this chapter, we have focused mainly on both NO and $ONOO^-$, as they are the most recognized biologically active molecules in both animal and plant cells.

10.2 Properties of Nitric Oxide

Basic properties of a signaling molecule such as a simple structure, small dimensions, and high diffusivity are endowed in NO molecule. Nitric oxide or nitrogen monoxide is a gaseous free radical with a relatively long half-life, estimated in biological systems to be 3–5 s (Henry et al. 1997; Tuteja et al. 2004). NO is one of the smallest diatomic molecules with a high diffusivity (4.8×10^{-5} cm^2 s^{-1} in H_2O), exhibiting hydrophobic properties. Thus, NO may not only easily migrate in the cytoplasm, but also freely diffuse through the lipid phase of biomembranes, exerting a broad range of interactions with various biological targets (Arasimowicz and Floryszak-Wieczorek 2007; Kovacs and Lindermayr 2013). As NO is extremely susceptible to both oxidation and reduction the free radical form of NO may be transformed into other redox forms, under physiological conditions. One-electron oxidation of NO leads to the formation of a nitrosonium cation (NO^+), while the product of one-electron reduction of NO is a nitroxyl radical (NO^-) (Stamler et al.

Table 10.1 Reactive nitrogen species including radicals and nonradical molecules

Radicals	Non-radicals
Nitric oxide, $NO^{\cdot}$	Nitrous acid: HNO_2
Nitrogen dioxide, $NO_2^{\cdot}$	Nitrosyl cation: NO^+
	Nitroxyl anion: NO^-
	Dinitrogen pentoxide: N_2O_5
	Dinitrogen tetroxide: N_2O_4
	Dinitrogen trioxide: N_2O_3
	Peroxynitrite: $ONOO^-$
	Peroxynitrous acid: ONOOH
	Nitronium (nitryl) cation: NO_2^+
	Alkyl peroxynitrites: ROONO
	Peroxyacetyl nitrate: $CH_3C(O)OONO_2$

1992). As a highly reactive species, NO in the presence of atmospheric oxygen forms other oxides, including nitrogen dioxide ($NO_2^{\cdot}$), dinitrogen trioxide (N_2O_3), and dinitrogen tetraoxide (N_2O_4), which may further react with cellular amines and thiols, or hydrolyse to NO_2^- and NO_3^- (Wendehenne et al. 2001). NO readily reacts with the superoxide anion-radical ($O_2^{-\cdot}$) forming $ONOO^-$.

10.3 Chemical Properties of Nitroxyl and Its Donors

It is well documented that nitroxyl or nitrosyl hydride (HNO) is the one-electron reduced and protonated sibling of NO. The chemistry of this triatomic molecule is surprisingly complex (Paolocci et al. 2007). As it was revealed in many studies, HNO is a weak acid (pKa > 11) (Bartberger et al. 2002; Shafirovich and Lymar 2002), which suggests that HNO, rather than the nitroxyl anion (NO^-), predominates at physiological pH. One of the most prevalent reactions of HNO is dimerization, generating a labile hyponitrous acid intermediate, which dehydrates to yield N_2O and H_2O (Eq. 10.1) (Kohout and Lampe 1965; Bazylinski and Hollocher 1985).

$$HNO + HNO \rightarrow [HONNOH] \rightarrow N_2O + H_2O \qquad (10.1)$$

Nitroxyl is highly reactive toward nucleophiles and it is resistant to scavenging by the superoxide anion. In turn, thermal degradation of Angeli's salt (HNO donor) in an aerobic solution did not produce detectable $ONOO^-$, instead, it generates an unidentified oxidant distinct from peroxynitrite (Miranda et al. 2002; Miranda 2005). Since nitroxyl cannot be stored, donor molecules have to be used in biological research (Irvine et al. 2008). Although there are several classes of HNO donors (Miranda 2005; Fukuto et al. 2008) but studies examining nitroxyl are generally performed using Angeli's salt and Piloty's acid, which have strikingly different biochemical designs. Piloty's acid (*N*-hydroxysulphenamide) generates

HNO at high alkaline pH (Seel and Bliefert 1972), while in the physiological pH the main degradation product is NO (Zamora et al. 1995).

Moreover, Angeli's salt (sodium trioxodinitrate; $Na_2N_2O_3$) produces nitroxyl by thermal decomposition of sodium oxyhyponitrite as the source of HNO/NO^- (King and Nagasawa 1999; Miranda 2005; Keefer 2005). Angeli's salt releases HNO with a half-life of 2.8 min at 37 °C over a broad pH range (4.4–8.1) (Eq. 10.2). Apart from Angeli's salt, Piloty's acid and its derivatives as well as cyanamide, diazenium diolate-derived compounds, acyl nitroso compounds, and acyloxy nitroso compounds are known as HNO donors (DuMond and King 2011).

$$Na_2N_2O_3 + H^+ \rightarrow [NaHN_2O_3] \rightarrow HNO + NaNO_2 \qquad (10.2)$$

10.3.1 Biological Reactivity of HNO

Although it is generally accepted that among RNS, a crucial role in biological systems is assigned to NO, an increasing amount of evidence highlights the importance of nitroxyl (HNO), the one-electron reduced and protonated congener of NO. Studies carried out on mammalian cells and utilizing the in vitro experimental approach reveal distinct or even opposite biological activities for HNO and NO (Wink et al. 2003; Miranda 2005; Paolocci et al. 2007; Fukuto et al. 2009; Switzer et al. 2009). Such effects include an ability of HNO to directly target thiols, complex ferric hemes, modify soluble guanylate cyclase (sGC) activity via an interaction at both the regulatory heme and cysteine thiols (Wong et al. 1998; Miranda et al. 2003; Farmer and Sulc 2005; Donzelli et al. 2006; Miller et al. 2009), and reveal resistance to scavenging by the superoxide anion (Miranda et al. 2002). Moreover, HNO and NO show differences in their ability to promote oxidative DNA damage (Wink et al. 1998; Chazotte-Aubert et al. 1999).

The complex redox chemistry of HNO makes this molecule suited for signaling in biological processes through a wide range of actions. Simultaneously, the chemical nature of nitroxyl explains how HNO and NO, despite common cellular targets, act so differently in biochemical systems (Fig. 10.1). The potential intracellular targets of HNO might involve thiols, transition metals, metalloproteins, iron-sulfur clusters, and redox active species, such as NADPH/NADH. Given the thiophilic nature of HNO, it is tempting to speculate that thiols and thiol proteins could be the primary targets of HNO in plants, similarly as in mammalian systems.

A direct interaction of HNO with thiols may result in either reversible or irreversible modification, depending on the nature and concentration of thiols in the cell environment (Fukuto et al. 2009). As shown in Fig. 10.2, in the presence of an excess of thiols, an intermediate, *N*-hydroxysulfenamide, reacts further to disulfide and hydroxylamine, which is considered to be a reversible process, since the reduction of disulfides back to their thiol oxidation state in biological systems is made possible in vivo by the thioredoxin system (DeMaster et al. 1997; Xian et al. 2000;

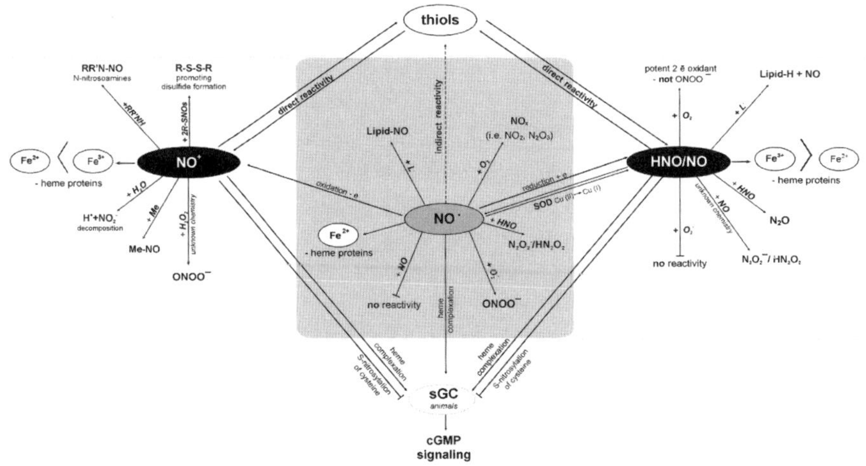

Fig. 10.1 The distinct chemical reactivity of $NO^{\cdot}$ versus NO^{+} and HNO/NO^{-} with potential biological targets. cGMP cyclic guanosine monophosphate, sGC soluble guanylate cyclase, H_2O_2 hydrogen peroxide, $L^{\cdot}$ lipid radical, Me metalloproteins, Me-NO metal nitrosylated proteins, $ONOO^{-}$ peroxynitrite, $O_2^{\cdot-}$ superoxide radical, SOD superoxide dismutase, *RR'*NH amines, *RR'R*NNO *N*-nitrosoamines, *R*-SNO *S*-nitrosothiols, *R-S-S-R* disulphide bridges, XO xanthine oxidase

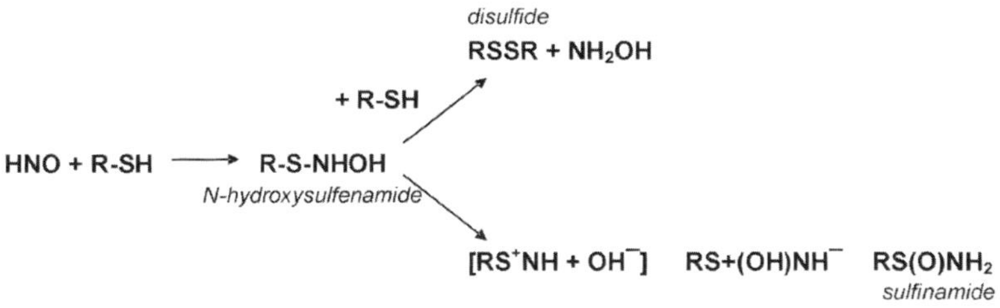

Fig. 10.2 The interaction of HNO with thiols

Lopez et al. 2005). On the other hand, in the low thiol environment an intramolecular rearrangement is observed in case of sulfinamide; however, conversion of thiols to sulfinamide is not reversed or at least much more difficult to reverse (Fukuto et al. 2009). Thus, the sulfinamide formation is a significant hallmark of HNO biological reactivity and may serve as a unique fingerprint for HNO presence, since at present no sensitive and specific assays are available for HNO detection. Only recently the capability of chemical systems of discerning HNO from NO have been reported (Rosenthal and Lippard 2010; Reisz et al. 2011; Zhou et al. 2012).

It is suggested that due to its relative nucleophilicity, HNO prefers an association with thiolates (RS-) over thiols (Lopez et al. 2007). Therefore, it is evident that HNO reacts very selectively with the thiol-containing proteins or specific compartmentalized "cysteine hot-spots," rather than modifying a "sea of cellular thiols" (Lopez et al. 2005). In confirmation, the HNO-mediated inhibition of glyceraldehyde 3-phosphate dehydrogenase (GADPH) activity observed in yeast cells was not correlated with the depletion of glutathione (GSH) pool and did not alter the GSH/GSSG redox status of the cell (Lopez et al. 2005). Moreover, in vitro experiments showed HNO targeted thiol residues on receptors and ion channels, including voltage-gated potassium and the calcium channel as well as sarcoplasmic ryanodine receptors (RyR1, RyR2) (Tocchetti et al. 2005; Cheong et al. 2005).

One of the most important reactions of HNO is reductive nitrosylation of metals such as Fe^{3+} and Cu^{2+} (Miranda 2005). The preferential targeting to ferric Fe^{3+} versus ferrous Fe^{2+} heme groups results in a stable Fe^{2+}-nitrosyl complex formation ($Fe^{2+}NO$) (Irvine et al. 2008). In turn, the interaction of HNO with Cu^{2+} of metalloprotein CuZnSOD produces free NO, which is suggested to be a reversible reaction (Niketic et al. 1999; Liochev and Fridovich 2002). Additionally, other biologically relevant molecules, including flavins, quinones, and cytochrome P450, seem to stimulate the oxidation of HNO to NO (Paolocci et al. 2007).

Nitroxyl might also involve the cGMP signaling cascade through the complex regulation of sGC activity. It has recently been reported by Miller et al. (2009) that HNO at low concentrations activates the enzyme via coordination to the ferrous heme, but not the ferric form of the enzyme, which seems unexpected, since HNO readily prefers ferric Fe^{3+}. Furthermore, at higher concentrations HNO may target regulatory thiols on sGC and attenuate this activity. Despite the fact that the identified potential of sGC as cGMP sources in plants still awaits demonstration, the functioning of the cGMP-dependent signaling pathway has been well documented (Leitner et al. 2009). It needs to be stressed here that the new class of plant-specific GC discovered in *Arabidopsis* does not possess the heme-binding motif and the activity in vitro is not dependent on the NO radical (Ludidi and Gehring 2003; Isner and Maathuis 2011).

Additionally, the indole nitrogen of tryptophan seems to be a significant biological candidate for the HNO target in plant tissue. In in vitro experiments Peyrot et al. (2006) observed the *N*-nitrosoindole formation, including 1-nitrosoindol-3-acetic acid as a result of the incubation of various indolic compounds (excluding indol-3-acetic acid, IAA) with an HNO releasing donor. Since tryptophan is a precursor of IAA, it is possible that *N*-nitrosoindoles might function as a missing link in the cross-talk between NO and IAA signaling.

The multiplicity of potential reactions of HNO has bidirectional consequences. Depending on the HNO concentration and the localization in the cellular milieu, the dual reactivity of nitroxyl leads to both pro- and antioxidant effects. Therefore, HNO at a high concentration may act as an oxidant and trigger the cytotoxicity effect mainly via double-stranded DNA breaks and depletion of intracellular GSH levels (Hewett et al. 2005). On the other hand, via donation of its hydrogen atom HNO possesses chemical properties, which predispose it to be also a one-electron reductant. It has been shown that HNO might serve as a potent chain-terminating antioxidant with an activity quantitatively similar to that of tocopherol (Lopez et al. 2007). Using yeast and in vitro model systems (Lopez et al. 2005, 2007) showed that HNO can function in both cytosol and within a lipid environment, carrying the potential to preserve membrane integrity from free radical damage (Paolocci et al. 2007).

10.4 Chemical Properties and Donors of Nitrosonium

Nitrosonium (NO^+) is a cation formed as a consequence of NO oxidation. NO can be oxidized by oxygen, the superoxide anion or transition metals to NO^+ by removal of single electron π^* when nitrogen oxidation state reaches +3. A relatively easy removal of an unpaired electron in NO is possible due to antibonding (ionization potential NO = 9.24 eV) (Bonner and Stedman 1996). In aqueous solutions NO^+ possesses a very short lifetime estimated at 3×10^{-10} s with the equilibrium constant $[NO^+]/([HNO_2][H^+]) \approx 10$–6.5 at 25 °C (Ridd 1979). The NO^+ cation is quickly hydrolyzed in aqueous solutions to nitrous acid (Eq. 10.3) (Hughes 1999). NO^+ may be transferred to thiolate of cysteine and form *S*-nitrosothiol (Arnelle and Stamler 1995) by *S*-nitrosylation (referred also as nitrosation).

$$NO^+ + H_2O \rightarrow H^+ + HNO_2 \tag{10.3}$$

Nitrosation is an electrophilic reaction involving the attack of NO^+ or a source of NO^+ at a nucleophilic center. In biological systems the thiol group of cysteine residues is subjected to *S*-nitrosylation after an NO^+ nucleophilic attack and deprotonation (Gaston et al. 2003). SNOs turned out to be unstable in solutions and preferably form disulfide bonds and release 2NO by homolysis (Collings et al. 1981), nevertheless, heterolytic decomposition to NO^- or NO^+ takes place, but in a less intensive way. Nitrosyl can react with nucleophilic centers such as *R*-OH, *R*-SH, or *RR'*NH to produce *R*-ONO, *R*-SNO, and *RR'*N-NO, respectively (Heck 2001; Stamler 1994). Under acidic conditions NO_2^- may be a source of NO^+ acting as an NO^+ donor (Stamler and Feelisch 1996). Furthermore, SNOs are donors of NO^+ equivalents in trans-nitrosylation reactions between -SNO and -SH groups of cysteine residues in amino acids and proteins. In trans-nitrosylation the transfer of NO^+ is more rapid than spontaneous NO^- release (Arnelle and Stamler 1995). Thus, SNOs may be perceived as NO^+, NO, NO^- donors in the homo- as well as heterolytic decomposition at physiological pH (Arnelle and Stamler 1995). SNO cleavage to NO radicals facilitates NO^+ bioactivity in the site of its release (Gaston et al. 2003). It was thought that NO may be converted into its congeners by metal-catalyzed oxidation or reduction (Stamler and Feelisch 1996; Radi 1996). Nevertheless, the mechanism may involve NO incorporation into dinitrosyl iron complexes (DNICs) containing a range of ligands. NOx as a part of a low mass dinitrosyl iron complex promotes the reaction of *S*-nitrosylation (Stojanović et al. 2004). Dinitrosyl iron complexes are organometallic-like compounds, which in vivo may play a role of an NO degradation structure in iron–sulfur clusters (Pulukkody et al. 2013). Moreover, their synthetic analogs exhibit a potential of NO donors. Cysteines in proteins and glutathione may be ligands of DNICs and might play a significant role in thiolate-dependent redox changes. It was demonstrated that MnSOD and FeSOD in *E. coli* orchestrated NO transition to NO^+ and HNO/NO^-, which in consequence indicates the role of metal-associated NO turnover in biological systems (Stojanović et al. 2004).

Sodium nitroprusside (SNP), $Na_2[Fe(CN)_5(NO^+)]$, is a nitrosyl complex formally carrying NO^+ and at neutral pH possessing the character of a nitrosonium donor (Hughes 1999). The mechanism of NO^+ release form SNP is not completely understood, although it is clear that induction by light (Floryszak-Wieczorek et al. 2006) or one electron reduction (Wang et al. 2002) is required. SNP irradiation by light in living systems is not sufficient to express NO^+ bioactivity, e.g., the vasodilation effect on vascular tissue, thus one electron reduction caused by thiols, ascorbate, and hemoglobins is more plausible. These abundant antioxidants might significantly affect NO donors used in animals and plants.

10.4.1 Biological Activity of NO^+

A neuroprotective effect is observed in animals when NO^+ realizing compounds such as *R*-SNOs are used. NO^+ may nitrosylate *N*-methyl-D-aspartate (NMDA) receptors in contrast to NO, which together with superoxide rather forms $ONOO^-$. NO release from SNOs depends on many factors including pH, O_2 concentration, buffer composition, light intensity, and the presence of chelators, and it often appears to be a result of metal contamination and redox state imbalance in in vivo experiments. Depending on the solution pH, -SNOs may be considered to be NO or NO^+ donors. NO^+ transfer to -SH groups modulates a wide range of proteins involved in a variety of metabolic and signaling pathways. Low mass SNOs such as GSNO and CysNO play a potential role of NO storage and carriers in living organisms (Wang et al. 2002). Inter-conversion between NO, NO^-, and NO^+ can take place in vitro and in vivo conditions, thereby NO effects should be considered in the experimental approach in relation to its congeners as well. Still less is known about NO bioactivity in plants in relation to animals, while, intensive studies are being conducted on the subject. The effects of NO^+ donors on plants and the kinetics of NO^+ release were determined by Floryszak-Wieczorek et al. (2006) and Ederli et al. (2009). In 0.2 mM SNP solution, the highest NO^+ concentration (6 μM) was recorded by the electrochemical method 2 h after constant illumination (120 μmol m^{-2} s^{-1}). Then a slight gradual decrease of NO^+ generation was observed, facilitating estimation of SNP half-life as $t½$ = ca. 12 h. Chopped illumination (30 min dark/30 min light) of 0.2 mM SNP solution indicated complete inhibition of NO^+ generation in the dark and a strong light-dependent manner of NO^+ release (Floryszak-Wieczorek et al. 2006). When considering the use of NO^+ donors, their concentration must be taken into account due to its dose-dependent bioactivity in plants.

10.5 Peroxynitrite

The nonenzymatic, one-to-one stoichiometry reaction between two free radicals, the superoxide anion, and NO leads to the formation of another RNS, i.e., $ONOO^-$. The rate constant of the reaction has been determined by several methodologies to be within the range of 4–16 × 109 M^{-1} s^{-1}. This very large value explains how $ONOO^-$ can be formed at all, considering that its precursors are very elusive species (Ferrer-Sueta and Radi 2009). The $ONOO^-$ at the physiological pH and temperature range is a relatively short-lived RNS, which may readily migrate through biological membranes and interact with target molecules also in the surrounding cells within the radius of one or two cells (~5–20 μm) (Szabó et al. 2007). Taking into account the half-life of the parent radicals the site of $ONOO^-$ generation seems to be spatially associated with the sources of the superoxide anion (NADPH-oxidase, xanthine oxidase, mitochondrial respiratory complexes), which half-life is much shorter than that of the NO molecule (Szabó et al. 2007). The concentrations of both $O_2^{\cdot -}$ and NO determine the rate of $ONOO^-$ formation. As it was found by Pacher et al. (2007), the rate of the reaction is at least 3–8 times faster than the rate of $O_2^{\cdot -}$ dismutation by superoxide dismutase (SOD).

Due to its chemical nature, $ONOO^-$ is an important biological oxidant and nitrated compound. Peroxynitrite may affect the target molecules either directly, through one or two-electron oxidation reactions, or indirectly through the formation of highly reactive radicals (Arasimowicz-Jelonek and Floryszak-Wieczorek 2011). In this regard, the rapid reaction with carbon dioxide generates carbonate ($CO_3^{\cdot -}$) and nitrogen dioxide ($^{\bullet}NO_2$) radicals. The peroxynitrite anion is in a pH-dependent protonation equilibrium with peroxynitrous acid (ONOOH), which further, through hemolytic cleavage, is a source of nitrogen dioxide ($^{\bullet}NO_2^-$) and a hydroxyl radical (HO^-) (Virag et al. 2003). However, in biological systems the reaction may become relevant mainly in hydrophobic phases to initiate lipid peroxidation and nitration processes (Radi et al. 1991; Szabó et al. 2007). Thus, peroxynitrite and its derived radicals may target tyrosine residues, thiols, DNA, and fatty-acid-containing phospholipids.

There is evidence that $ONOO^-$ is a more toxic molecule than NO or the superoxide (Szabó 2003), however, in contrast to animal cells, $ONOO^-$ itself seems to be less destructive for plant cells (Delledonne et al. 2001). As it was suggested earlier by Romero-Puertas et al. (2004), $ONOO^-$ may be continuously formed as an inevitable event in plant cell metabolism, adapted to detoxify its excess. Potentially, plants cope with $ONOO^-$ overproduction by a broad range of its decomposition mechanisms, including ascorbates, flavonoids, γ-tocopherol, thiols, and metalloporphyrins (Arasimowicz-Jelonek and Floryszak-Wieczorek 2011). Importantly, the peroxynitrite-detoxifying activity has been demonstrated for two plastid peroxiredoxins, i.e., PrxIIE and 2-Cys-Prx (Sakamoto et al. 2003; Romero-Puertas et al. 2007). A similar activity toward $ONOO^-$ inactivation was earlier shown in vitro by selenocysteine-containing glutathione-peroxidase (GPx) (Sies et al. 1997).

10.6 Biotargets of Reactive Nitrogen Species

The number of studies on the interaction between RNS and other biomolecules in living systems has substantially increased during the last decade. The chemical nature of NO and NO-derived molecules is directly or indirectly involved in redox-based posttranslational modification, including binding to metal centers, *S*-nitrosylation of thiol groups and nitration of tyrosine. As it has been indicated by recent research, nitration of unsaturated fatty acids might also constitute an important mode of RNS action in plant cells (Sánchez-Calvo et al. 2013).

10.6.1 Tyrosine Nitration

Tyrosine nitration is the reaction of a nitrating agent with a tyrosine residue of a target protein that leads to the formation of a stable product 3-nitrotyrosine, by the addition of a nitro group to the *ortho*-position, adjacent to the hydroxyl group of tyrosine (Dixit et al. 2009). Although nitrotyrosine was initially considered as a specific marker of peroxynitrite generation in vivo, factors other than $ONOO^-$ can also induce tyrosine nitration. The potential mechanism involves a hemoperoxidase that, in the presence of hydrogen peroxide (H_2O_2) and nitrite (NO_2^-), can generate the radical nitrogen dioxide ($^{\bullet}NO_2$) which acts as a nitrating agent (Souza et al. 2008).

This posttranslational modification is considered to be a selective process. In most proteins the number of tyrosine residues is around 3–4 % out of the primary structure, but only few of these tyrosine may become preferentially nitrated (Corpas et al. 2013). Therefore, an increase of the total protein pool undergoing tyrosine nitration phenomena in response to stress stimuli could be a footprint of homeostasis misbalance (Arasimowicz-Jelonek and Floryszak-Wieczorek 2011).

It is important to note that Tyr-nitration occurs in plant cells under optimal, physiological conditions, which might indicate an unknown physiological function of plant nitration via $ONOO^-$. Among the identified nitrated proteins from sunflower hypocotyls there were proteins associated with signal transduction, i.e., putative serine/threonine-protein kinase, 14-3-3-like protein, Calmodulin-like protein, thus providing evidence for the possible signaling role of Tyr-nitration in plant cells (Chaki et al. 2009a). More recently Begara-Morales et al. (2013) documented in a pea plant experimental model that each organ has a specific protein nitration pattern. Moreover, the intensity of the proteins which undergo this post-translation modification increases during the senescence of roots, stems, and leaves.

Tyrosine nitration has been assumed as a reliable marker of nitro-oxidative stress, since this posttranslational modification is frequently associated with plant pathophysiological states (Corpas et al. 2007). A rise in the protein pool undergoing tyrosine nitration was demonstrated in response to abiotic stresses, including shear stress (Gong and Yuan 2006), salt stress (Valderrama et al. 2007; Corpas et al. 2009; Tanou et al. 2012), extreme temperature (Chaki et al. 2011; Airaki et al. 2012), continuous light and high light intensity (Corpas et al. 2008), high bicarbonate and high pH (Cellini et al. 2011), water stress (Signorelli et al. 2013), and arsenic (Leterrier et al. 2012). Moreover, modifications of the nitrated protein pattern were also detected in plant responses to biotic stimuli (Saito et al. 2006; Chaki et al. 2009b; Cecconi et al. 2009).

Detailed studies on tyrosine-nitrated proteins in sunflower seedlings exposed to high temperature showed that the activity of ferredoxin-NADP reductase (FNR) and carbonic anhydrase (CA) is inhibited by 31 and 43 %, respectively. This stress-mediated enzyme inhibition was confirmed under in vitro conditions, where their respective activities were determined in the presence of a peroxynitrite donor—SIN-1 (Chaki et al. 2011, 2013). Moreover, an in silico analysis of the pea CA protein sequence suggests that Tyr(205) is the most likely potential target for nitration (Chaki et al. 2013). In *A. thaliana* plants tyrosine nitration provoked the inhibition of *O*-acetylserine(thiol)lyase A1(OASA1), a crucial enzyme for cysteine homeostasis (Álvarez et al. 2011). The authors demonstrated that this protein undergoes Tyr nitration selectively on its Tyr(302) residue in vivo after SIN-1 treatment. Also in relation to *S*-adenosylhomocysteine hydrolase (SAHH), an inhibition of enzyme activity after tyrosine modification was well-documented in an in vitro experiment. In this case, an in silico analysis of the barley SAHH sequence revealed Tyr(448) as the target for the potential nitration reaction (Chaki et al. 2009a).

10.6.2 Nitration of Unsaturated Fatty Acids

The complexity of RNS signaling in living cells includes also the formation of nitro-fatty acids (NO_2-FA). The products of the interaction between RNS and lipids possess biochemical properties distinct from the precursor lipids and might function as mediators that regulate various signal transduction pathways (Trostchansky et al. 2013).

A hypothetical model of NO_2-FA action in plant cells was recently proposed by Sánchez-Calvo et al. (2013). Based on pharmacological experiments the authors found that nitrolinolenate ($LnNO_2$) induced NO production in 30-day-old *Arabidopsis* plants. Thus, $LnNO_2$ may serve as an NO donor. The released NO may interact with GSH to form mobile signal GSNO or may affect cysteine residues in proteins through a reversible *S*-nitrosylation. Alternatively, similar to animal systems, $LnNO_2$ can form reversible covalent adducts on nucleophilic sites of proteins, such as cysteine residues in a process known as nitroalkylation (Sánchez-Calvo et al. 2013).

10.6.3 Protein S-Nitrosylation

The reversible covalent binding of an NO moiety to the thiol side chain of protein cysteine leads to the formation of *S*-nitrosothiols. Currently, this is the most studied and the best described NO-dependent, redox-based posttranslational modification in plants. Kovacs and Lindermayr (2013) summarized four potential mechanisms of *S*-nitrosylation occurring in the biological milieu including an oxidative pathway with NO in a higher oxidation status, a radical-mediated pathway with ·NO and thiyl (RS·) radicals, metal-catalyzed RSNO formation in the presence of transition metals, and trans-nitrosylation. The transfer of the NO moiety from one *S*-nitrosylated protein to another in a trans-nitrosylation reaction creates an important mechanism of cell signaling (Astier and Lindermayr 2012). Moreover, the reversibility of *S*-nitrosylation constitutes a feedback mechanism controlling NO availability in living cells. The removal of the NO group from proteins might occur through different enzyme systems, of which especially *S*-nitrosoglutathione (GSNO) reductase and thioredoxin play a prime role (Benhar et al. 2009).

Based on the Biotin Switch method numerous putative protein targets for *S*-nitrosylation have been found in plants, including cytoskeleton, metabolic, redox-related, stress-related, and signaling/regulating proteins (Astier and Lindermayr 2012; Kovacs and Lindermayr 2013). However, the number of experimentally confirmed functional modifications is much lower. At present, around 20 different candidate proteins have been characterized in more detail (Astier and Lindermayr 2012). These include peroxiredoxin II E displaying an $ONOO^-$ reductase activity (Romero-Puertas et al. 2007), the nonexpressor of pathogenesis-related gene 1 (NPR1; Tada et al. 2008). Methionine adenosyltransferase (MAT) (Lindermayr et al. 2006), metacaspase 9 (Belenghi et al. 2007), glyceraldehyde 3-phosphate dehydrogenase (GADPH) (Holtgrefe et al. 2008), salicylic acid binding protein 3 (SABP3) (Wang et al. 2009), transcription factor (TGA1) (Lindermayr et al. 2010), NADPH oxidase (Yun et al. 2011), cytosolic fructose 1,6-bisphosphate aldolase (cALD2) (van der Linde et al. 2011), transport inhibitor response 1 (TIR1) (Terrile et al. 2012) and cell division cycle 48 (CDC48) (Astier et al. 2012).

10.6.4 Metal Nitrosylation

The chemical nature of NO determines the reactive metal centers of proteins, next to cysteine residues, to be NO targets in the biological milieu. Similarly to *S*-nitrosylation, metal nitrosylation is also a precisely targeted and reversible posttranslational modification that allows living cells to flexibly and specifically react to changes in their environment (Mannick and Schonhoff 2004).

Nitrosylation of the crucial metal centers has been widely studied in living organisms, particularly in the case of mammalian sGC. NO binding to the ferrous state of the heme group of sGC severs the bond between the heme iron and the histidine residue via an axial dislocation of iron. This triggers a conformational change resulting in cGMP synthesis. The basal activity of the enzyme can be increased up to 200 times by binding NO, however, the lifetime of the NO–heme complex is very short, with half-life as low as 0.2 s (Bruckdorfer 2005).

Both cGMP and cGMP-mediated processes were reported in plants many years ago; however, no sequenced higher plant genome contains homologs of the nucleotidyl cyclase genes that are recognizable in diverse eukaryotic kingdoms (Ashton 2011). However, a guanylyl cyclase (GC) domain was identified within the cytosolic kinase region of *Arabidopsis* AtBRI1 and it was shown to have catalytic activity in vitro (Kwezi et al. 2007). More recently, the cytosolic kinase region of the leucine-rich repeat receptor-like kinase receptor AtPepR1 also contains a putative GC domain with the same functionally assigned residues in the catalytic center as AtBRI1 (Qi et al. 2010). Still it needs to be stressed that NO-sensitive GC have not yet been unraveled in the plant kingdom.

Analogous to NO–sGC interaction, NO may interact with iron present in other heme proteins. In this way cytochrome c is nitrosylated in vivo on its heme iron during apoptosis toward the proapoptotic activity of cytochrome c (Schonhoff et al. 2003). Nitrosylation may also be an allosteric regulator of cytochrome c oxidase (COX) function. In this case, the interaction of NO with COX results in an inhibition of the enzyme activity by NO binding to the iron/copper binuclear center (Schonhoff et al. 2003).

Experimental data have revealed that heme-containing plant NO sensors involve catalase (CAT) and ascorbate peroxidase (APX). According to Clark et al. (2000), a reversible inhibition of CAT and APX is possible by the formation of an iron–nitrosyl complex between NO and the iron atom of the heme group. Interestingly, pharmacological approaches using NO donors showed that the duration of this inhibition is combined with the time of donor compound decomposition and is usually limited to the first 24 h (Floryszak-Wieczorek et al. 2006, 2007). Since plant cells contain several isoforms of the enzyme, it is possible that only some isoforms are inhibited by NO (Clark et al. 2000). The periodical inhibition of CAT and APX via NO may potentially regulate ROS level in the cell, e.g., during PCD in xylem formation or in wound healing mechanisms (Ferrer and Barcelo 1999; Clarke et al. 2000; Arasimowicz-Jelonek et al. 2009).

The major NO sensor in legume nodules is leghemoglobin (Lb) (Meilhoc et al. 2011). It has been demonstrated that NO can bind this symbiotic hemoglobin to form stable nitrosylleghemoglobin (LbFeIINO) complexes in nodules of soybean (Mathieu et al. 1998; Meakin et al. 2007; Sánchez-Calvo et al. 2010), cowpea (Maskall et al. 1977) and pea (Kanayama and Yamamoto 1991). Herold and Puppo (2005) demonstrated that Lb-NO complexes, including in vivo formation of oxyLb and ferrylLb, are engaged in scavenging of NO and $ONOO^-$. Interestingly, Lb-NO complexes are most abundant in the youngest nodules, suggesting a beneficial role of Lb in the protection of nitrogenase activity, which is rapidly inactivated by RNS, and in consequence they prevent rejection of symbiotic rhizobia (Becana et al. 2010). NO targets also nonsymbiotic class-1 hemoglobins, resulting in the formation of a nitrosyl complex (Perazzolli et al. 2004).

Non-heme iron nitrosylation of the iron–sulfur cluster protein aconitase leads to the loss of an iron atom from the iron–sulfur cluster and inhibition of aconitase activity (Pieper et al. 2003). As it was found by Navarre et al. (2000), tobacco cytosolic and mitochondrial aconitase activities are inhibited by NO. Based on in vitro experiments it was found that lipoxygenase-1 could serve as a non-heme iron plant NO sensor too, resulting in the metal-nitrosyl complex formation (Nelson 1987). It should be stressed that at the physiological pH range complexes of NO, Fe^{2+} and low-molecular thiols, referred to as dinitroso-iron complexes (DNICs) or mononitroso-iron complexes (MNICs), may also be formed (Graziano and Lamattina 2005).

10.7 Conclusion

At present, the current state of knowledge on the nitroxyl anion and the nitrosonium cation as chemically distinct redox siblings of NO constitutes a mysterious topic in plant biology. Interest in this molecule has recently been renewed with data that these compounds might be produced endogenously and that they reveal unique effects compared with NO. In the physiological state or under low metabolic stress conditions a balance between NO and peroxynitrite as quenching or resetting signals following stimulation results in nitration of target proteins or plays a role as a modulator of diverse patho- and physiological processes in plants. Biotargeting by various forms of RNS cellular compounds appears to be highly selective and localized, which is extremely important in developmental and stress responses of living organisms. Finally, the functional cooperation of NO with RNS in a cellular milieu may induce numerous pathways of signaling networks in biological systems. New noninvasive technologies of RNS monitoring and further progress in the identification of different targets of protein *S*-nitrosylation and Tyr nitration might provide opportunities to unravel their physiological implications in plants.

Acknowledgments This work was supported by funds from Ministry of Science and Higher Education project IUVENTUS PLUS no. 2011 000671 and National Science Centre project PRELUDIUM no. 2011/03/N/NZ9/00290.

References

Airaki M, Leterrier M, Mateos RM et al (2012) Metabolism of reactive oxygen species and reactive nitrogen species in pepper (*Capsicum annuum* L.) plants under low temperature stress. Plant Cell Environ 35:281–295

Álvarez C, Lozano-Juste J, Romero LC et al (2011) Inhibition of *Arabidopsis O*-acetylserine(thiol)lyase A1 by tyrosine nitration. J Biol Chem 86:578–586

Arasimowicz M, Floryszak-Wieczorek J (2007) Nitric oxide as a bioactive signalling molecule in plant stress responses. Plant Sci 172:876–887

Arasimowicz-Jelonek M, Floryszak-Wieczorek J, Milczarek G (2009) Nitric oxide induced by wounding mediates redox regulation in pelargonium leaves. Plant Biol 11:650–663

Arasimowicz-Jelonek M, Floryszak-Wieczorek J (2011) Understanding the fate of peroxynitrite in plant cells - from physiology to pathophysiology. Phytochemistry 72:681–688

Arnelle DR, Stamler JS (1995) NO^+, $NO^•$, NO^- donation by *S*-nitrosothiols: implications for regulation of physiological functions by *S*-nitrosylation and acceleration of disulphide formation. Arch Biochem Biophys 318(2):79–285

Ashton AR (2011) Guanylyl cyclase activity in plants. Proc Natl Acad Sci USA 108:E96

Astier J, Besson-Bard A, Lamotte O et al (2012) Nitric oxide inhibits the ATPase activity of the chaperone-like AAA^+ ATPase CDC48, a target for *S*-nitrosylation in cryptogein signaling in tobacco cells. Biochem J 447:249–260

Astier J, Lindermayr C (2012) Nitric oxide-dependent posttranslational modification in plants: an update. Int J Mol Sci 13:15193–15208

Bartberger MD, Liu W, Ford E et al (2002) The reduction potential of nitric oxide (NO) and its importance to NO biochemistry. Proc Natl Acad Sci USA 99:10958–10963

Bazylinski DA, Hollocher TC (1985) Evidence from the reaction between trioxodinitrate (II) and 15NO that trioxodinitrate (II) decomposes into nitrosyl hydride and nitrite in neutral aqueous solution. Inorg Chem 24:4285–4288

Becana M, Matamoros MA, Udvardi M, Dalton DA (2010) Recent insights into antioxidant defenses of legume root nodules. New Phytol 188:960–976

Begara-Morales JC, Chaki M, Sánchez-Calvo B et al (2013) Protein tyrosine nitration in pea roots during development and senescence. J Exp Bot 64:1121–1134

Belenghi B, Romero-Puertas MC, Vercammen D et al (2007) Metacaspase activity of *Arabidopsis thaliana* is regulated by *S*-nitrosylation of a critical cysteine residue. J Biol Chem 282:1352–1358

Benhar M, Forrester MT, Stamler JS (2009) Protein denitrosylation: enzymatic mechanisms and cellular functions. Nat Rev Mol Cell Biol 10:721–732

Bonner FT, Stedman G (1996) The chemistry of nitric oxide and redox-related species. In: Feelisch M, Stamler JS (eds) Methods in nitric oxide research. Wiley, Chichester, pp 3–18

Bruckdorfer KR (2005) Cardiovascular disease: diet, nutrition and emerging risk factors (Report of the British nutrition foundation task force). In: Stanner S (ed) Cardiovascular disease: diet, nutrition and emerging risk factors. Oxford, Blackwell, pp 78–99

Cecconi D, Orzetti S, Vandelle E et al (2009) Protein nitration during defense response in *Arabidopsis thaliana*. Electrophoresis 30:2460–2468

Cellini A, Corpas FJ, Barroso JB, Masia A (2011) Nitric oxide content is associated with tolerance to bicarbonate-induced chlorosis in micropropagated *Prunus* explants. J Plant Physiol 168:1543–1549

Chaki M, Carreras A, López-Jaramillo J et al (2013) Tyrosine nitration provokes inhibition of carbonic anhydrase (β-CA) activity under high temperature stress. Nitric Oxide 29:30–33

Chaki M, Valderrama R, Fernandez-Ocana AM et al (2011) High temperature triggers the metabolism of *S*-nitrosothiols in sunflower mediating a process of nitrosative stress which provokes the inhibition of ferredoxin-NADP reductase by tyrosine nitration. Plant Cell Environ 34:1803–1818

Chaki M, Valderrama R, Fernandez-Ocana AM et al (2009a) Protein targets of tyrosine nitration in sunflower (*Helianthus annuus* L.) hypocotyls. J Exp Bot 60:4221–4234

Chaki M, Fernandez-Ocana A, Valderrama R et al (2009b) Involvement of reactive nitrogen and oxygen species (RNS and ROS) in sunflower-mildew interaction. Plant Cell Physiol 50:265–279

Chazotte-Aubert L, Oikawa S, Gilibert I et al (1999) Cytotoxicity and site-specific DNA damage induced by nitroxyl anion (NO^-) in the presence of hydrogen peroxide. J Biol Chem 274:20909–20915

Cheong E, Tumbev V, Abramson J et al (2005) Nitroxyl triggers Ca^{2+} release from skeletal and cardiac sarcoplasmic reticulum by oxidizing ryanodine receptors. Cell Calcium 37:87–96

Clark D, Durner J, Navarre DA, Klessig DF (2000) Nitric oxide inhibition of tobacco catalase and ascorbate peroxidase. Mol Plant Microbe Interact 13:1380–1384

Clarke A, Desikan R, Hurst RD et al (2000) NO way back: nitric oxide and programmed cell death in *Arabidopsis thaliana* suspension cultures. Plant J 4:667–677

Collings P, Garley MS, Stedman G (1981) Kinetics and mechanism of the decomposition of $[(NH_2)_2CSNO]^+$. J Chem Soc Dalton Trans 2:31–335

Corpas FJ, Chaki M, Fernández-Ocaña A et al (2008) Metabolism of reactive nitrogen species in pea plants under abiotic stress conditions. Plant Cell Physiol 49:1711–1722

Corpas F, del Río LA, Barroso JB (2007) Need of biomarkers of nitrosative stress in plants. Trends Plant Sci 12:436–438

Corpas FJ, Hayashi M, Mano S et al (2009) Peroxisomes are required for in vivo nitric oxide accumulation in the cytosol following salinity stress of *Arabidopsis* plants. Plant Physiol 151:2083–2094

Corpas FJ, Palma JM, del Río LA, Barroso JB (2013) Protein tyrosine nitration in higher plants grown under natural and stress conditions. Front Plant Sci 4:29

Delledonne M, Xia Y, Dixon RA, Lamb C (1998) Nitric oxide functions as a signal in plant disease resistance. Nature 394:585–588

Delledonne M, Zeier J, Marocco A, Lamb C (2001) Signal interactions between nitric oxide and reactive oxygen intermediates in the plant hypersensitive disease response. Proc Natl Acad Sci USA 98:13454–13459

DeMaster EG, Redfer B, Quast BJ et al (1997) Mechanism for the inhibition of aldehyde dehydrogenase by nitric oxide. Alcohol 14:181–189

Dixit K, Khan MA, Sharma ID et al (2009) Physicochemical studies on peroxynitrite-modified H3 histone. Int J Biol Macromol 46:20–26

Donzelli S, Espey MG, Thomas DD et al (2006) Discriminating formation of HNO from other reactive nitrogen oxide species. Free Radic Biol Med 40:1056–1066

DuMond JF, King SB (2011) The chemistry of nitroxyl-releasing compounds. Antioxid Redox Signal 14:1637–1648

Durner J, Wendenhenne D, Klessig DF (1998) Defence gene induction in tobacco by nitric oxide, cyclic GMP, and cyclic ADP-ribose. Proc Natl Acad Sci USA 95:10328–10333

Ederli L, Reale L, Madeo L, Ferranti F, Gehring C, Fornaciari M, Romano B, Pasqualini S (2009) NO release by nitric oxide donors in vitro and in planta. Plant Physiol Biochem 47:42–48

Farmer PJ, Sulc F (2005) Coordination chemistry of the HNO ligand with hemes and synthetic coordination complexes. J Inorg Biochem 99:166–184

Ferrer MA, Barcelo AR (1999) Differential effects of nitric oxide on peroxidase and H_2O_2 production by the xylem of *Zinnia elegans*. Plant, Cell Environ 22:891–897

Ferrer-Sueta G, Radi R (2009) Chemical biology of peroxynitrite: kinetics, diffusion, and radicals. ACS Chem Biol 4:161–177

Floryszak-Wieczorek J, Milczarek G, Arasimowicz M, Ciszewski A (2006) Do nitric oxide donors mimic endogenous NO-related response in plants? Planta 224:1363–1372

Floryszak-Wieczorek J, Arasimowicz M, Milczarek G et al (2007) Only an early nitric oxide burst and the following wave of secondary nitric oxide generation enhanced effective defence responses of pelargonium to a necrotrophic pathogen. New Phytol 175:718–730

Fukuto JM, Bianco CL, Chavez TA (2009) Nitroxyl (HNO) signaling. Free Radical Biol Med 47:1318–1324

Fukuto JM, Jackson MI, Kaludercic N, Paolocci N (2008) Examining nitroxyl in biological systems. Meth Enzymol 440:411–437

Graziano M, Lamattina L (2005) Nitric oxide and iron in plants: an emerging and converging story. Trends Plant Sci 1:4–8

Gaston BM, Carver J, Doctor A, Palmer LA (2003) *S*-nitrosylation signalling in cell biology. Mol Interv 3:253–263

Gong YW, Yuan YJ (2006) Nitric oxide mediates inactivation of glutathione *S*-transferase in suspension culture of Taxus cuspidata during shear stress. J Biotechnol 123:85–192

Heck DE (2001) ·NO, RSNO, $ONOO^-$, NO^+, ·NOO, NOx: dynamic regulation of oxidant scavenging, nitric oxide stores, and cyclic GMP-independent cell signaling. Antioxid Redox Signal 3:249–260

Henry YA, Ducastel B, Guissani A (1997) Basic chemistry of nitric oxide and related nitrogen oxides. In: Henry YA, Guissani A, Ducastel B (eds) Nitric oxide research from chemistry to biology. Austin, TX: Landes, Co. Biomed Publ pp 15–46

Herold S, Puppo A (2005) Oxyleghemoglobin scavengers nitrogen monoxide and peroxynitrite: a possible role in functioning nodules. J Biol Inorg Chem 10:935–945

Hewett SJ, Espey MG, Uliasza TF, Wink DA (2005) Neurotoxicity of nitroxyl: insights into HNO and NO biochemical imbalance. Free Radic Biol Med 39:1478–1488

Holtgrefe S, Gohlke J, Starmann J et al (2008) Regulation of plant cytosolic glyceraldehyde 3-phosphate dehydrogenase isoforms by thiol modifications. Physiol Plant 133:211–228

Hughes MN (1999) Relationship between nitric oxide, nitrosyl ion, nitrosonium cation and peroxynitrite. Biochim Biophys Acta 1411:263–272

Irvine JC, Ritchie RH, Favaloro JL et al (2008) Nitroxyl (HNO): the Cinderella of the nitric oxide story. Trends Pharmacol Sc. 29:601–608

Isner JC, Maathuis FJM (2011) Measurement of cellular cGMP in plant cells and tissues using the endogenous fluorescent reporter Flinc G. Plant J 65:329–334

Kanayama Y, Yamamoto Y (1991) Formation of nitrosylleghemoglobin in nodules of nitrate-treated cowpea and pea plants. Plant Cell Physiol 32:19–24

Keefer LK (2005) Nitric oxide (NO)- and nitroxyl (HNO)-generating diazeniumdiolates (NONOates): emerging commercial opportunities. Curr Top Med Chem 5:625–636

King SB, Nagasawa HT (1999) Chemical approaches towards generation of nitroxyl. Meth Enzymol 301:211–221

Kohout FC, Lampe FW (1965) On the role of nitroxyl molecule in the reaction of hydrogen atoms with nitric oxide. J Am Chem Soc 87:5795–5796

Kovacs I, Lindermayr C (2013) Nitric oxide-based protein modification: formation and site-specificity of protein *S*-nitrosylation. Front. Plant Sci 4:137

Kwezi L, Meier S, Mungur L et al (2007) The *Arabidopsis thaliana* brassinosteroid receptor (AtBRI1) contains a domain that functions as a guanylyl cyclase in vitro. PLoS ONE 2:449

Leitner M, Vandelle E, Gaupels F et al (2009) NO signals in the haze nitric oxide signalling in plant defence. Curr Opin Plant Biol 12:1–8

Leterrier M, Barroso JB, Valderrama R et al (2012) NADP-dependent isocitrate dehydrogenase from *Arabidopsis* roots contributes in the mechanism of defence against the nitro-oxidative stress induced by salinity. Sci World J 2012:694–740

Lindermayr C, Saalbach G, Bahnweg G, Durner J (2006) Differential inhibition of *Arabidopsis* methionine adenosyltransferases by protein *S*-nitrosylation. J Biol Chem 281:4285–4291

Lindermayr C, Sell S, Muller B et al (2010) Redox regulation of the NPR1-TGA1 system of *Arabidopsis thaliana* by nitric oxide. Plant Cell 22:2894–2907

Liochev SI, Fridovich I (2002) Nitroxyl (NO^-): substrate for superoxide dismutase. Arch Biochem Biophys 402:166–171

Lopez BE, Rodriguez CE, Pribadi M et al (2005) Inhibition of yeast glycolysis by nitroxyl (HNO): mechanism of HNO toxicity and implications to HNO biology. Arch Biochem Biophys 442:140–148

Lopez BE, Shinyashiki M, Han TH, Fukuto JM (2007) Antioxidant actions of nitroxyl (HNO). Free Radical Biol Med 42:482–491

Ludidi N, Gehring C (2003) Identification of a novel protein with guanylyl cyclase activity in Arabidopsis thaliana. J Biol Chem 278:6490–6494

Mannick JB, Schonhoff CM (2004) NO means no and yes: Regulation of cell signaling by protein nitrosylation. Free Radic Res 38:1–7

Maskall CS, Gibson JF, Dart PJ (1977) Electron-paramagnetic-resonance studies of leghaemoglobins from soya-bean and cowpea root nodules. Identification of nitrosyl-leghaemoglobin in crude leghaemoglobin preparations. Biochem J 167:435–445

Mathieu C, Moreau S, Frendo P et al (1998) Direct detection of radicals in intact soybean nodules: presence of nitric oxide-leghemoglobin complexes. Free Radical Biol Med 24:1242–1249

Meakin GE, Bueno E, Jepson B et al (2007) The contribution of bacteroidal nitrate and nitrite reduction to the formation of nitrosylleghaemoglobin complexes in soybean root nodules. Microbiology 153:411–419

Meilhoc E, Boscari A, Bruand C et al (2011) Nitric oxide in legume-rhizobium symbiosis. Plant Sci 181:573–581

Miller TW, Cherney MM, Lee AJ et al (2009) The effects of nitroxyl (HNO) on soluble guanylate cyclase activity: interactions at ferrous heme and cysteine thiols. J Biol Chem 284:21788–21796

Miranda KM (2005) The chemistry of nitroxyl (HNO) and implications in biology. Coord Chem Rev 249:433–455

Miranda KM, Nims RW, Thomas DD et al (2003) Comparison of the reactivity of nitric oxide and nitroxyl with heme proteins. A chemical discussion of the differential biological effects of these redox related products of NOS. J Inorg Biochem 93:52–60

Miranda KM, Yamada K, Espey DD et al (2002) Further evidence for distinct reactive intermediates from nitroxyl and peroxynitrite: effects of buffer composition on the chemistry of Angeli's salt and synthetic peroxynitrite. Arch Biochem Biophys 401:134–144

Navarre DA, Wendehenne D, Durner J et al (2000) Nitric oxide modulates the activity of tobacco aconitase. Plant Physiol 122:573–582

Nelson MT (1987) Effects of permeant ions and blockers on properties of single calcium channels from brains. Soc Gen Physiol Ser 41:215–230

Niketic V, Stojanovic S, Nikolic A et al (1999) Exposure of Mn and FeSODs, but not Cu/ZnSOD, to NO leads to nitrosonium and nitroxyl ions generation which cause enzyme modification and inactivation: an in vitro study. Free Radical Biol Med 27:992–996

Pacher P, Beckman JS, Liaudet L (2007) Nitric oxide and peroxynitrite in health and disease. Physiol Rev 87:315–424

Paolocci N, Jackson MI, Lopez BE et al (2007) The pharmacology of nitroxyl (HNO) and its therapeutic potential: not just the janus face of NO. Pharmacol Therap 113:442–458

Perazzolli M, Dominici P, Romero-Puertas MC et al (2004) *Arabidopsis* nonsymbiotic hemoglobin AHb1 modulates nitric oxide bioactivity. Plant Cell 16:2785–2794

Peyrot F, Fernandez BO, Bryan NS et al (2006) *N*-Nitroso products from the reaction of indoles with Angeli's salt. Chem Res Toxicol 19:58–67

Pieper GM, Halligan NLN, Hilton G et al (2003) Non-heme iron protein: a potential target of nitric oxide in acute cardiac allograft rejection. Proc Natl Acad Sci USA 100:3125–3130

Pulukkody R, Kyran SJ, Bethel RD et al (2013) Carbon monoxide induced reductive elimination of disulfide in an *N*-heterocyclic carbene (NHC)/thiolate dinitrosyl iron complex (DNIC). J Am Chem Soc 135:8423–8430

Qi Z, Verma R, Gehring C et al (2010) Ca^{2+} signaling by plant *Arabidopsis thaliana* Pep peptides depends on AtPepR1, a receptor with guanylyl cyclase activity, and cGMP-activated Ca^{2+} channels. Proc Natl Acad Sci USA 107:21193–21198

Radi R, Beckman JS, Bush KM, Freeman BA (1991) Peroxynitrite oxidation of sulfhydryls. The cytotoxic potential of superoxide and nitric oxide. J Biol Chem 266:4244–4250

Radi R (1996) Reaction of nitric oxide with metalloproteins. Chem Res Toxicol 9:828–835

Reisz JA, Zink CN, King SB (2011) Rapid and selective nitroxyl (HNO) trapping by phosphines: kinetics and new aqueous ligations for HNO detection and quantitation. J Am Chem Soc 133:11675–11685

Ridd JH (1979) Diffusion control and pre-association in nitrosation, nitration and halogenation. In: Gold V, Bethell D (eds) Advances in physical organic chemistry. Academic Press, New York, pp 1–51

Romero-Puertas MC, Laxa M, Matte A et al (2007) *S*-nitrosylation of peroxiredoxin II E promotes peroxynitrite-mediated tyrosine nitration. Plant Cell 19:4120–4130

Romero-Puertas MC, Perazzolli M, Zago ED, Delledonne M (2004) Nitric oxide signalling functions in plant–pathogen interactions. Cell Microbiol 6:795–803

Rosenthal J, Lippard SJ (2010) Direct detection of nitroxyl in aqueous solution using a tripodal copper (II) BODIPY complex. J Am Chem Soc 132:5536–5537
Saito S, Yamamoto-Katou A, Yoshioka H et al (2006) Peroxynitrite generation and tyrosine nitration in defense responses in tobacco BY-2 cells. Plant Cell Physiol 47:689–697
Sakamoto A, Tsukamoto S, Yamamoto H et al (2003) Functional complementation in yeast reveals a protective role of chloroplast 2-Cys peroxiredoxin against reactive nitrogen species. Plant J 33:841–851
Sánchez-Calvo Gates AJ, Meakin GE et al (2010) Production of nitric oxide and nitrosylleghemoglobin complexes in soybean nodules in response to flooding. Mol Plant Microbe Interact 23:702–711
Sánchez-Calvo B, Barroso JB, Corpas FJ (2013) Hypothesis: Nitro-fatty acids play a role in plant metabolism. Plant Sci 199–200:1–6
Schonhoff CM, Gaston B, Mannick JB (2003) Nitrosylation of cytochrome c during apoptosis. J Biol Chem 278:18265–18270
Seel F, Bliefert C (1972) Der Mechanismus der Zersetzung des Natriumsalzes der Benzolsulfhydroxamsäure in wäßriger Lösung. J Inorg Gen Chem 394:187–196
Shafirovich V, Lymar SV (2002) Nitroxyl and its anion in aqueous solutions: spin states, protic equilibria, and reactivities toward oxygen and nitric oxide. Proc Natl Acad Sci USA 99:7340–7345
Sies H, Sharov VS, Klotz LO, Briviba K (1997) Glutathione peroxidase protects against peroxynitrite-mediated oxidations. A new function for selenoproteins as peroxynitrite reductase. J Biol Chem 272:27812–27817
Signorelli S, Corpas FJ, Omar Borsani O et al (2013) Water stress induces a differential and spatially distributed nitro-oxidative stress response in roots and leaves of *Lotus japonicus*. Plant Sci 201–202:137–146
Souza JM, Peluffo G, Radi R (2008) Protein tyrosine nitration-Functional alteration or just a biomarker? Free Radical Biol Med 45:357–366
Stamler JS (1994) Redox signaling: nitrosylation and related target interactions of nitric oxide. Cell 78:931–936
Stamler JS, Feelisch M (1996) Biochemistry of nitric oxide and redox related species. In: Feelisch M, Stamler JS (eds) Methods in nitric oxide research. Wiley, Chichester, pp 19–27
Stamler JS, Singel DJ, Loscalzo J (1992) Biochemistry of nitric oxide and its redox-activated forms. Scince 258:1898–1902
Stojanović S, Stanić D, Nikolić M et al (2004) Iron catalyzed conversion f NO into nitrosonium (NO^+) and nitroksyl (HNO/NO^-) species. Nitric Oxide 11:256–262
Switzer CH, Flores-Santana W, Mancardi D et al (2009) The emergence of nitroxyl (HNO) as a pharmacological agent. Bioch Biophys Acta 1787:35–840
Szabó C (2003) Multiple pathways of peroxynitrite cytotoxicity. Toxicol Lett 140–141:105–112
Szabó C, Ischiropoulos H, Radi R (2007) Peroxynitrite: biochemistry, pathophysiology and development of therapeutics. Nat Rev Drug Discov 6:662–680
Tada Y, Spoel SH, Pajerowska-Mukhtar K et al (2008) Plant immunity requires conformational changes [corrected] of NPR1 via *S*-nitrosylation and thioredoxins. Science 321:952–956
Tanou G, Filippou P, Belghazi M et al (2012) Oxidative and nitrosative-based signaling and associated post-translational modifications orchestrate the acclimation of citrus plants to salinity stress. Plant J 72:585–595
Terrile MC, Paris R, Calderon-Villalobos LI et al (2012) Nitric oxide influences auxin signaling through *S*-nitrosylation of the *Arabidopsis* transport inhibitor response 1 auxin receptor. Plant J 70:492–500
Tocchetti CG, Wang W, Valdivia HH et al (2005) Nitroxyl anion: a novel thiol-sensitive positive inotrope that enhances SR Ca^{2+} release and uptake in murine cardiomyocytes. Circulation 112:U363 1553 Suppl S
Trostchansky A, Bonilla L, González-Perilli L, Rubbo H (2013) Nitro-fatty acids: formation, redox signaling, and therapeutic potential. Antioxid Redox Signal 19:1257–1265

Tuteja N, Chandra M, Tuteja R, Misra MK (2004) Nitric oxide as a unique bioactive signaling messenger in physiology and pathophysiology. J Biomed Biotech 4:227–237

Valderrama R, Corpas FJ, Carreras A et al (2007) Nitrosative stress in plants. FEBS Lett 581:453–461

van der Linde K, Gutsche N, Leffers HM et al (2011) Regulation of plant cytosolic aldolase functions by redox-modifications. Plant Physiol Biochem 49:946–957

Virag L, Szabó E, Gergely P, Szabó C (2003) Peroxynitrite-induced cytotoxicity: mechanism and opportunities for intervention. Toxicol Lett 140–141:113–124

Wang PG, Xian M, Tang X et al (2002) Nitric oxide donors: chemical activities and biological application. Chem Rev 102:1091–1134

Wang YQ, Feechan A, Yun BW et al (2009) *S*-nitrosylation of AtSABP3 antagonises the expression of plant immunity. J Biol Chem 284:2131–2137

Wendehenne D, Pugin A, Klessig D, Durner J (2001) Nitric oxide: comparative synthesis and signaling in animal and plant cells. Trends Plant Sci 6:177–183

Wink DA, Feelisch M, Fukuto J et al (1998) The cytotoxicity of nitroxyl: possible implications for the pathophysiological role of NO. Arch Biochem Biophys 351:66–74

Wink DA, Miranda KM, Katori T et al (2003) Orthogonal properties of the redox siblings nitroxyl and nitric oxide in the cardiovascular system: a novel redox paradigm. Am J Physiol Heart Circ Physiol 285:H2264–H2276

Wong PSY, Hyuan J, Fukuto JM et al (1998) Reaction between *S*-nitrosothiols and thiols: generation of nitroxyl (HNO) and subsequent chemistry. Biochemistry 37:5362–5371

Xian M, Chen X, Liu Z et al (2000) Inhibition of papain by *S*-nitrosothiols. Formation of mixed disulfides. J Biol Chem 275:20467–20473

Yun BW, Feechan A, Yin M et al (2011) *S*-nitrosylation of NADPH oxidase regulates cell death in plant immunity. Nature 478:264–268

Zamora R, Grzesiok A, Weber H, Feelish M (1995) Oxidative release of nitric oxide accounts for guanylyl cyclase stimulating, vasodilator and anti-platelet activity of Piloty's acid: a comparison with Angeli's salt. Biochem J 312:333–339

Zhou Y, Yao YW, Li JY et al (2012) Nitroxyl induced fluorescence enhancement via reduction of copper (II) coumarin-ester complex: its application for bioimaging in vivo. Sens Actuators B Chem 174:414–420

Chapter 11
Nitric Oxide and Other Signaling Molecules: A Cross Talk in Response to Abiotic Stress

Wei-Biao Liao and Ji-Hua Yu

Abstract Nitric oxide (NO), an easily diffusible bioactive molecule, has emerged as a biological messenger in plants. The study of NO has contributed to a better knowledge of many mechanisms and functions that were not well understood until very recently. NO may act as a signal molecule in multiple physiological processes in plants such as seed germination, plant maturation and senescence, floral transition, stomatal movement, lateral and adventitious root development. Depending upon the concentration and location in the plant cells, potential roles of NO as a regulator of many abiotic stresses have been identified. NO functions as a signaling molecule that mediates plant responses to various stimuli. Intracellular signaling responses to NO under stresses involve synthesis of cyclic guanosine monophosphate (cGMP), cyclic ADP ribose (cADPR), hydrogen peroxide (H_2O_2), elevation of cytosolic calcium (Ca^{2+}), and so on. In this chapter, our goal is to highlight the recent advances in NO signal transduction and its interactions with other signaling molecules in response to abiotic stress.

Keywords Abscisic acid · Calcium · Hydrogen peroxide · Mitogen-activated protein kinase · Signal transduction

11.1 Introduction

Nitric oxide (NO) regulates many physiological and biochemical processes in plants, including seed germination or dormancy, plant maturation and senescence, leaf extension, chlorophyll biosynthesis, pollen tubes growth, root organogenesis, programmed cell death, respiration, photosynthesis, stomatal movement, apoptosis,

W.-B. Liao (✉) · J.-H. Yu
College of Agronomy, Gansu Agricultural University, No. 1 Yinmen Village, Anning District, Lanzhou 730070, China
e-mail: liaowb@gsau.edu.cn

M. N. Khan et al. (eds.), *Nitric Oxide in Plants: Metabolism and Role in Stress Physiology*, DOI: 10.1007/978-3-319-06710-0_11,

hypersensitive responses, and many others (Neill et al. 2002; Courtois et al. 2008). In the enzymatic synthesis of NO in animal tissues, the NOS converts L-arginine to L-citrulline and NO. In addition to L-arginine as substrate, the reaction catalyzed by NOS requires molecular oxygen, NADPH, and other cofactors such as tetrahydrobiopterin (BH_4), flavin mononucleotide (FMN), flavin adenine dinucleotide (FAD), heme, calcium (Ca^{2+}), and calmodulin (CaM) (Knowles and Moncada 1994, Alderton et al. 2001). There are numerous enzymatic sources of NO biosynthesis in plants, for example, NO synthase (NOS), nitrate reductase (NR), xanthine oxidase/dehydrogenase (XDH), and nitrite-NO oxidoreductase (Ni-NOR) (Rockel et al. 2002; Neill et al. 2008). However, NOS found in animal is still elusive in plants. On the other hand, there are many reports in which NOS inhibitors have been shown to repress various processes in plants (Zhao et al. 2004; Crawford et al. 2006; Jasid et al. 2006; Hao et al. 2008; Liao et al. 2009, 2013).

Table 11.1 indicates that NO may be involved in plant abiotic and biotic stress responses such as drought, low and high temperatures, UV, ozone exposure, and disease infection (Neill et al. 2008; Liao et al. 2012a). Although extensive research has been carried out on the functioning of NO as a signaling molecule in plants, the exact cellular mechanism of NO signaling in plants is still incomprehensible. The present chapter is focused on the recent advances in the role of NO in signal transduction and its interaction with other signaling molecules during abiotic stress.

11.2 NO Signal Transduction

To function as a signaling molecule, it has to possess certain properties facilitating its direct influence on secondary messengers. NO is highly reactive due to the presence of an unpaired electron, which explains its existence as three interchangeable species such as the radical ($NO\cdot$), the nitrosonium cation (NO^+), and the nitroxyl radical (NO^-) (Stamler et al. 1992; Wojtaszek 2000). NO, which is synthesized in roots, operates downstream of indole acetic acid (IAA) promoting adventitious root development in cucumber through the guanylyl cyclase (GC)-catalyzed synthesis of cyclic guanosine monophosphate (cGMP) (Pagnussat et al. 2003). NO seems to possess the properties of a signaling molecule, such as a simple structure, small dimensions, and high diffusivity (Arasimowicz and Floryszak-Wieczorek 2007). The evidence that the signaling function of NO may be executed via a cGMP-dependent or cGMP-independent pathway has been obtained in several systems (Neill et al. 2008). NO treatment induced dramatic and transient increases in endogenous cGMP levels in tobacco (Durner et al. 1998). The authors also found that NO-induced activation of phenylalanine ammonia lyase (PAL) was blocked by two inhibitors of GC, 6-anilino-5,8-quinolinedione and 1*H*-(1,2,4)-oxadiazole[4,3-a]quinoxalin-1-one. It has also been observed that cGMP levels in *Arabidopsis thaliana* seedlings increased rapidly and to different degrees after salt and osmotic stress (Donaldson et al. 2004). ABA- and NO-induced stomatal closure in pea required the synthesis and action of cGMP and cyclic ADP ribose (cADPR)

Table 11.1 Reports of nitric oxide (NO) induction by abiotic stress and NO-mediated effect

Stressor	NO-mediated effect	Species of induced NO	References
Drought/osmotic	Involved in ABA signaling, stomatal closure	*Nicotiana tabacum*	Gould et al. (2003)
	Induction of ABA synthesis, LEA expression	*Pisum sativum*	Leshem and Haramaty (1996)
	Maintained higher RWC and lower leaf water loss	*Triticum aestivum* L.	Tan et al. (2008)
	Improved seed germination	*Antiaris toxicaria*	Bai et al. (2011)
Salinity	Increased osmotic tolerance	*N. tabacum*	Gould et al. (2003)
	Induce expression of Na^{+}/H^{+} antiporter gene	*Zea mays* L.	Zhang et al. (2006)
	Enhanced seedling growth	*Oryza sativa* L.	Uchida et al. (2002)
	Increased seed germination and seed respiration rate and ATP synthesis	*T. aestivum* L.	Zheng et al. (2009)
	Increased seedling growth, photosynthetic pigment content, proline accumulation, net photosynthetic	*Cucumis sativus* *Lycopersicom esculentum* Mill.	Fan et al. (2007) Wu et al. (2011)
UV-B radiation	Increased shoot and root dry weight	*Arabidopsis thaliana*	Mackerness et al. (2001)
	Induced the expression of CHS gene	*Z. mays* L.	An et al. (2005)
	Increased leaf area and biomass	*P. sativum*	Qu et al. (2006)
	Increased stem length	*Z. mays* L.	Kim et al. (2010)
	Prevented chlorophyll content reduction and of higher quantum yield for photosystem II		
Heavy metal	Increased the root elongation; reduced the NOS activity	*Hibiscus moscheutos* (Al^{3+})	Tian et al. (2007) Zhang et al. (2008)
	Increased chlorophyll content and proline accumulation and soluble protein	*T. aestivum* L. (Al^{3+})	
	Increased root length	*A. thaliana* (Pb^{2+})	Phang et al. (2011)
	Enhanced root growth	*T. aestivum* L. (Cd^{2+})	Groppa et al. (2008)

(continued)

Table 11.1 (continued)

Stressor	NO-mediated effect	Species of induced NO	References
High temperature	Increased chlorophyll a fluorescence parameters, membrane integrity	*Phaseolus radiates*	Yang et al. (2006)
	Decreased electrolyte leakage and relative ion leakage	*Phragmites communis*	Song et al. (2006)
	Improved survival rate of seedlings and quantum yield for photosystem II	*O. sativa* L.	Uchida et al. (2002)
Low temperature	Decline in the ROS level	*Scenedesmus obliquus*	Mallick et al. (2000)
	Increased soluble sugar and chlorophyll content	*C. sativus*	Liu et al. (2011)
	Increases in membrane permeability	*C. sativus*	Yang et al. (2011)
Ozone	Decreased cell death	*A. thaliana*	Ahlfors et al. (2009)
Mechanical injury	NO burst result in cell death	*A. thaliana*	Garces et al. (2001)
		Taxus brevifolia	Pedroso et al. (2000)

(Neill et al. 2002). The cyclic nucleotide cGMP has also been shown to be involved in plant hormone signaling and alters phosphorylation of *A. thaliana* root proteins (Isner et al. 2012). In plants, NO, cGMP, and cADPR have also been suggested to mediate stomatal closure induced by ABA (Neill et al. 2003; Garcia-Mata and Lamattina 2002). cGMP has been clearly identified in various plant tissues, but the mechanisms by which cGMP might be turned over in plant cells under abiotic stress remain unknown until now. Pharmacological studies using inhibitors of NO-sensitive GC have implicated cGMP downstream of NO and ABA signaling in guard cells (Neill et al. 2003). Recently, Joudoi et al. (2013) investigated the nitrated cGMP derivative 8-nitro-cGMP functions in guard cell signaling. They found that 8-nitro-cGMP may act as a guard cell signaling molecule and that a NO/8-nitro-cGMP signaling cascade may be involved in guard cells.

In addition to cGMP-dependent signaling, NO may also signal its presence through other mechanisms such as direct *S*-nitrosylation. *S*-nitrosylation has been shown to be the reversible covalent attachment of NO to the thiol group of cysteine residues forming an *S*-nitrosothiol (SNO) and may be a highly conserved cell signaling mechanism (Wang et al. 2006). Recently, Lin et al. (2012) identified *nitric oxide excess1* (*noe1*) (NO accumulation mutant), in *Oryza sativa*. They isolated the corresponding gene and analyzed its role in NO-mediated leaf cell death. Their results suggested that both NO and SNOs are important mediators in the process of H_2O_2-induced leaf cell death in *O. sativa*. Up to date, there are relatively little endogenous *S*-nitrosylated proteins have been characterized in plants. The development of a sensitive proteomic approach will be very useful that may identify endogenous *S*-nitrosylated proteins.

11.3 NO Interaction with Other Signaling Molecules in Response to Abiotic Stress

11.3.1 Interaction of NO with Ca^{2+}

A number of studies have concluded that NO regulated the signaling cascade through cADPR and Ca^{2+} mobilization. Nicotinamide, a potential inhibitor of cADPR synthesis, inhibited ABA- and NO-induced stomatal closure (Neill et al. 2002), suggesting that inhibition of ABA responses by nicotinamide is, at least partly, due to inhibition of cADPR biosynthesis following NO production. In addition, Ca^{2+} has been shown to be the downstream target of NO. NO-induced intracellular Ca^{2+} release and regulation of guard cell plasma membrane K^+ and Cl^- channels are mediated by a cGMP- and cADPR-dependent pathway (Garcia-Mata et al. 2003). Thus, NO might act firmly within one branch of the Ca^{2+}-signaling pathways engaged by ABA and define the boundaries of parallel signaling events in the control of guard cell movements. The data also suggested that NO may act through cGMP and cADPR to modulate intracellular Ca^{2+}-permeable channels to elevate free cytosolic calcium levels in cells. Cell suspensions of *Nicotiana plumbaginifolia* expressing the calcium reporter aequorin provided evidence that NO participated in the elevation of free Ca^{2+} in plant cells exposed to abiotic stresses such as high temperatures, hyperosmotic conditions, and salinity stress (Gould et al. 2003). The potential roles for NO as an endogenous regulator of Ca^{2+} mobilization in physiological contexts have been recognized widely (Fig. 11.1).

Rodríguez-Serrano et al. (2009) studied the effect of cadmium on the production of reactive oxygen species (ROS) and NO in growing pea leaves in vivo by confocal laser microscopy. They observed an induction of ROS production by cadmium mainly in mesophyll cells, probably associated with chloroplasts, mitochondria, and peroxisomes, and in plasma membrane from epidermal cells. In contrast with ROS, the production of NO was strongly reduced by cadmium. This result suggests the existence of cross talk between NO, ROS, and Ca^{2+} under cadmium toxicity (Rodríguez-Serrano et al. 2009). Ca^{2+}/CaM-dependent protein kinase (CCaMK) is a strong candidate for the decoder of Ca^{2+} signals. Recently, we investigated Ca^{2+} and CaM in NO- and H_2O_2-induced adventitious rooting in marigold (Liao et al. 2012b). Ca^{2+} chelators and CaM antagonists both prevented NO- and H_2O_2-induced adventitious rooting. Ca^{2+} might be involved as an upstream signaling molecule for CaM during NO- and H_2O_2-induced rooting (Liao et al. 2012b). González et al. (2012) also noted a copper-induced cross talk among Ca^{2+}, H_2O_2, and NO in *Ulva compressa* and a calcium-dependent activation of gene expression involving CaMs and calcium-dependent protein kinases.

11.3.2 Interaction of NO with H_2O_2 and ABA

It has conclusively been shown that both NO and H_2O_2 are generated under similar stress situations and with similar kinetics (Fig. 11.1). For example, H_2O_2 can react with NO to form the reactive molecule peroxynitrite, which may have unique signaling properties. Thus, it is suggested that they could both influence the same or related signaling pathways and thereby lead to additive and possibly synergistic responses (Neill et al. 2007, 2008). Recently, there are much data on the cellular processes by which NO and H_2O_2 act either negatively or positively in guard and mesophyll cells (She et al. 2004; Bright et al. 2006; Zhang et al. 2007). For example, Zhang et al. (2007) investigated the role of NO and the relationship between NO, H_2O_2, and mitogen-activated protein kinase (MAPK) in ABA-induced antioxidant defense in leaves of maize plants. They found that ABA-induced H_2O_2 production mediates NO generation that activates MAPK and results in the up-regulation in the expression and the activities of antioxidant enzymes in ABA signaling. The signal interactions of NO and H_2O_2 in disease, salt and drought resistance (Li et al. 2009), heat shock-induced hypericin production (Xu et al. 2008), stomatal movement (She et al. 2004; Bright et al. 2006), and adventitious root development (Liao et al. 2009, 2010) have been commonly observed. H_2O_2 might induce NO synthesis and accumulation in *Vicia faba* guard

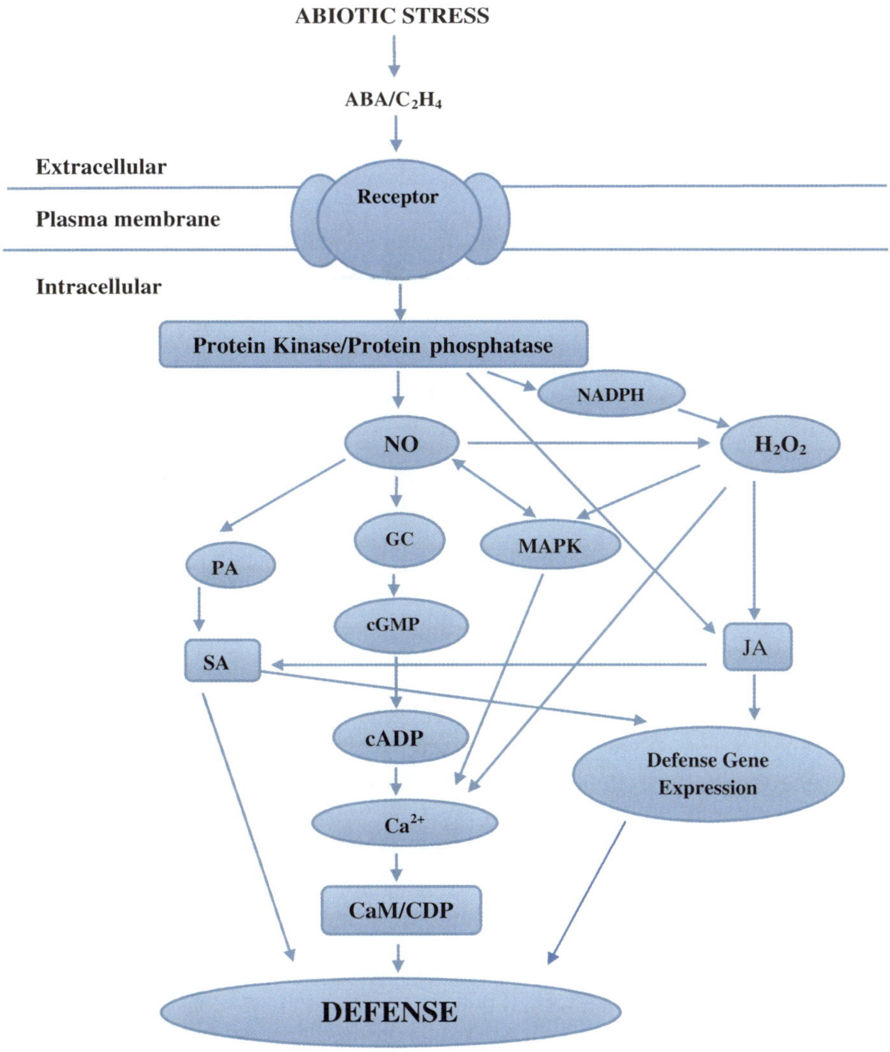

Fig. 11.1 Schematic model of the signaling networks involving NO and its interactions with other signaling molecules in responses to abiotic stress. ABA, abscisic acid; NOS, nitric oxide synthase; NR, nitrate reductase; NADPH, nicotinamide adenine dinucleotide phosphate; NO, nitric oxide; PAL, phenylalanine ammonia lyase; GC, guanylate cyclase; MAPK, mitogen-activated protein kinase; H_2O_2, hydrogen peroxide; SA, salicylic acid; cGMP, cyclic guanosine monophosphate; cADPR, cyclic ADP ribose; JA, jasmonic acid; CaM, calmodulin; CDPKs, calcium-dependent protein kinases

cells, and this accumulation could be reduced by an NOS inhibitor (She et al. 2004). On the other hand, there have also been evidences that points toward the modulation of H_2O_2 levels by NO in *V. faba* guard cells (She et al. 2004). Contrarily, other experiments did not observe this phenomenon (Bright et al. 2006;

Zhang et al. 2007). Our experiment showed that NO and H_2O_2 may be downstream signal molecules in the auxin signaling cascade, and NO may be involved as an upstream signaling molecule for H_2O_2 production (Liao et al. 2011).

Recently, Ma et al. (2012) reported that ZmCCaMK might be required for ABA-induced antioxidant defense in maize. H_2O_2-dependent NO production played an important role in the ABA-induced activation of ZmCCaMK. NO, ROS, and cell death and the possible cross talk between them in microspore embryogenesis in barley under specific stress treatments (cold, starvation) were investigated recently (Rodríguez-Serrano et al. 2012). ROS increase was involved in the stress-induced programmed cell death at early stages in both non-induced microspores and embryogenic suspension cultures of barley, whereas NO played a dual role after stress in the two in vitro systems, one involved in programmed cell death in embryogenic suspension cultures and the other in the initiation of cell division leading to embryogenesis in reprogrammed microspores (Rodríguez-Serrano et al. 2012).

11.3.3 Interactions of NO with MAPK, cGMP, and Ethylene

The interactions of NO with other signaling molecules such as MAPK, cGMP, and ethylene have also been noted (Fig. 11.1).

Zhang et al. (2007), using pharmacological and biochemical approaches, studied the role of NO and the relationship between NO, H_2O_2, and MAPK in ABA-induced up-regulation in the expression of several antioxidant genes in maize leaves. They found that MAPK activation was similarly targeted by H_2O_2 and NO in mesophyll cells which might be required for downstream signaling to enhance antioxidant gene expression and enzyme activity. Both ABA and H_2O_2 activated an MAPK enzyme in maize leaves, which was largely inhibited by the removal of NO. Thus, the results show that NO may be involved in the ABA-induced up-regulation in the expression and the activities of antioxidant enzymes (Zhang et al. 2007). She and Song (2008) investigated the roles of MAPKK/CDPK and their effects on NO levels of guard cells during dark-induced stomatal closure in broad bean. Two specific CDPK inhibitors 20-amino-30-methoxyflavone (PD98059) and trifluoperazine (TFP) reduced NO content in guard cells and significantly reversed dark-induced stomatal closure, suggesting that MAPKK/CDPK may mediate dark-induced stomatal closure by enhancing NO levels in guard cells. PD98059 and TFP also reversed stomatal closure by sodium nitroprusside (SNP), an NO donor, and by dark, indicating MAPKK and CDPK may be related to restraining the NO scavenging to elevate NO levels. The authors noted that there may be a causal and interdependent relationship between MAPKK/CDPK and NO in dark-induced stomatal closure, and in the process this cross talk may lead to the formation of a self-amplification loop (She and Song 2008). Further research should be done to investigate whether the MAPK lies upstream and regulates NO production in plants under abiotic stresses.

It has been demonstrated that NO action requires the synthesis of cGMP in animal cells. In plant cells, cGMP has been shown to be synthesized which may be enhanced by NO (Newton et al. 1999; Neill et al. 2003). cGMP may be an important component of NO-mediated signaling pathways, and GC, together with phosphodiesterase (PDE), regulated the endogenous level of cellular messenger cGMP (Suita et al. 2009). Previous study has indicated that NO operated downstream of indole-3-acetic acid (IAA) promoting cucumber adventitious root development through the GC-catalyzed synthesis of cGMP (Pagnussat et al. 2003). cGMP synthesis was involved in, but not sufficient, for stomatal closure in pea (Neill 2002). However, H_2O_2-induced closure was not inhibited by ODQ, suggesting that H_2O_2 and NO may be in separate signaling pathways in terms of cGMP signaling (Desikan et al. 2004). Interestingly, specific GC inhibitor LY83583 was able to reduce adventitious root number and length in SNP-treated marigold explants, and this inhibition could be reserved by 8-Br-cGMP. However, H_2O_2-mediated signaling pathways were not involved in cGMP. Thus, cGMP was involved in NO-induced root formation of marigold, but it was not involved in H_2O_2-mediated rooting process (Liao et al. 2009).

It has been shown that ethylene reduced NO levels in *V. faba* guard cells via a pattern of NO scavenging, then induced stomatal opening in the dark (Song et al. 2011). An earlier study has reported that exogenously applied NO delayed the senescence of postharvest horticultural produce by decreasing the sensitivity of ethylene and suppression of ethylene production (Eum et al. 2009). NO may function as a signal molecule involved in the senescence of cut rose regulated by ethylene. Exogenous NO decreased 1-aminocyclopropane-1-carboxylate oxidase (ACO) activity and ethylene production, and cPTIO [2-(4-carboxyphenyl)-4,4,5,5-tetramethylimidazoline-1-oxyl-3-oxide] was able to prevent the negative effects of NO, implying that NO may act as an antagonist to ethylene in the senescence of cut rose flowers (Liao et al. 2013). The *AOX1a* gene was used as a molecular probe to investigate its regulation by signal molecules such as H_2O_2, NO, ethylene, salicylic acid, and jasmonic acid, all of them reported to be involved in the ozone response. Ederli et al. (2006) reported that both NO- and ethylene-dependent pathways were required for ozone-induced up-regulation of *AOX1a* in tobacco. However, only NO was indispensable for the activation of *AOX1a* gene expression. Thus, that in ozone-fumigated tobacco plants NO was the preferred signaling molecule involved in AOX gene expression, which was coordinately activated by ethylene (Ederli et al. 2006).

11.4 Conclusions and Perspectives

Rapidly increasing evidences indicate that NO is actively involved in several plant abiotic stresses. To protect plants under stress, NO may act as an antioxidant, eliminating the superoxide radicals and compromising the toxicity caused by ROS. However, there has been much disagreement regarding the mechanism(s) by which

NO reduces abiotic stress. In addition, with a small sample size, caution must be applied, as most studies of NO synthesis and signaling are based on pharmacological studies that use NO donors, NO scavengers, and NO inhibitors. One major drawback of this approach is that the exogenous NO response functions are not uniquely identified.

Although several NO signaling transduction pathways in plants have been suggested, biochemical and molecular details of each pathway remained obscure. For example, there is a clear need for the development of techniques to identify, visualize, and quantify cGMP and cADPR in plant cells, and to clone the genes required for their synthesis and degradation. Moreover, it is unclear how these identified pathways cooperate with each other in plants, and which pathway operates in each particular tissue or organ or at a specific time. It has been suggested in previous studies that there is a connection among NO, H_2O_2, Ca^{2+}, MAPK, cGMP, and ethylene in plants. To date, the relationships between them have become more apparent. At present, however, the exact cross talk among them remains to be elucidated. Therefore, more research will be required to determine the complex NO signaling network. Also, in future studies, it will be important to know more about the mechanisms of NO signaling transduction and to identify and characterize its direct targets and their functions.

Acknowledgements This work was supported by the National Natural Science Foundation of China (No. 31160398), the Post Doctoral Foundation of China (Nos. 20100470887, 2012T50828), the Key Project of Chinese Ministry of Education (No. 211182), and the Research Fund for the Doctoral Program of Higher Education (No. 20116202120005).

References

Ahlfors R, Brosché M, Kollist H, Kangasjärvi J (2009) Nitric oxide modulates ozone-induced cell death, hormone biosynthesis and gene expression in *Arabidopsis thaliana*. Plant J 58:1–12

Alderton WK, Cooper CE, Knowles RG (2001) Nitric oxide synthases: structure, function and inhibition. Biochem J 357:593–615

An LZ, Liu YH, Zhang MX et al (2005) Effects of nitric oxide on growth of maize seedling leaves in the presence or absence of ultraviolet-B radiation. J Plant Physiol 162:317–326

Arasimowicz M, Floryszak-Wieczorek J (2007) Nitric oxide as a bioactive signalling molecule in plant stress responses. Plant Sci 172:876–887

Bai X, Yang L, Tian M et al (2011) Nitric oxide enhances desiccation tolerance of recalcitrant *Antiaris toxicaria* seeds via protein *S*-nitrosylation and carbonylation. PLoS ONE 6:e20714

Bright J, Desikan R, Hancock JT et al (2006) ABA-induced NO generation and stomatal closure in Arabidopsis are dependent on H_2O_2 synthesis. Plant J 45:113–122

Courtois C, Besson A, Dahan J et al (2008) Nitric oxide signalling in plants: Interplays with Ca^{2+} and protein kinases. J Exp Bot 59:155–163

Crawford NM, Galli M, Tischner R et al (2006) Response to Zemojtel et al.: plant nitric oxide synthase: back to square one. Trends Plant Sci 11:526–527

Desikan R, Cheung MK, Bright J et al (2004) ABA, hydrogen peroxide and nitric oxide signalling in stomatal guard cells. J Exp Bot 55:205–212

Donaldson L, Ludidi N, Knight MR et al (2004) Salt and osmotic stress cause rapid increases in *Arabidopsis thaliana* cGMP levels. FEBS Lett 569:317–320

Durner J, Wendehenne D, Klessig DF (1998) Defense gene induction in tobacco by nitric oxide, cyclic GMP, and cyclic ADP-ribose. P Natl Acad Sci USA 95:10328–10333

Ederli L, Morettini R, Borgogni A et al (2006) Interaction between nitric oxide and ethylene in the induction of alternative oxidase in ozone-treated tobacco plants. Plant Physiol 142:595–608

Eum HL, Kim HB, Choi SB, Lee SK (2009) Regulation of ethylene biosynthesis by nitric oxide in tomato (*Solanum lycopersicum* L.) fruit harvested at different ripening stages. Eur Food Res Technol 228:331–338

Fan H, Guo S, Jiao Y et al (2007) Effects of exogenous nitric oxide on growth, active oxygen species metabolism, and photosynthetic characteristics in cucumber seedlings under NaCl stress. Front Agric China 1:308–314

Garces H, Durzan DJ, Pedroso MC (2001) Mechanical stress elicits nitric oxide formation and DNA fragmentation in *Arabidopsis thaliana*. Ann Bot 87:567–574

Garcia-Mata C, Gay R, Sokolovski S et al (2003) Nitric oxide regulates K^+ and Cl^- channels in guard cells through a subset of abscisic acid-evoked signaling pathways. Proc Natl Acad Sci USA 100:11116–11121

Garcia-Mata C, Lamattina L (2002) Nitric oxide and abscisic acid cross talk in guard cells. Plant Physiol 128:790–792

González A, Cabrera M de los Á, Henríquez MJ et al (2012) Cross talk among calcium, hydrogen peroxide, and nitric oxide and activation of gene expression involving calmodulins and calcium-dependent protein kinases in *Ulva compressa* exposed to copper excess. Plant Physiol 158: 1451–1462

Gould KS, Lamotte O, Klinguer A et al (2003) Nitric oxide production in tobacco leaf cells: a generalized stress response? Plant Cell Environ 26:1851–1862

Groppa D, Rosales EP, Iannone MF, Benavides MP (2008) Nitric oxide, polyamines and Cd-induced phytotoxicity in wheat roots. Phytochemistry 69:2609–2615

Hao GP, Xing Y, Zhang JH (2008) Role of nitric oxide dependence on nitric oxide synthase-like activity in the water stress signaling of maize seedling. J Integr Plant Biol 50:435–442

Isner JC, Nühse T, Maathuis FJM (2012) The cyclic nucleotide cGMP is involved in plant hormone signalling and alters phosphorylation of *Arabidopsis thaliana* root proteins. J Exp Bot 63:3199–3205

Jasid S, Simontacchi M, Bartoli CG, Puntarulo S (2006) Chloroplasts as a nitric oxide cellular source. Effect of reactive nitrogen species on chloroplastic lipids and proteins. Plant Physiol 142:1246–1255

Joudoi T, Shichiri Y, Kamizono N et al (2013) Nitrated cyclic GMP modulates guard cell signaling in Arabidopsis. Plant Cell 25:558–5571

Kim T-Y, Jo M-H, Hong J-H (2010) Protective effect of nitric oxide against oxidative stress under UV-B radiation in maize leaves. J Environ Sci 19:1323–1334

Knowles RG, Moncada S (1994) Nitric oxide synthases in mammals. Biochem J 298:249–258

Leshem YY, Haramaty E (1996) The characterization and contrasting effects of the nitric oxide free radical in vegetative stress of *Pisum sativum* Linn. foliage. J Plant Physiol 148:258–263

Li Y, Yin H, Wang Q et al (2009) Oligochitosan induced *Brassica napus* L. production of NO and H_2O_2 and their physiological function. Carbohyd Polym 4:612–617

Liao WB, Huang GB, Yu JH et al (2011) Nitric oxide and hydrogen peroxide are involved in indole-3-butyric acid-induced adventitious root development in marigold. J Hortic Sci Biotechnol 86:159–165

Liao WB, Xiao HL, Zhang ML (2009) Role and relationship of nitric oxide and hydrogen peroxide in adventitious root development of marigold. Acta Physiol Plant 31:1279–1289

Liao WB, Xiao HL, Zhang ML (2010) Effect of nitric oxide and hydrogen peroxide on adventitious root development from cuttings of ground-cover chrysanthemum and associated biochemical changes. J Plant Growth Regul 29:338–348

Liao WB, Zhang ML, Huang GB, Yu JH (2012a) Hydrogen peroxide in the vase solution increases vase life and keeping quality of cut Oriental × Trumpet hybrid lily 'Manissa'. Sci Hortic 139:32–38

Liao WB, Zhang ML, Huang GB, Yu JH (2012b) Ca^{2+} and CaM are involved in NO- and H_2O_2-induced adventitious root development in marigold. J Plant Growth Regul 31:253–264
Liao WB, Zhang ML, Yu JH (2013) Role of nitric oxide in delaying senescence of cut rose flowers and its interaction with ethylene. Sci Hortic 155:30–38
Lin A, Wang Y, Tang J et al (2012) Nitric oxide and protein *S*-nitrosylation are integral to hydrogen peroxide-induced leaf cell death in rice. Plant Physiol 158:451–464
Liu X, Wang L, Liu L et al (2011) Alleviating effect of exogenous nitric oxide in cucumber seedling against chilling stress. Afr J Biotechnol 10:4380–4386
Ma F, Lu R, Liu H et al (2012) Nitric oxide-activated calcium/calmodulin-dependent protein kinase regulates the abscisic acid-induced antioxidant defence in maize. J Exp Bot 63:4835–4847
Mackerness SAH, John CF, Jordan B, Thomas B (2001) Early signaling components in ultraviolet-B responses: distinct roles for different reactive oxygen species and nitric oxide. FEBS Lett 489:237–242
Mallick N, Mohn FH, Rai L, Soeder CJ (2000) Impact of physiological stresses on nitric oxide formation by green alga, *Scenedesmus obliquus*. J Microbiol Biotechnol 10:300–306
Neill S (2007) Interactions between abscisic acid, hydrogen peroxide and nitric oxide mediate survival responses during water stress. New Phytol 175:4–6
Neill S, Barros R, Bright J, Desikan R et al (2008) Nitric oxide, stomatal closure, and abiotic stress. J Exp Bot 59:165–176
Neill SJ, Desikan D, Clarke A, Hancock JT (2002) Nitric oxide is a novel component of abscisic acid signaling in stomatal guard cells. Plant Physiol 128:13–16
Neill SJ, Desikan R, Hancock JT (2003) Nitric oxide signalling in plants. New Phytol 159:11–35
Newton RP, Roef L, Witters E, Van Onckelen H (1999) Cyclic nucleotides in higher plants: the enduring paradox. New Phytol 143:427–455
Pagnussat GC, Lanteri ML, Lamattina L (2003) Nitric oxide and cyclic GMP are messengers in the indole acetic acid-induced adventitious rooting process. Plant Physiol 132:1241–1248
Pedroso MC, Magalhaes JR, Durzan D (2000) Nitric oxide induces cell death in Taxus cells. Plant Sci 157:173–180
Phang IC, Leung DW, Taylor HH, Burritt DJ (2011) The protective effect of sodium nitroprusside (SNP) treatment on *Arabidopsis thaliana* seedlings exposed to toxic level of Pb is not linked to avoidance of Pb uptake. Ecotoxicol Environ Saf 74:1310–1315
Qu Y, Feng HY, Wang YB et al (2006) Nitric oxide functions as a signal in UV-B inhibition of pea stems elongation. Plant Sci 170:994–1000
Rockel P, Strube F, Rockel A et al (2002) Regulation of nitric oxide (NO) production by plant nitrate reductase in vivo and in vitro. J Exp Bot 53:103–110
Rodríguez-Serrano M, Bárány I, Prem D et al (2012) NO, ROS, and cell death associated with caspase-like activity increase in stress-induced microspore embryogenesis of barley. J Exp Bot 63:2007–2024
Rodríguez-Serrano M, Romero-Puertas MC, Pazmiño DM et al (2009) Cellular response of pea plants to cadmium toxicity: cross talk between reactive oxygen species, nitric oxide, and calcium. Plant Physiol 150:229–243
She XP, Song XG (2008) Pharmacological evidence indicates that MAPKK/CDPK modulate NO levels in darkness-induced stomatal closure of broad bean. Aust J Bot 56:347–357
She XP, Song XG, He JM (2004) The role and relationship of nitric oxide and hydrogen peroxide in light/dark-regulated stomatal movement in *Vicia faba*. Acta Bot Sin 46:1292–1300
Song L, Ding W, Zhao M et al (2006) Nitric oxide protects against oxidative stress under heat stress in the calluses from two ecotypes of reed. Plant Sci 171:449–458
Song XG, She XP, Wang J, Sun YC (2011) Ethylene inhibits darkness-induced stomatal closure by scavenging nitric oxide in guard cells of *Vicia faba*. Funct Plant Biol 38:767–777
Stamler JS, Singel DJ, Loscalzo J (1992) Biochemistry of nitric oxide and its redox-activated forms. Science 258:1898–1902
Suita K, Kiryu T, Sawada M et al (2009) Cyclic GMP acts as a common regulator for the transcriptional activation of the flavonoid biosynthetic pathway in soybean. Planta 229:402–413

Tan J, Zhao H, Hong J et al (2008) Effects of exogenous nitric oxide on photosynthesis, antioxidant capacity and proline accumulation in wheat seedlings subjected to osmotic stress. World J Agric Sci 4:307–313

Tian QY, Sun DH, Zhao MG, Zhang WH (2007) Inhibition of nitric oxide synthase (NOS) underlies aluminum-induced inhibition of root elongation in *Hibiscus moscheutos*. New Phytol 174:322–331

Uchida A, Jagendorf AT, Hibino T, Takabe T, Takabe T (2002) Effects of hydrogen peroxide and nitric oxide on both salt and heat stress tolerance in rice. Plant Sci 163:515–523

Wang Y, Yun BW, Kwon EJ et al (2006) *S*-nitrosylation: an emerging redox-based post-translational modification in plants. J Exp Bot 57:1777–1784

Wojtaszek P (2000) Nitric oxide in plants: to NO or not to. Phytochemistry 54:1–4

Wu X, Zhu W, Zhang H et al (2011) Exogenous nitric oxide protects against salt-induced oxidative stress in the leaves from two genotypes of tomato (*Lycopersicom esculentum* Mill.). Acta Physiol Plant 33:1199–1209

Xu MJ, Dong JF, Zhang XB (2008) Signal interaction between nitric oxide and hydrogen peroxide in heat shock-induced hypericin production of *Hypericum perforatum* suspension cells. Sci China C Life Sci 51:676–686

Yang H, Wu F, Cheng J (2011) Reduced chilling injury in cucumber by nitric oxide and the antioxidant response. Food Chem 127:1237–1242

Yang JD, Yun JY, Zhang TH, Zhao HL (2006) Presoaking with nitric oxide donor SNP alleviates heat shock damages in mung bean leaf discs. Bot Stud 47:129–136

Zhang AY, Jiang MY, Zhang JH et al (2007) Nitric oxide induced by hydrogen peroxide mediates abscisic acid-induced activation of the mitogen-activated protein kinase cascade involved in antioxidant defense in maize leaves. New Phytol 175:36–50

Zhang H, Li YH, Hu LY et al (2008) Effects of exogenous nitric oxide donor on antioxidant metabolism in wheat leaves under aluminum stress. Russ J Plant Physiol 55:469–474

Zhang Y, Wang L, Liu Y et al (2006) Nitric oxide enhances salt tolerance in maize seedlings through increasing activities of proton-pump and Na^+/H^+ antiport in the tonoplast. Planta 224:545–555

Zhao LQ, Zhang F, Guo JK et al (2004) Nitric oxide functions as a signal in salt resistance in the calluses from two ecotypes of reed. Plant Physiol 134:849–857

Zheng C, Jiang D, Liu F et al (2009) Exogenous nitric oxide improves seed germination in wheat against mitochondrial oxidative damage induced by high salinity. Environ Exp Bot 67:222–227

Chapter 12
Cytoprotective Role of Nitric Oxide Under Oxidative Stress

Y. S. Bakakina, E. V. Kolesneva, L. V. Dubovskaya and I. D. Volotovski

Abstract NO is a known important effector in living organisms. Currently, it is obvious that NO as a second messenger participates in various plant physiological reactions including stress responses. NO due to its physical and chemical properties (small size, high diffusion coefficient, and no charge) can also react with a variety of intracellular targets directly. One of the fastest reactions of NO in biological systems is interaction with ROS. Formation of ROS is a common feature of plant cell responses to multiple stresses. Excessive ROS production causes oxidative stress resulting in damage to all classes of biomolecules. In this chapter the data concerning protective effects of NO under oxidative stress in plant cell is summarized. In particular, we present the results of our experiments on the effect of exogenously applied NO at high and low concentrations on structural and functional parameters of plant cells under H_2O_2-induced oxidative stress. The mechanism of synergistic action of NO and H_2O_2 is also discussed.

Keywords Apoptosis · DNA fragmentation · Lipid peroxidation · Nitric oxide · Oxidative stress · Programmed cell death

12.1 Introduction

Plants are frequently subjected to diverse stresses, and in the process of evolution the plants have developed various defense mechanisms to survive under acute stress conditions. The generation of reactive oxygen species (ROS) is a common feature of majority of plant stress responses. The oxidative damage can result from such processes as pathogen infections, drought, herbicide treatment, and other

Y. S. Bakakina · E. V. Kolesneva · L. V. Dubovskaya (✉) · I. D. Volotovski
Institute of Biophysics and Cell Engineering, National Academy of Sciences of Belarus, Akademicheskaya ul. 27, 220072 Minsk, Belarus
e-mail: dubovsk@mail.ru

M. N. Khan et al. (eds.), *Nitric Oxide in Plants: Metabolism and Role in Stress Physiology*, DOI: 10.1007/978-3-319-06710-0_12,

stresses. The possibility exists that nitric oxide (NO) production occurs under natural conditions as a generalized stress response (Gould et al. 2003). It is shown that NO can function to diminish the adverse effects from stress effects in different plant species (Leshem and Kuiper 1996). Two interrelated mechanisms by which NO may reduce stress have been proposed. Firstly, NO might function as an antioxidant, directly scavenging ROS that is generated by most of stresses (Dubovskaya et al. 2007). NO can interact rapidly with superoxide radical to form peroxynitrite ($ONOO^-$) (Radi et al. 1991). Although peroxynitrite and its protonated form ONOOH are themselves oxidizing agents, they are considered to be less toxic than peroxides and may therefore minimize cell damage (Wink et al. 1993). Secondly, NO may function as a signaling molecule in the cascade of events leading to gene expression (Wendehenne et al. 2001). The chemical properties of NO (small molecule, short life time, absence of charge, and high diffusivity) suggest that it would be an ideal inter- and intramolecular signaling molecule in plant stress responses (Foissner et al. 2000).

Plants are considered to be subjected to oxidative stress when ROS are produced in uncontrollable toxic amounts (Beligni and Lamattina 2002). Generation of superoxides, hydrogen peroxide, hydroxyl radicals, and other free radicals can result from the involvement of oxygen in normal respiratory processes and the production of oxygen during photosynthesis (Bowler et al. 1992). The protective mechanisms against oxidative damage are operating with the involvement of some enzymes (superoxide dismutase, catalases, and peroxidases) and free radical scavengers (carotenoids, ascorbate, tocopherols, oxidized and reduced glutathione, GSSG, and GSH, respectively) (Halliwel and Gutteridge 2007). In animal cells the GSH/GSSG ratio is crucial for cell stress reaction. The GSH/GSSG ratio influences cytosolic calcium homeostasis resulting from oxidative changes in sensitive thiols of Ca^{2+}-ATPases (Nicoterra et al. 1992).

High concentrations of oxidants induce lipid peroxidation, DNA and RNA fragmentation, protein degradation, and the ions leakage from the intracellular compartments (Heath 1987). NO can break off the chain reaction of oxidation, minimizing oxidative damage to cells (Dubovskaya et al. 2007). NO can also bind superoxide radicals generated in the electron transport chain of chloroplasts and mitochondria, preventing the formation of highly toxic hydroxyl radicals. However, high doses of NO can inhibit the electron flow through cytochrome *c* oxidase, leading to increased superoxide production in mitochondria. Moreover, the interaction between NO and superoxide leads to the generation of greatly cytotoxic agent peroxynitrite (Radi et al. 1991). Thus, NO and ROS interaction can result in both cytoprotective and cytotoxic effects depending on the balance of reacting molecules content (Dubovskaya et al. 2007). This chapter will focus on the relationship between oxidative stress, cytoprotective role of NO, and structural and functional parameters of plant cells.

12.2 The Generation of Reactive Oxygen Species

ROS in plant cell is generated in response to multiple abiotic and biotic stresses, including different chemical pollutants, drought, intense light, extreme temperatures, UV radiation, ozone, excess electron excitation by absorption of light by pigments, phytohormones, mechanical damage, herbicides, pathogens, elicitors, etc. (Miller et al. 2008; Torres 2010). ROS includes free radical molecules (superoxide radical, hydroperoxyl, hydroxyl radical) and neutral molecules (H_2O_2, singlet oxygen). ROS is produced in different organelles of plant cell which may be due to the enzymatic as well as with nonenzymatic processes (Vranová et al. 2002). For example, in a cell wall H_2O_2 generated due to the activity of amine oxidases and peroxidases, in a plasma membrane and peroxisomal membrane—due to the oxidation of reduced pyridine nucleotides by NADPH oxidase.

Moreover, in membrane structures of plant cell, ROS may be formed by carbon dioxide fixation by ribulose-1,5-biphosphate carboxylase (Rubisco). It is known that Rubisco catalyzes two reactions: the carboxylation and oxygenation. The second one ultimately leads to the formation of H_2O_2 (Mahalingam and Fedoroff 2003).

Recently enzyme xanthine oxidase was detected in plant peroxisomes, which in certain conditions can produce large amounts of superoxide anion radical ($O_2^{\bullet-}$) (del Río et al. 2006). Appearance of oxygen free radicals in plants is often the result of oxidation of various substrates by peroxidases localized in the cytosol and other compartments. In the cytosol, ROS generation may be associated with some biochemical reactions, for example, catalyzed by flavin oxidases, peroxisomal enzyme systems, and respiratory electron transport chain in mitochondria (Neill et al. 2002).

Under physiological conditions, ROS is predominantly produced during electron transport in mitochondria and chloroplasts (Wise and Naylor 1987). Therefore, the term ROS is often used to describe the products generated during the successive one-electron reduction of molecular oxygen to $O_2^{\bullet-}$, H_2O_2, hydroxyl radical ($OH^{\bullet}$), and water.

Therefore, in plant cells the ROS production is confined to chloroplasts, mitochondria, peroxisomes, and apoplast and normally maintained at a low level. Under adverse conditions, intensive production and accumulation of ROS that can damage all biological molecules are occurred. The ability of cells to resist oxidative damage is determined by the potential of antioxidant system, the operation of which will be discussed below.

12.3 Physiological Consequences of Oxidative Stress in Plants

Oxidative stress induces damage to almost all of the structural components of cells. High doses of oxidants result in the oxidation of lipids and proteins, DNA fragmentation, degradation of RNA, leakage of ions that may lead to cell death

(Dubovskaya et al. 2007). Lipid peroxidation, in turn, can result in damage of cell membranes. In addition, the products of lipid peroxidation (malondialdehyde (MDA), 4-hydroxyalkenals, etc.) are able to exert mutagenic and cytotoxic activity (Dubovskaya et al. 2007).

ROS causes the modification of the native structure of proteins, which makes them more sensitive to the action of proteases (Mehta et al. 1992). Oxidative modification of proteins leads to chemical changes in the amino acid residues and disruption of the protein tertiary structure and, as a consequence, aggregation, denaturation, and loss of function. Some proteins such as Rubisco is directly fragmented by ROS (Ishida et al. 1999).

Necessary to emphasize that only $OH^{\bullet}$ directly causes DNA damage (oxidation of nitrogenous bases, their modification, breaks the sugar phosphate bones, chromosome damage). DNA mutations can lead to pathological conditions and cell death. It is known that high levels of ROS inhibit DNA synthesis and cell division, which also leads to their death (Arora et al. 2002).

12.4 NO and Oxidative Stress

NO is an uncharged lipophilic molecule, and thus has an ability to diffuse within the cell and can cross plant membranes (Leshem 2001). NO is an unstable molecule. Its half-life is less than 6 s, and it can diffuse a distance of about 30 microns (Bethke et al. 2004). There are several potential precursors of NO in plants (Fig. 12.1), and, apparently, the importance of each of them depends on a type of plant, a type of tissue/cells, plant growth conditions, and a structure of signaling pathways that are active under these conditions. NO can be generated in plants by nitrate-/nitrite-dependent pathway with the action of nitrate reductase and nitrite-NO reductase, L-arginine-dependent pathway involving nitric oxide synthase (NOS)-like enzymes, as well as nonenzymatic synthesis. Several plant systems produce NO by xanthine dehydrogenase/oxidase. Other enzymes may also be involved in NO production (Gupta et al. 2011).

Drought, extreme temperatures, ozone, UV irradiation, and other abiotic stresses lead to the excessive ROS production in the cell (Vranová et al. 2002; Neill et al. 2002). On the other hand, various chemical and mechanical stresses, environmental signals induce rapid and significant increase in NO concentration in plants (Dubovskaya et al. 2011; Mur et al. 2013).

To date, two possible mechanisms of NO action under adverse environmental factors have been described for plants. First, NO possesses antioxidant properties, it scavenges ROS produced by different abiotic and biotic stress conditions, thus protecting the cell from injuring action of ROS. This mechanism was observed, for example, in potato plants treated with an herbicide Diquat (Beligni and Lamattina 2002). It is assumed that NO can directly interact with the superoxide radical or H_2O_2, which leads to production of peroxynitrite, a substance less toxic than peroxides, thereby minimizes the cell injury (Scheel 1998).

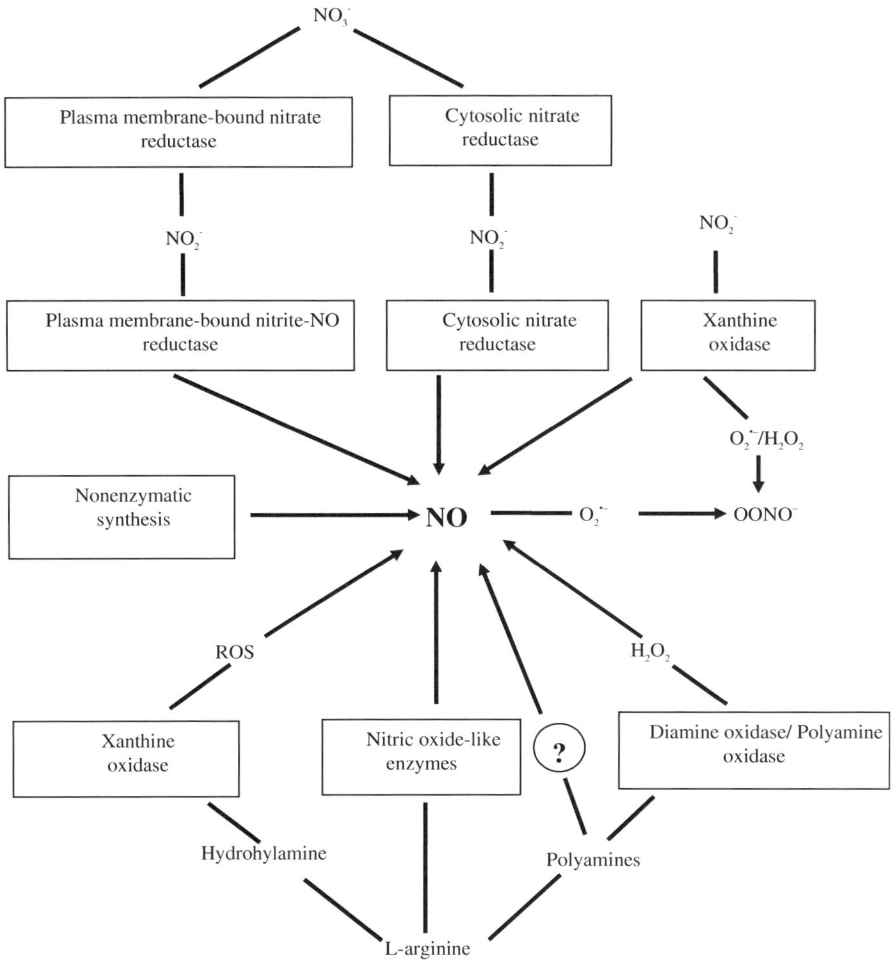

Fig. 12.1 Pathways of NO biosynthesis in plants

Second, NO functions as a signaling molecule and triggers the cascade of events leading to expression of numerous genes (Lamotte et al. 2004). NO has been emerging to be a key signaling molecule in plant signal transduction pathways, where cyclic guanosine 3′,5′-monophosphate (cGMP) may be its downstream mediator (Dubovskaya et al. 2011). NO may directly or indirectly interact with other signaling molecules. It was shown that the expression of tobacco genes responsible for the synthesis of protective metabolites is induced by NO, cGMP, and cyclic adenosine diphosphate ribose (cADPR) (Durner et al. 1998). Established that in plants, as well as in animals, cGMP and cADPR participate as signaling mediators between NO and the genome (Delledonne et al. 1998; Durner et al. 1998). In tobacco cells, cGMP, cADPR, as well as NO, induced the expression of pathogenesis-related protein 1 (*PR-1*) and phenylalanine ammonia lyase (*PAL*) (Durner et al. 1998). Furthermore, the antagonist cADPR 8-bromo-cADPR inhibited the accumulation of *PR-1* transcripts (Klessig et al. 2000). Moreover, the expression and the activity of PAL increased in the presence of 8-bromo-cGMP, a membrane-permeable analog of cGMP (Durner et al. 1998).

Changes in cytosolic free calcium concentration ($[Ca^{2+}]_{cyt}$) are known to occur in response to diverse environmental stimuli, including oxidative stress (Kolesneva et al. 2006). We have demonstrated that the membrane-permeable cGMP analog 8-bromo-cGMP elevated the cytosolic $[Ca^{2+}]_{cyt}$ in *Nicotiana plumbaginofolia* protoplasts (Volotovski et al. 1998). In animal cells, NO can regulate $[Ca^{2+}]_{cyt}$ directly interacting with Ca^{2+} channels and leading to their activation or inhibition both in vivo and in vitro, by *S*-nitrosylation or tyrosine nitrosylation. Besides, NO can open the voltage-dependent Ca^{2+} channel by depolarization of the plasma membrane too (Willmott et al. 2000). NO can also modulate the activity of Ca^{2+} channels through the activation of cGMP-dependent protein kinase G (PKG) in several ways: (1) PKG directly phosphorylates voltage-dependent Ca^{2+} channel or phosphorylates proteins associated with plasma membrane leading to the activation of Ca^{2+} channels; (2) PKG triggers signaling reactions leading to the synthesis of inositol 1,4,5-trisphosphate (IP_3) and subsequent activation of IP_3-dependent Ca^{2+} channels; (3) PKG activates ADPR cyclase, responsible for the synthesis of cADPR, a specific modulator of ryanodine receptor calcium release channels (RyR-channels) (Wang et al. 2000; Courtois et al. 2008). Now it is evident that NO-mediated signal transduction in plant cells involves the same key structural elements as in animal systems.

We have shown that pretreatment of plants with exogenous NO decreased $[Ca^{2+}]_{cyt}$ in response to oxidative stress, indicating a possible protective function of NO (Kolesneva et al. 2006). The participation of endogenous NO, being produced with the involvement of NOS-like enzyme, in H_2O_2-induced Ca^{2+} response was also confirmed using the inhibitory analysis. We also found that NO mediates its effects on $[Ca^{2+}]_{cyt}$ by activating plasma membrane Ca^{2+} channels and RyR-like intracellular Ca^{2+}-channels regulated by cADPR and with the involvement of cGMP.

There are also cGMP-independent mechanisms of NO action in cells. NO can trigger biological effects directly without the participation of cGMP and/or cADPR (Besson-Bard et al. 2008). Biological effects of NO are based on the chemical modification of biomolecules by binding to the metals with the variable valence in metalloproteins (metal nitrosylation), covalent modification of cysteine residues (*S*-nitrosylation) and tyrosine residues (tyrosine nitrosylation) in proteins. These processes are considered as a specific post-translational modification of proteins. There are more than 100 proteins that have been identified as targets for NO (Besson-Bard et al. 2008).

Using the plants with over-expression or lack of expression of hemoglobin, it was found that a functional interaction between NO and hemoglobin reduces the intracellular concentration of NO during hypoxia and pathogen infection

(Perazzolli et al. 2004; Seregelyes et al. 2004). Over-expression of hemoglobin in transgenic plants led to the increased stability to hypoxia and reduced necrosis symptoms in response to virulent pathogens. Thus, one of the functions of hemoglobin may be involved in stress adaptation to protect plants from nitrosative stress.

NO can regulate the activity of cytosolic and mitochondrial aconitase, catalase, ascorbate peroxidase, cytochrome *c* oxidase inactivating them by nitrosylation (Nelson 1987; Millar and Day 1996; Clark et al. 2000). Besides, NO can transduce a signal by tyrosine nitrosylation. It was found that nitration of tyrosine residues in proteins occurred in olive leaves under salt stress and in tobacco mutants over-producing NO (Morot-Gaudry-Talarmain et al. 2002; Valderrama et al. 2007).

Existence of *S*-nitrosylation as physiologically significant transduction mechanisms in plants involves the reversibility of the process. De-*S*-nitrosylation could be mediated by GSH to form *S*-nitrosoglutathione (GSNO), an endogenous donor of NO pool in cells. It is shown that GSNO-reductase plays an important role in triggering *S*-nitrosothiols-mediated effects (in particular, the development of systemic acquired resistance) in plants infected by pathogens (Feechan et al. 2005; Rusterucci et al. 2007).

Physical and chemical properties (small molecule, fast metabolism, the lack of electric charge, and high diffusion coefficient) allow NO to react with a variety of intracellular targets. One of the fastest reactions of NO in biological systems is its interaction with ROS. Furthermore, NO can break down chain reactions of oxidation and, thus, minimize oxidative damage to cells (Wink et al. 1993). NO can also bind superoxide radicals generated in the electron transport chains of chloroplasts and mitochondria, preventing the formation of highly toxic hydroxyl radicals.

Beligni and Lamattina (2002) have found that NO interferes with plant photo-oxidative stress induced by bipyridinium herbicide Diquat. They demonstrated that two NO donors, sodium nitroprusside (SNP) and *S*-nitroso-*N*-acetylpenicillamine (SNAP), strongly reduced lipid peroxidation and the protein loss caused by the application to potato leaf pieces or isolated chloroplasts of high doses of Diquat. NO donors also protected the RNA against oxidative damage. Their results have provided the evidence that NO is a potent antioxidant in plants and that its action may, at least in part, be explained by its ability to directly scavenge ROS. NO was demonstrated to confer a water-deficit tolerance to both detached wheat leaves and wheat seedlings under drought stress condition, as NO availability to induces stomatal closure (Beligni and Lamattina 2001; Dubovskaya et al. 2011). NO was postulated to act as an antioxidant and protect membranes and lipoproteins from oxidation either directly by inactivating ROS such as lipid hydroxyl radical or indirectly by inhibiting lipoxygenase activity (Beligni and Lamattina 1999; Beligni et al. 2002).

Lipid peroxidation is the consequence of oxidative damage of the cell. NO rapidly reacts with lipid alcoxyl or lipid peroxyl radicals and breaks the self-perpetuating chain reaction during lipid peroxidation (Beligni and Lamattina 1999, 2002). We have demonstrated that NO applied at low nanomolar concentrations diminished the impacts of H_2O_2-induced oxidative stress in tobacco leaves

(Dubovskaya et al. 2007). For example, NO acts as an antioxidant in plants by preventing the formation of MDA, a lipid peroxidation product. These data are similar to the results obtained by Beligni and Lamattina (2002) and Hung et al. (2002) who demonstrated that NO performs protective role during oxidative stress induced in potato plants treated with Diquat (Beligni and Lamattina 2002) and in rice plants treated with Paraquat (Hung et al. 2002). High concentrations of ROS can activate the programmed cell death (PCD), which is characterized by a fragmentation of chromatin and the release of low molecular DNA fragments (Wendehenne et al. 2001). We found that H_2O_2-induced fragmentation of total DNA was entirely suppressed by NO applied in the form of SNP (100 μM, equivalent to nanomolar concentrations of NO) (Dubovskaya et al. 2007). It was supposed that H_2O_2 induced the synthesis of new soluble proteins, which was manifested in the increase in total protein content in the cytosol. This increase in protein content in tobacco leaves was fully prevented by the addition of 100 μM SNP, even though SNP applied alone had no effect on protein content. Experiments with SNP in the presence of an NO antagonist, 2-(4-carboxyphenyl)-4, 5-dihydro-4,4,5,5-tetramethyl-1H-imidazole-1-oxyl-3-oxide potassium salt (cPTIO) confirm that the protective effect is caused directly by NO and not by the source substance (SNP itself) (Dubovskaya et al. 2007).

NO can trigger the expression of certain genes and promote the protective oxidative stress-induced responses of plants due to its signaling nature. Besides defense genes (*PR-1, PAL*), NO activates the expression of antioxidant genes such as glutathione-*S*-transferase (*GST*), chalcone synthase (*CHS*), glutathione peroxidase (*GPX*), and alternative oxidase (*AOX1a*) genes and inhibits gene expression of thylakoid ascorbate peroxidase (*tAPX*) modulating oxidative status of plant cell (Murgia et al. 2004).

NO plays a crucial role in plant immune response. For example, in soybean cell suspensions inoculated with avirulent *Pseudomonas syringae* pv. *glycinea,* NO participates in cooperation with other ROS and activated hypersensitive response and PCD (Delledonne et al. 1998). At the same time, NO functioned independently of other ROS to induce defense genes for the synthesis of protective natural products. Moreover, the authors have found that the inhibitors of NO synthesis potentiated the hypersensitive disease-resistance response of *Arabidopsis* leaves to avirulent *Pseudomonas syringae* pv. *maculicola*, promoting disease and bacterial growth (Delledonne et al. 1998).

In our work it was demonstrated that NO applied at high concentrations in the form of 5 mM SNP (equivalent to micromolar concentrations of NO) may function as signaling inductor of programmed cascade of events leading to activation of caspase-like proteases, fragmentation of total DNA, protein degradation, and decrease in ATP content in the cytosol of H_2O_2-treated tobacco leaves (Dubovskaya et al. 2007). The ATP content in animal cells is a reliable parameter to distinguish the necrotic cells from apoptotic ones (Lemasters 1999). Similar results were obtained on tobacco BY-2 cells treated with benzyladenosine (Mlejnek et al. 2003), a powerful inductor of apoptosis in animal cells, and *Arabidopsis* cells exposed to oxidative stress (Tiwari et al. 2002).

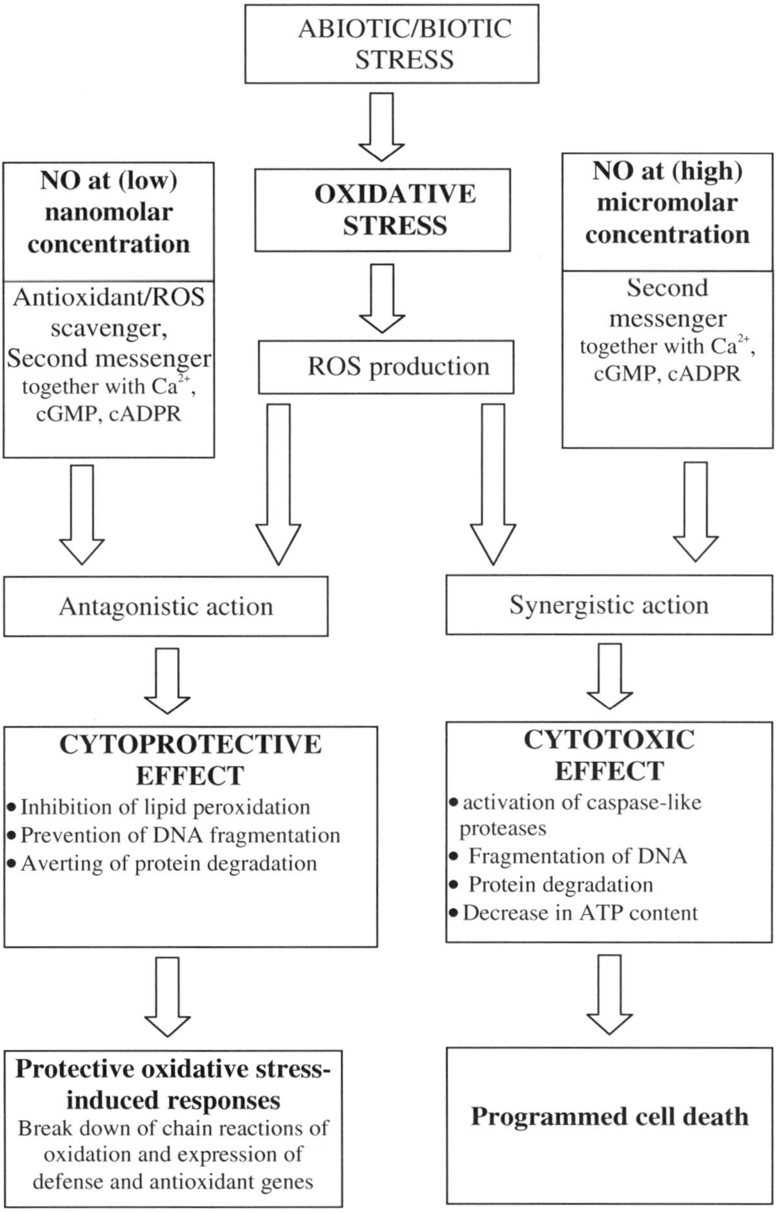

Fig. 12.2 Cytoprotective and cytotoxic role of NO depending on its concentration in tobacco cells under oxidative stress

Therefore, we have shown that NO and H_2O_2 exerted antagonistic effects on tobacco leaves at NO nanomolar concentrations but induced synergistic effects at NO micromolar concentrations (Fig. 12.2). So interaction between NO and H_2O_2

can resulted in cytoprotective and cytotoxic effects, respectively. During H_2O_2-induced oxidative stress, low concentrations of NO inhibited lipid peroxidation, prevented the fragmentation of DNA, and prevented accumulation of soluble proteins in tobacco cells. When applied at high concentrations, NO induced the caspase-like activity, promoted degradation of DNA and soluble proteins, and reduced ATP synthesis. The results are consistent with the hypothesis that NO performs a dual role in plants, acting as an antioxidant (scavenger of ROS) and as a signaling molecule.

Hence, NO performs its protective role during oxidative stress in tobacco leaves irrespective of underlying mechanisms, because NO, even at high concentrations, exerts prolonged rather than immediate action, and its effect is realized through the activation of caspase-like proteases leading to PCD, a physiological process of the self-destruction of the cell. The PCD may serve, for example, to prevent infection and subsequent spreading of a pathogen in functionally active plant organs.

12.5 Conclusion

Consequently, it can be postulated that under stress conditions NO can function as a cytoprotective molecule. Protective role of NO could be due to its signaling nature as well as direct interaction with ROS. Thus, our results confirmed the hypothesis on the dual role of NO in plants as it was described in animals and determines the importance of studying the role of NO in plants and its participation in signaling mechanisms at various stress conditions. Cytoprotective and cytotoxic effects of NO depend on its location and concentration, the balance between ROS production and scavenging and a type of cell.

References

Arora A, Sairam RK, Srivastava GC (2002) Oxidative stress and antioxidative system in plants. Cur Sci 82:1227–1238

Beligni MV, Lamattina L (1999) Is nitric oxide toxic or protective? Trends Plant Sci 4:299–300

Beligni MV, Lamattina L (2001) Nitric oxide induces stomatal closure and enhances the adaptive plant responses against drought stress. Plant Physiol 126:1196–1204

Beligni MV, Lamattina L (2002) Nitric oxide interferes with plant photo-oxidative stress by detoxifying reactive oxygen species. Plant Cell Env 25:737–748

Beligni MV, Fath A, Bethke PC et al (2002) Nitric oxide acts an antioxidant and delays programmed cell death in barley aleurone layers. Plant Physiol 129:1642–1650

Besson-Bard A, Pugin A, Wendehenne D (2008) New insights into nitric oxide signaling in plants. Annu Rev Plant Biol 59:21–39

Bethkea PC, Badgerb MR, Jones RL (2004) Apoplastic synthesis of nitric oxide by plant tissues. Plant Cell 16:332–341

Bowler C, Van Montagu M, Inze D (1992) Superoxide dismutase and stress tolerance. Annu Rev Plant Physiol Plant Mol Biol 43:83–116

Clark D, Durner J, Navarre DA, Klessig DF (2000) Nitric oxide inhibition of tobacco catalase and ascorbate peroxidases. Mol Plant-Microbe Interact 13:1380–1384

Courtois C, Besson A, Dahan J et al (2008) Nitric oxide signalling in plants: interplays with Ca^{2+} and protein kinases. J Exp Bot 59:155–163

del Río LA, Sandalio LM, Corpas FJ et al (2006) Reactive oxygen species and reactive nitrogen species in peroxisomes. Production, scavenging, and role in cell signaling. Plant Physiol 141:330–335

Delledonne M, Xia Y, Dixon RA, Lamb C (1998) Nitric oxide functions as a signal in plant disease resistance. Nature 394:585–588

Dubovskaya LV, Kolesneva EV, Knyazev DM, Volotovski ID (2007) Protective role of nitric oxide during hydrogen peroxide-induced oxidative stress in tobacco plants. Rus J Plant Physiol 54:755–762

Dubovskaya LV, Bakakina YS, Kolesneva EV et al (2011) cGMP-dependent ABA-induced stomatal closure in the ABA-insensitive *Arabidopsis* mutant *abi1-1*. New Phytol 191:57–69

Durner J, Wendehenne D, Klessig DF (1998) Defense gene induction in tobacco by nitric oxide, cyclic GMP and cyclic ADP-ribose. Proc Natl Acad Sci USA 95:10328–10333

Feechan A, Kwon E, Yun BW et al (2005) A central role for *S*-nitrosothiols in plant disease resistance. Proc Natl Acad Sci USA 102:8054–8059

Foissner I, Wendehenne D, Langebartels C, Durner J (2000) In vivo imaging of an elicitor-induced nitric oxide burst in tobacco. Plant J 23:817–824

Gould KS, Lamotte O, Klinguer A et al (2003) Nitric oxide production in tobacco leaf cells: a generalized stress response? Plant Cell Env 26:1851–1862

Gupta KJ, Fernie AR, Kaiser WM, van Dongen JT (2011) On the origins of nitric oxide. Trends Plant Sci 16:160–168

Halliwell B, Gutteridge JMC (2007) Free radicals in biology and medicine. Oxford University Press, Oxford, UK, p 704

Heath RL (1987) The biochemistry of ozone attack on the plasma membrane of plant cells. Adv Phytochem 21:29–54

Hung KT, Chang CJ, Kao CH (2002) Paraquat toxicity is reduced by nitric oxide in rice leaves. J Plant Physiol 159:159–166

Ishida H, Makino A, Mae T (1999) Fragmentation of the large subunit of ribulose-1, 5-bisphosphate carboxylase by reactive oxygen species occurs near Gly-329. J Biol Chem 274:5222–5226

Klessig DF, Durner J, Noad R et al (2000) Nitric oxide and salicylic acid signaling in plant defense. Proc Natl Acad Sci USA 97:8849–8855

Kolesneva EV, Dubovskaya LV, Volotovski ID (2006) Effect of nitric oxide on concentration of cytosolic free Ca^{2+} in transgenic *Arabidopsis thaliana* plants under oxidative stress. Cell biology and instrumentation: UV radiation, nitric oxide and cell death in plants 371:199–207

Lamotte O, Gould K, Lecourieux D et al (2004) Analysis of nitric oxide signaling functions in tobacco cells challenged by the elicitor cryptogein. Plant Physiol 135:516–529

Lemasters JL (1999) Mechanisms of hepatic toxicity. v. necrapoptosis and mitochondrial permeability transition: shared pathways to necrosis and apoptosis. Am J Physiol 276:1–6

Leshem Y (2001) Nitric oxide in plants. Kluwer Academic Publishers, London, UK, p 180

Leshem YY, Kuiper PJC (1996) Is there a GAS (general adaptation syndrome) response to various types of environmental stress? Biol Plant 38:1–18

Mahalingam R, Fedoroff N (2003) Stress response, cell death and signaling: the many faces of ROS. Physiol Plantarum 119:56–68

Mehta RA, Fawcett TW, Porath D, Mattoo AK (1992) Oxidative stress causes rapid membrane translocation and in vivo degradation of ribulose-1,5-bisphosphate carboxylase/oxygenase. J Biol Chem 267:2810–2816

Millar AH, Day DA (1996) Nitric oxide inhibits the cytochrome oxidase but not the alternative oxidase of plant mitochondria. FEBS Lett 398:155–158

Miller G, Shulaev V, Mittler R (2008) Reactive oxygen signaling and abiotic stress. Physiol Plant 133:481–489

Mlejnek P, Dolezjel P, Prochazka S (2003) Intracellular phosphorylation of benzyladenosine is related to apoptosis induction in tobacco BY-2 cells. Plant Cell Env 26:1723–1735

Morot-Gaudry-Talarmain Y, Rockel P, Moureaux T et al (2002) Nitrite accumulation and nitric oxide emission in relation to cellular signalling in nitrite reductase antisense tobacco. Planta 215:708–715

Mur LA, Mandon J, Persijn S et al (2013) Nitric oxide in plants: an assessment of the current state of knowledge. AoB Plants 5:1–17

Murgia I, de Pinto MC, Delledonne M et al (2004) Comparative effects of various nitric oxide donors on ferritin regulation, programmed cell death, and cell redox state in plant cells. J Plant Physiol 161:777–783

Neill SJ, Desikan R, Clarke A et al (2002) Hydrogen peroxide and nitric oxide as signalling molecules in plants. J Exp Bot 53:1237–1247

Nelson MJ (1987) The nitric oxide complex of ferrous soybean lipoxygenase-1. Substrate, pH, and ethanol effects on the active-site iron. J Biol Chem 262:12137–12142

Nicoterra P, Kass GEN, Orrenius S (1992) Calcium mediated mechanisms in chemically-induced cell death. Annu Rev Pharmacol Toxicol 32:449–470

Perazzolli M, Dominici P, Romero-Puertas MC et al (2004) *Arabidopsis* nonsymbiotic hemoglobin AHb1 modulates nitric oxide bioactivity. Plant Cell 16:2785–2794

Radi R, Beckman JS, Bash KM, Freeman RA (1991) Peroxynitrite induced membrane lipid peroxidation: cytotoxic potential of superoxide and nitric oxide. Arch Biochem Biophys 228:481–487

Rustérucci C, Espunya MC, Díaz M et al (2007) *S*-nitrosoglutathione reductase affords protection against pathogens in *Arabidopsis*, both locally and systemically. Plant Physiol 143:1282–1292

Scheel D (1998) Resistance response physiology and signal transduction. Curr Opin Plant Biol 1:305–310

Seregélyes C, Igamberdiev AU, Maassen A et al (2004) NO degradation by alfalfa class 1 hemoglobin (Mhb1): a possible link to PR-1a gene expression in Mhb1-overproducing tobacco plants. FEBS Lett 571:61–66

Tiwari BS, Belenghi B, Levine A (2002) Oxidative stress increased respiration and generation of reactive oxygen species, resulting in ATP depletion, opening of mitochondrial permeability transition, and programmed cell death. Plant Physiol 128:1271–1281

Torres MA (2010) ROS in biotic interactions. Physiol Plant 138:414–429

Valderrama R, Corpas FJ, Carreras A et al (2007) Nitrosative stress in plants. FEBS Lett 581:453–461

Volotovski ID, Sokolovsky SG, Molchan OV, Knight MR (1998) Second messengers mediate increases in cytosolic calcium in tobacco protoplasts. Plant Physiol 117:1023–1030

Vranová E, Inzé D, Van Breusegem F (2002) *Signal transduction* during oxidative stress. J Exp Bot 53:1227–1236

Wang Y, Wagner MB, Joyner RW, Kumar R (2000) cGMP-dependent protein kinase mediates stimulation of L-type calcium current by cGMP in rabbit atrial cells. Cardiovasc Res 48:310–322

Wendehenne D, Pugin A, Klessig DF, Durner J (2001) Nitric oxide: comparative synthesis and signaling in animal and plant cells. Trends Plant Sci 6:177–183

Willmott NJ, Wong K, Strong AJ (2000) Intercellular Ca^{2+}-waves in rat hippocampal slice and dissociated glial-neuron cultures mediated by nitric oxide. FEBS Lett 487:239–247

Wink DA, Hanbauer I, Krishna MC et al (1993) Nitric oxide protects against cellular damage and cytotoxicity from reactive oxygen species. Proc Natl Acad Sci USA 90:9813–9817

Wise RR, Naylor AW (1987) Chilling-enhanced peroxidation: the peroxidative destruction of lipids during chilling injury to photosynthesis and ultrastructure. Plant Physiol 83:272–277

Chapter 13
Phytohormones and Nitric Oxide Interactions During Abiotic Stress Responses

Paulo T. Mioto, Luciano Freschi and Helenice Mercier

Abstract Although nitric oxide (NO) is considered as an important signaling molecule in plants, very little is known regarding its mechanisms of action, synthesis, and possible interactions with other molecules. In the last years, NO was related to a wide array of processes in plants, which are also under the influence of the five major hormonal classes. Despite some recent advances in the knowledge of how the interaction between hormones and NO occurs, this is still a vast field for research, as many questions have yet to be answered. Therefore, this chapter focuses on the current knowledge of possible interactions between NO and auxins, gibberellins, cytokinins, abscisic acid, and ethylene during plant abiotic stress responses.

Keywords Abscisic acid · Auxin · Cytokinin · Gibberellins · Nitric oxide

13.1 Introduction

In a constantly changing environment, plant hormones play a crucial role by integrating a multitude of endogenous and exogenous stimuli into common signaling pathways, which ultimately lead to coordinated developmental and biochemical responses and, therefore, a maximization of plant fitness either under optimal or unfavorable conditions. Naturally, to develop such a critical function, plant hormones need to interact among themselves and also with other endogenous signals, such as second messengers, transcription factors, and other nonhormonal signal molecules.

P. T. Mioto · L. Freschi · H. Mercier (✉)
Department of Botany, Institute of Biosciences, University of São Paulo,
São Paulo, SP CEP 05508-090, Brazil
e-mail: hmercier@usp.br

M. N. Khan et al. (eds.), *Nitric Oxide in Plants: Metabolism and Role in Stress Physiology*, DOI: 10.1007/978-3-319-06710-0_13,

In this context, the gaseous free radical nitric oxide (NO) has been the focus of increasing interest during the last two decades since this molecule has been described to be part of innumerous signaling cascades controlling both plant development and stress responses. Together with other reactive nitrogen species (RNS), NO intermediates metabolic, growth, and organogenetic adjustments of plants to environmental challenges, such as heat, cold, drought, high light, heavy metals, pathogen attack, among others. In this chapter, we will specifically address the interplay between these signaling molecules in plant stress responses. Naturally, only plant stress events whose signaling cascade implicates both phytohormones and NO will be discussed, and, therefore, not all NO-mediated plant responses to environmental challenges will be considered in this chapter.

13.2 Phytohormones and Nitric Oxide Interactions Under Abiotic Stress

13.2.1 Temperature Stress

Low temperature is a common challenge that plants, mainly from temperate or cold regions, must overcome in order to survive. Low temperature can cause several adverse effects in plants, such as protein denaturation, oxidative and osmotic stresses, and membrane rigidity, which may lead to ion leakage (Ruelland et al. 2009). In order to resist these deleterious effects, plants must perceive the stress and trigger the adequate responses. It is interesting to note that since cold may also trigger some degree of osmotic stress, there might be an overlap in the responses to these conditions, such as osmolyte production and stomatal closure.

NO is likely involved in the response to low temperature since, among other stresses, cold stress is one of the most effective in inducing nitric oxide synthase (NOS)-like and *S*-nitrosoglutathione reductase (GSNOR) activities, as well as NO production. *S*-nitrosothiols and nitration of tyrosine residues have also been observed in *Pisum sativum* (Corpas et al. 2008). This may reflect an important role of NO in this stress condition. For *Brassica juncea* plants, it is possible that at least part of Rubisco inhibition due to low temperatures may be a consequence of its *S*-nitrosylation (Abat and Deswal 2009). In *Arabidopsis thaliana* plants, nitrate reductase (NR)-dependent NO generation was detected when kept under 4 °C (Zhao et al. 2009). Cantrel et al. (2011) also detected an increase of NO in response to cold in *A. thaliana*.

One of the most important responses to NO production might be the activation of *C*-repeat binding factors (CBF) genes. Cantrel et al. (2011) noted remarkable reductions in the levels of CBF in *A. thaliana* defective in NR (*nia1nia2* mutant) under chilled environment. CBF genes appear to be important regulators in cold response as a constitutive expression of *CBF2* was capable of inducing 85 of the 302 surveyed genes that were up-regulated under cold stress (Vogel et al. 2005). Apparently, the activation of CBF genes leads to the accumulation of DELLA proteins by up-regulating GA-2-oxidase, an enzyme responsible for the inactivation of gibberellins (GA) (Achard et al. 2008a, b). These results indicate a possible link between NO and GA in response to cold, with NO acting through expression of CBF proteins to activate GA-2-oxidase and, consequently, inhibit GA response. GA inhibition resulted in an accumulation of DELLA proteins. Apparently, DELLAs confer a degree of stress tolerance associated with a decrease in growth, resulting in enhanced survivability (Achard et al. 2006, 2008a, b). If NO does, in fact, influence DELLA proteins, this could indicate a hub of interactions between NO and several hormonal classes.

Abscisic acid (ABA) is also involved in cold stress response. Production of ABA is an initial response to the cold stress that leads to stomatal closure or osmolyte production (Ruelland et al. 2009; Kosová et al. 2012). There is a well-established interaction between ABA and NO in stomatal closure and perhaps other drought responses, as addressed in Sect. 16.2.2 of this chapter. Kosová et al. (2012) also noted a decrease in bioactive gibberellins (GAs), indole-3-acetic acid (IAA), and cytokinins (CKs) during initial days of cold exposure in a cold-resistant wheat cultivar. It is conceivable that the observed decrease in GA levels was due to NO-mediated expression of CBF genes and subsequent activation of GA-2-oxidase.

In both spring and winter wheat, an increase in NO emission and in the contents of IAA and 1-aminocyclopropanecarboxylic acid (ACC) was detected after cold hardening at 5 °C for 12 days, also indicating some degree of interaction between these molecules (Majláth et al. 2012). Ethylene production is enhanced by cold in tomato plants and, apparently, triggers a mitogen-activated protein kinase (MAPK) cascade to induce some aspects of cold resistance, like proline production and expression of *CBF* genes (Zhao et al. 2013).

Liu et al. (2012) compared the transcriptome of *Solanum habrochaites* (cold resistant) and *S. lycopersicum* (cold susceptible). They noted a significant change in gene expression related to hormones, mainly auxin (Aux) and ABA, calcium, and reactive oxygen species (ROS) signaling. Since NO has antioxidant properties depending on its concentration, it is possible that it indirectly regulates ROS concentration and, consequently, its signaling (Gupta et al. 2011). Accumulation of glutathione (GSH) under cold stress is also a commonly observed response, probably as a mechanism to enhance antioxidant defense under adverse condition (Tomashow 2010). However, in addition to its antioxidant function, GSH also has an important role in NO scavenging and/or transport mechanism through the formation of *S*-nitrosoglutathione (GSNO). Therefore, there must be a fine-tuned balance in GSH/GSNO, NO, and ROS to regulate the responses of plants under cold stress. A tentative summary of the interactions between hormones and NO in cold signaling is provided in Fig. 13.1.

Under heat stress, an influx of calcium in the cytosol and a subsequent activation of calmodulins (CaMs) and, possibly, several other protein kinases promote perception by the membrane (Saidi et al. 2011). In *A. thaliana*, NO seems to be involved in the activation of a specific calmodulin (*CaM3*), which triggers heat-shock protein (HSP) accumulation under high temperature. Xuan et al. (2010) demonstrated that *noa1* and *nia1nia2* mutants had impaired heat resistance and lower levels of

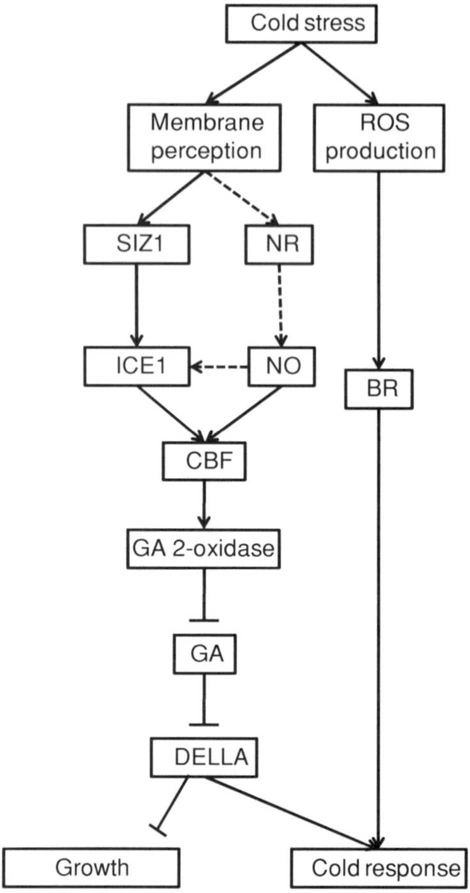

Fig. 13.1 A summary of possible interactions between elements from cold signaling involving NO and hormones. Dashed arrows indicate possible relations between the components that were not directly described in the literature. Cold stress leads to ROS production, which modulates cold response through brassinosteroids (BR). Cold stress also leads to perception, possibly through changes in membrane fluidity, which may trigger nitrate reductase (NR)-dependent NO synthesis. Along with high expression of osmotically responsive genes 1 (HOS1) and inducer of Cbf expression 1 (ICE1), NO modulates the expression of (CBF) genes. CBF genes enhance gibberellin (GA) degradation through activation of GA-2-oxidase leading to an increase of DELLA proteins, which then promotes the cold response and inhibits growth. This diagram was constructed based on several species; therefore, it is highly hypothetical

expression of *CaM3* and this phenotype was reversed by the application of NO. By using a pharmacological approach, Song et al. (2008) showed that NO mediates ABA responses to high temperature in calluses of two ecotypes of reeds (*Phragmites communis*). The production of NO in this case seems to be, at least in part, via NOS pathway. It is interesting to note that a similar interaction between NO and Ca^{2+} was also found under drought stress, showing that both possibly originated from ABA signaling (Freschi et al. 2010a). However, there are some genes exclusive to heat that are not expressed under drought or osmotic stresses (Saidi et al. 2009, 2011). Another possible role of NO in heat stress is ferredoxin-NADPH reductase inactivation by tyrosine nitration, which could indicate a possible consequence of protein–NO interaction (Chaki et al. 2011).

Studying *Nicotiana glauca* guard cell protoplasts, Beard et al. (2012) proposed an interesting interplay between NO, Aux, and ethylene. When these protoplasts were cultivated under 32 °C, NO production increased, possibly nitrosylating transport inhibitor response 1 (TIR1) Aux receptor and, consequently, increasing Aux responses. One of these responses is the transcription of ACS (ACC synthase), an enzyme responsible for ethylene biosynthesis. The activation of this pathway inhibits cell death signaling and reactivates the cell cycle. Interestingly, when these same cells were cultivated at 38 °C, the cell cycle was not activated, although cell death was inhibited. This effect of temperature can be mimicked by the application of L-NMMA (NG-monomethyl-L-arginine), an inhibitor of NOS activity.

It is important to note that factors, such as cold, light, NO, and hormones, play important roles in plant developmental processes, such as flowering and germination. The study of such processes may reveal further information about interactions between all these factors, which perhaps are not observed when taking into account only stress responses.

13.2.2 Drought Stress

Nitric oxide is often associated with drought stress, interacting mainly with ABA (Hancock et al. 2012). There are several papers showing that the application of NO enhances plant tolerance to drought (Lei et al. 2007; Hao et al. 2008; Shao et al. 2012; Tian and Lei 2006; Xing et al. 2004). However, the specific role of NO in drought response is still under discussion.

One of the most important responses to drought is stomatal closure, which diminishes water loss to the environment. The signaling pathways for stomatal control are well studied, and a number of interesting reviews are currently available (Chen et al. 2012; Zhu et al. 2012). Therefore, this chapter will only focus on the pathways that may involve the participation of NO and its interaction with hormones.

Available evidence shows an involvement of NO in stomatal closure and ABA. García-Mata and Lamattina (2001) verified that the application of NO donors significantly reduced the transpiration and, therefore, increased water retention of detached wheat leaves. Later on, studies conducted by Neill et al. (2002) with ABA and NO donors or scavengers substantiated the claim that NO might be indispensable to the ABA-mediated stomatal closure.

Another key component of stomatal closure seems to be H_2O_2. The production of this ROS is enhanced by ABA and controls the influx of Ca^{2+} into the cytosol, causing the closure of stomata (Desikan et al. 2004; Pei et al. 2000). Apparently, both NO and H_2O_2 result in a Ca^{2+} outburst that precedes stomatal closure and seem to act synergistically (Desikan et al. 2004). Recently, Puli and Raghavendra (2012) verified that the application of pyrabactin, an ABA agonist, is capable of inducing NO and H_2O_2 production in *P. sativum*. Similarly, methyl jasmonate (MeJA) is also capable of inducing stomatal closure and stimulates the production of NO and ROS (Saito et al. 2009). Apparently, MeJA and ABA are both capable of activating protein phosphatase 2A (PP2A) through its regulatory subunit ROOTS CURL IN NPA (RCN1), resulting in H_2O_2 and NO production (Saito et al. 2008).

Ethylene was also found to be involved in stomatal closure, possibly by stimulating the production of H_2O_2, as seen by impaired stomatal closure and H_2O_2 production in ethylene-insensitive mutants or wild-type plants treated with the ethylene perception inhibitors 1-methylcyclopropene (1-MCP) and silver (Desikan et al. 2006). This ethylene-mediated production of H_2O_2 appears to be a result of the activation of an NADPH oxidase present in chloroplasts called AtrbohF (Desikan et al. 2006). Auxs and CKs are involved in stomatal control too. Both hormones promoted stomatal opening in *Vicia faba* epidermal strips while causing a reduction in NO levels (Xiao-Ping and Xi-Gui 2006).

Phosphatidic acid (PA) seems to be required for ABA-induced stomatal closure, possibly generated by an NO/H_2O_2-induced phospholipase Dδ (Distéfano et al. 2012). PA is known to impair blue light-mediated stomatal opening, indicating a possible point of interaction between ABA-mediated stomatal closure and blue light-dependent stomatal opening. A scheme highlighting the interactions between NO and hormones in stomatal closure is given in Fig. 13.2. It is important to highlight that this figure was constructed based on the available information from several species submitted to drought and, therefore, it is highly speculative.

Although very important, stomatal closure is only one of many plant responses to drought. Another significant aspect of drought was highlighted by Freschi et al. (2010a) in their investigation of Crassulacean acid metabolism (CAM) up-regulation in young pineapple plants. By using mixed strategies of donating/scavenging compounds and endogenous quantification of hormones, they found that ABA was capable of up-regulating CAM through NO and Ca^{2+}. Alternatively, CKs repressed CAM, indicating that *Ananas comosus* CAM up-regulation is dependent on a balance between ABA and CKs. Later, Mioto and Mercier (2013) investigated CAM up-regulation in the C_3-CAM facultative bromeliad *Guzmania monostachia*. As determined by the previous work of Freschi et al. (2010b), CAM equally expressed along the length of the leaves—the apical portion shows a remarkable up-regulation of CAM, while the basal portion does not. Using detached leaves, it was found that after 7 days of exposure to polyethylene glycol (PEG), ABA content was significantly higher in both the basal and apical portions, but NO showed an increase exclusively in the apical part of the leaf (Mioto and Mercier 2013). These results indicate that NO is possibly a signal produced in response to ABA and controls only processes that are happening in the apical portion of the leaf, such as CAM up-regulation and stomatal closure.

Since drought generates oxidative stress, another important response is the up-regulation of the antioxidant system in response to NO. The ascorbate-glutathione

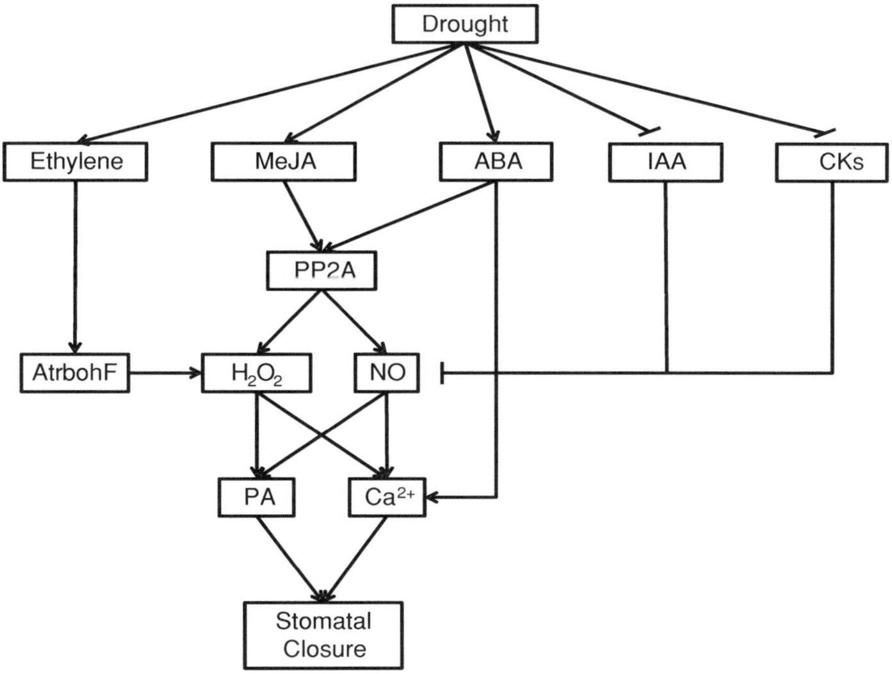

Fig. 13.2 A summary of possible interactions between elements from the stomatal closure pathway involving NO and hormones. Drought stimulates the production of ethylene, methyl jasmonate (MeJA), and abscisic acid (ABA) while reducing the amounts of indole-3-acetic acid (IAA) and cytokinins (CKs). Both IAA and CKs reduce the production of NO, while MeJA and ABA enhance it through the activation of protein phosphatase 2A, along with the production of H_2O_2. Ethylene, through AtrbohF, also increases H_2O_2 amounts. Both NO and H_2O_2 increase the amounts of phosphatidic acid (PA) and cytosolic calcium (Ca^{2+}), leading to stomatal closure. ABA is also capable of increasing Ca^{2+} directly. This diagram was constructed based on several species; therefore, it is highly hypothetical

cycle is an important component of such a system and is enhanced by NO, due to a possible *S*-nitrosylation of its enzymes (Bai et al. 2011). In fact, results obtained in *G. monostachia* showed that in the apical portion of the leaf, where NO is present in drought stressed plants, H_2O_2 levels remain low. The opposite is observed in the basal region (Mioto and Mercier 2013).

13.2.3 Salt Stress

Salt stress leads to many responses similar to drought. *B. juncea* plants treated with the NO donor sodium nitroprusside (SNP) fared better than those left untreated and showed lower levels of H_2O_2, suggesting that NO may act in resistance to salinity by increasing activity of tonoplast H^+-ATPases, which is necessary for driving the Na^+/H^+ exchange (Khan et al. 2012).

Achard et al. (2006) tested an *A. thaliana* mutant deficient in four of the five DELLAs possessed by this species. DELLAs play an important role in restricting growth and regulating development during salt stress but seem to be less effective in metabolic responses, such as expression of proteins with chaperone function. Based on the study of mutants lacking DELLAs which were insensitive to ethylene or ABA, they concluded that DELLA is a convergence point between these two pathways. Salt stress results in DELLA accumulation due to a drop in GA content by enhanced GA-2-oxidase activity (Achard et al. 2008b). Therefore, in response to salt stress, DELLAs could regulate the inhibition of growth in response to ABA, GA, and ethylene. To date, no interaction between DELLAs and NO has been shown during salt stress, but since it is likely to occur in other stresses (Achard et al. 2008a, b; Cantrel et al. 2011), this may also happen during salt stress.

Salicylic acid (SA) also appears to be involved in salt stress response. Application of high concentrations of SA increased H_2O_2 and NO amounts but decreased viability of *S. lycopersicum* root tips (Gémes et al. 2011). The mechanisms underlying these responses, however, are still undefined. Szepesi et al. (2009) detected an increase in aldehyde oxidase and a subsequent increase in ABA production after the application of SA. The authors could not conclude whether this was a direct effect of SA or a consequence of the lowering in the water potential of the tissues. SA application, however, improved the growth under NaCl stress. A schematic representation of possible pathways leading to salt stress response is presented in Fig. 13.3. It is important to highlight that this figure was constructed based on the available information from several species under salt stress and, therefore, it is highly speculative.

Interestingly, Alonso-Ramírez et al. (2009) found strong evidence that GAs can increase SA contents, possibly through degradation of DELLAs, and increase viability of *A. thaliana* seeds exposed to NaCl and paraquat. This work appears to be in conflict with the other works shown, but since GAs and ABA are very important in regulating germination, the stress signaling in seeds could be somewhat different from germinated plants.

13.2.4 Heavy Metal Stress

Heavy metal stresses tend to change the hormonal balance of the roots, resulting in alterations of the root structure. Since NO seems to participate in root branching and elongation, many of the possible interactions between NO and hormones under heavy metal stress are perhaps related to these structural changes of the roots. Also, a common feature of stress originating from all heavy metals is the production of ROS (He et al. 2012a). Again, NO could act directly in ROS scavenging or stimulate enzymes, such as superoxide dismutase and catalase (He et al. 2012a). To date, cadmium (Cd) and aluminum (Al) are the heavy metals with the most

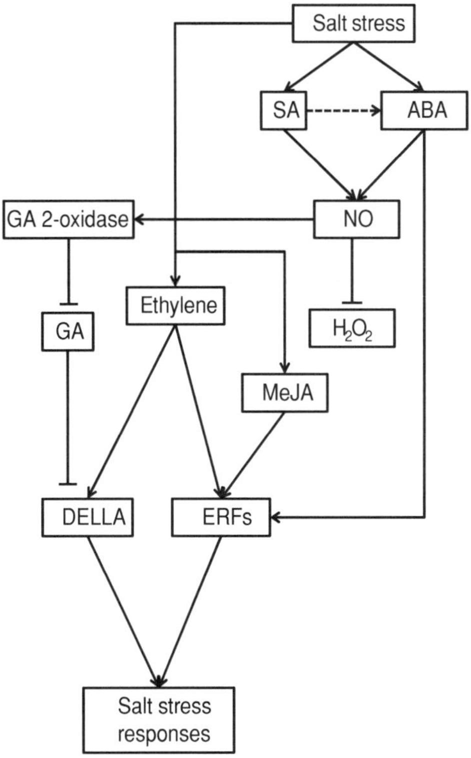

Fig. 13.3 A summary of possible interactions between elements from the salt stress response pathway involving NO and hormones. Dashed arrows indicate possible relations between the components that were not directly described in the literature. Salt stress leads to the production of salicylic acid (SA) and abscisic acid (ABA), both stimulating the production of NO. NO decreases the amounts of H_2O_2 and, through activation of GA-2-oxidase. A reduction in gibberellin (GA) content leads to the accumulation of DELLA proteins, also regulated positively by ethylene. Ethylene response factors (ERFs) are produced by salt stress-induced ABA, ethylene, and methyl jasmonate (MeJA). Accumulation of DELLA and ERFs stimulates salt stress responses. This diagram was constructed based on several species; therefore, it is highly hypothetical

publications related to NO. However, some of the data are controversial, sometimes showing increases and sometimes decreases in NO content in response to heavy metals, indicating that the responses may be much more complex, thereby deserving further analysis.

In pea (*P. sativum*) and *A. thaliana* plants, Cd seems to reduce NO production (Barroso et al. 2006; Rodríguez-Serrano et al. 2009; Zhu et al. 2012; Xu et al. 2011). The level of detected NO depleted with a consequent decrease in GSH, GSNO and in the activity of GSNOR in *P. sativum* (Barroso et al. 2006). Moreover, Cd increased the ROS production, MeJA, and ethylene (Rodríguez-Serrano et al. 2009). Treatment with exogenous GA reduced NO levels in *A. thaliana* and suppressed iron-responsive transporter (IRT1), that transports Cd, resulting in the

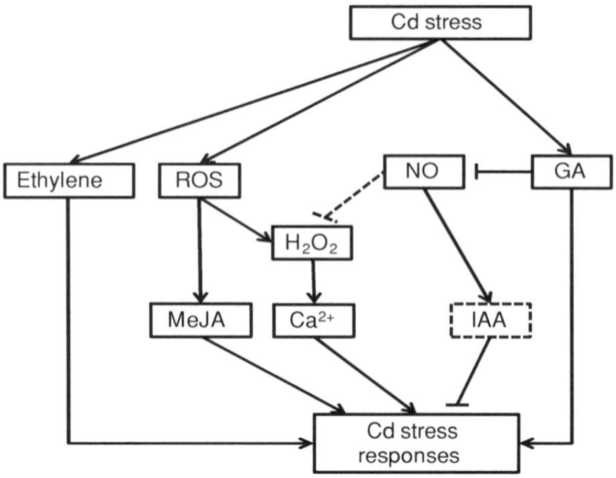

Fig. 13.4 A summary of possible interactions between elements from the cadmium signaling pathway involving NO and hormones. Dashed arrows indicate possible relations between the components that were not directly described in the literature. Cd stress leads to increased amounts of ethylene, gibberellins (GA), and reactive oxygen species (ROS). GA decreases the production of NO, causing a subsequent decrease in indole-3-acetic acid (IAA), therefore leading to Cd stress responses. ROS increases the concentration of cytosolic calcium (Ca^{2+}) through H_2O_2 and the amounts of methyl jasmonate (MeJA), also leading to Cd stress response. NO might reduce the amount of H_2O_2, providing a possible interaction between GA and ROS pathways. This diagram was constructed based on several species; therefore, it is highly hypothetical

reduction of the uptake of this heavy metal (Zhu et al. 2012). In this case, however, the reduction in NO levels could be the consequence of the stress alleviation caused by GA and may not represent a more direct interaction between these molecules. However, it is not difficult to imagine DELLAs as a point of interaction between NO and GA, as already postulated in temperature and salt stresses (Achard et al. 2006). *Medicago truncatula* seedlings also showed a decrease in NO production in response to Cd, and application of NO inactivated IAA-oxidase, possibly reducing IAA degradation (Xu et al. 2011). A possible pathway of Cd responses taking into account these studies is provided in Fig. 13.4. It is important to highlight that this figure was constructed based on the available information from several species submitted to heavy metal stress, and, therefore, it is highly speculative.

Some results, however, show that Cd exposure results in greater NO production. Xu et al. (2011) detected an increase in NO production with a concomitant decrease in Aux-responsive genes in *A. thaliana* seedlings exposed to Cd. De Michele et al. (2009) showed that Cd induces NO production and results in programmed cell death (PCD), and application of L-NMMA, an inhibitor of NOS activity, almost completely inhibited this response, indicating that NO seems to be very important in PCD. Similarly, NO also participates in PCD triggered by different agents, such as pathogen attack, and is often associated with ABA production (Mur et al. 2006).

PCD also seems to be controlled, in some cases, by ethylene, such as during lysigenous aerenchyma formation in rice (Rzewuski and Sauter 2008). Whether NO–ethylene interaction actually does occur in this process remains to be seen, but a synergistic interaction between NO and ethylene is not very commonly observed in plant stress responses.

Regarding Al stress, effects on NO production still need to be substantiated, but it is possible that there is an initial drop in NO during the first hour of exposure followed by an increase in this free radical in the next hours up to a few days (He et al. 2012a). Application of NO reduced aluminum accumulation in both wheat (*Triticum aestivum*) and rye (*Secale cereale*) roots exposed to this heavy metal (He et al. 2012b). The same study also showed that NO is involved in root elongation and, strangely, SNP application increased the levels of both "growth" hormones (e.g., IAA, Zeatin riboside—ZR, and GA) and stress-related ABA. Therefore, in roots, NO may interact with several hormones to promote root growth and also signal stress, but how this interaction occurs is still unclear.

Cu^{2+} excess provokes several changes in *A. thaliana* morphology, and these changes appear to be related to a reduction in IAA-controlled gene expression and a concomitant increase in NO production, possibly via NR (Kolbert et al. 2012). In fact, application of NO resulted in reduced IAA levels, and inhibition of Aux transport (Petó et al. 2011). In the alga *Ulva compressa*, NO is produced in response to high levels of cadmium, leading to Ca^{2+} release in the cytosol and activation of calcium-dependent protein kinases (CDPKs) and calmodulins that lead to gene regulation responses. H_2O_2 is also involved in this event, stimulating both Ca^{2+} and NO production (Gonzáles et al. 2012).

13.3 Concluding Remarks

Although we treated each stress separately in this chapter, it is important to keep in mind that plants are often challenged simultaneously with several stresses. Moreover, there is a wide range of responses common to several stresses (stomatal closure and proline production, for example), indicating a possible common pathway among them. As indicated by Potters et al. (2009), at least in the long term, different stresses may result in very similar results in plant morphology, leading to the assumption that they could all have a common core of responses.

Some of the figures presented in this chapter were put together based on several species, and even among the same species there are differences regarding the details of how each study applied the stresses to the plants. Based on this, it is not possible to ascertain where all the interactions indicated in the figures actually happen. In order to clarify this, a model plant and standardized experimental conditions are necessary so that each study can be directly compared to others and the signaling pathway of each stress can be determined more easily and accurately. On the other hand, it is also important to study all the biodiversity and the uniqueness of each species, as the behavior found in the model plant may not be true for all species. However, it might be easier to discuss results based on a well-established signaling pathway and comparing the pathway to the peculiarities of each species.

References

Abat JK, Deswal R (2009) Differential modulation of *S*-nitrosoproteome of *Brassica juncea* by low temperature: Change in *S*-nitrosylation of Rubisco is responsible for the inactivation of its carboxylase activity. Proteomics 9:4368–4380

Achard P, Cheng H, De Grauwe L et al (2006) Integration of plant responses to environmentally activated phytohormonal signals. Science 311:91–94

Achard P, Gong F, Cheminant S et al (2008a) The cold-inducible CBF1 factor–dependent signaling pathway modulates the accumulation of the growth-repressing DELLA proteins via its effect on gibberellin metabolism. Plant Cell 20:2117–2129

Achard P, Renou JP, Berthomé R et al (2008b) Plant DELLAs restrain growth and promote survival of adversity by reducing the levels of reactive oxygen species. Curr Biol 18:656–660

Alonso-Ramírez A, Rodríguez D, Reyes D et al (2009) Evidence for a role of gibberellins in salicylic acid-modulated early plant responses to abiotic stress in *Arabidopsis* seeds. Plant Physiol 150:1335–1344

Bai X, Yang L, Tian M et al (2011) Nitric oxide enhances desiccation tolerance of recalcitrant *Antiaris toxicaria* seeds via protein *S*-nitrosylation and carbonylation. PLoS ONE 6:1–11

Barroso J, Corpas FJ, Carreras A et al (2006) Localization of *S*-nitrosoglutathione and expression of *S*-nitrosoglutathione reductase in pea plants under cadmium stress. J Exp Bot 57:1785–1793

Beard RA, Anderson DJ, Bufford JL, Tallman G (2012) Heat reduces nitric oxide production required for auxin-mediated gene expression and fate determination in tree tobacco guard cell protoplasts. Plant Physiol 159:1608–1623

Cantrel C, Vazquez T, Puyaubert J et al (2011) Nitric oxide participates in cold-responsive phosphosphingolipid formation and gene expression in *Arabidopsis thaliana*. New Phytol 189:415–427

Chaki M, Valderrama R, Fernández-Ocaña A et al (2011) High temperature triggers the metabolism of *S*-nitrosothiols in sunflower mediating a process of nitrosative stress which provokes the inhibition of ferredoxin–NADP reductase by tyrosine nitration. Plant, Cell Environ 34:1803–1818

Chen C, Xiao YG, Li X, Ni M (2012) Light-regulated stomatal aperture in arabidopsis. Mol Plant 5:556–572

Corpas FJ, Chaki M, Fernández-Ocaña A et al (2008) Metabolism of reactive nitrogen species in pea plants under abiotic stress conditions. Plant Cell Physiol 49:1711–1722

De Michele R, Vurro E, Rigo C et al (2009) Nitric oxide is involved in cadmium-1 induced programmed cell death in *Arabidopsis* suspension cultures. Plant Physiol 150:217–228

Desikan R, Cheung MK, Bright J et al (2004) ABA, hydrogen peroxide and nitric oxide signalling in stomatal guard cells. J Exp Bot 55:205–212

Desikan R, Last K, Harrett-Williams R et al (2006) Ethylene-induced stomatal closure in Arabidopsis occurs via AtrbohF-mediated hydrogen peroxide synthesis. Plant J 47:907–916

Distéfano AM, Scuffi D, García-Mata C et al (2012) Phospholipase Dδ is involved in nitric oxide-induced stomatal closure. Planta 236:1899–1907

Freschi L, Rodrigues MA, Domingues DS et al (2010a) Nitric oxide mediates the hormonal control of Crassulacean acid metabolism expression in young pineapple plants. Plant Physiol 152:1971–1985

Freschi L, Takahashi CA, Cambui C et al (2010b) Specific leaf areas of the tank bromeliad *Guzmania monostachia* perform distinct functions in response to water shortage. J Plant Physiol 167:526–533

Garcia-Mata C, Lamattina L (2001) Nitric oxide induces stomatal closure and enhances the adaptive plant responses against drought stress. Plant Physiol 126:1196–1204

Gémes K, Poór P, Horváth E et al (2011) Cross-talk between salicylic acid and NaCl-generated reactive oxygen species and nitric oxide in tomato during acclimation to high salinity. Physiol Plant 142:179–192

Gonzáles A, Cabrera MA, Henríquez MJ et al (2012) Cross talk among calcium, hydrogen peroxide, and nitric oxide and activation of gene expression involving calmodulins and calcium-dependent protein kinases in *Ulva compressa* exposed to copper excess. Plant Physiol 158:1451–1462

Gupta K, Hincha DK, Mur LAJ (2011) NO way to treat a cold. New Phytol 189:360–363

Hancock JT, Neill SJ, Wilson ID (2012) Nitric oxide and ABA in the control of plant function. Plant Sci 181:555–559

Hao GP, Xing Y, Zhang JH (2008) Role of nitric oxide dependence on nitric oxide synthase-like activity in the water stress signaling of maize seedling. J Integr Plant Biol 50:435–442

He H, Zhan J, He L, Minghua G (2012a) Nitric oxide signaling in aluminum stress in plants. Protoplasma 249:483–492

He HY, He LF, Gu MH, Li XF (2012b) Nitric oxide improves aluminum tolerance by regulating hormonal equilibrium in the root apices of rye and wheat. Plant Sci 183:123–130

Khan MN, Siddiqui MH, Mohammad F, Naeem M (2012) Interactive role of nitric oxide and calcium chloride in enhancing tolerance to salt stress. Nitric Oxide Biol Chem 27:210–218

Kolbert Z, Petö A, Lehotai N et al (2012) Long-term copper (Cu^{2+}) exposure impacts on auxin, nitric oxide (NO) metabolism and morphology of *Arabidopsis thaliana* L. Plant Growth Regul 68:151–159

Kosová K, Prásil IT, Vítámvás P et al (2012) Complex phytohormone responses during the cold acclimation of two wheat cultivars differing in cold tolerance, winter samanta and spring Sandra. J Plant Physiol 169:567–576

Lei Y, Yin C, Li C (2007) Adaptive responses of *Populus przewalskii* to drought stress and SNP application. Acta Physiol Plant 29:519–526

Liu H, Ouyang B, Zhang J et al (2012) Differential modulation of photosynthesis, signaling, and transcriptional regulation between tolerant and sensitive tomato genotypes under cold stress. PLoS ONE 7:1–16

Majláth I, Szalai G, Soós V (2012) Effect of light on the gene expression and hormonal status of winter and spring wheat plants during cold hardening. Physiol Plant 145:296–314

Mioto PT, Mercier H (2013) Abscisic acid and nitric oxide signaling in two different portions of detached leaves of *Guzmania monostachia* with CAM up-regulated by drought. J Plant Physiol 170:996–1002

Mur LA, Kenton P, Atzorn R et al (2006) The outcomes of concentration-specific interactions between salicylate and jasmonate signaling include synergy, antagonism, and oxidative stress leading to cell death. Plant Physiol 140:249–262

Neill SJ, Desikian R, Clarke A, Hancock JT (2002) Nitric oxide is a novel component of abscisic acid signaling in stomatal guard cells. Plant Physiol 128:13–16

Pei ZM, Murata Y, Benning G et al (2000) Calcium channels activated by hydrogen peroxide mediate abscisic acid signalling in guard cells. Nature 406:731–734

Petó A, Lehotai N, Lozano-Juste J et al (2011) Involvement of nitric oxide and auxin in signal transduction of copper-induced morphological responses in *Arabidopsis* seedlings. Ann Bot 108:449–457

Potters G, Pasternak TP, Guisez Y et al (2009) Different stresses, similar morphogenic responses: integrating a plethora of pathways. Plant Cell Environ 32:158–169

Puli MR, Raghavendra AS (2012) Pyrabactin, an ABA agonist, induced stomatal closure and changes in signalling components of guard cells in abaxial epidermis of *Pisum sativum*. J Exp Bot 63:1349–1356

Rodríguez-Serrano M, Romero-Puertas MC, Pazmiño DM et al (2009) Cellular response of pea plants to cadmium toxicity: cross talk between reactive oxygen species, nitric oxide, and calcium. Plant Physiol 150:229–243
Ruelland E, Vaultier MN, Zachowski A, Hurry V (2009) Cold signalling and cold acclimation in plants. Adv Bot Res 49:35–146
Rzewuski G, Sauter M (2008) Ethylene biosynthesis and signaling in rice. Plant Sci 175:32–42
Saidi Y, Finka A, Muriset M (2009) The heat shock response in moss plants is regulated by specific calcium-permeable channels in the plasma membrane. Plant Cell 21:2829–2843
Saidi Y, Finka A, Goloubinoff P (2011) Heat perception and signalling in plants: a tortuous path to thermotolerance. New Phytol 190:556–565
Saito N, Munemasa S, Nakamura Y et al (2008) Roles of RCN1, regulatory a subunit of protein phosphatase 2A, in methyl jasmonate signaling and signal crosstalk between methyl jasmonate and abscisic acid. Plant Cell Physiol 49:1396–1401
Saito N, Nakamura Y, Mori IC, Murata Yoshiyuki (2009) Nitric oxide functions in both methyl jasmonate signaling and abscisic acid signaling in *Arabidopsis* guard cells. Plant Signal Behav 4:119–120
Shao RX, Yang QH, Xin LF, Shangguan ZP (2012) Effects of exogenous nitric oxide and cytokinin on the growth and photosynthesis of wheat seedlings under water deficit. J Food Agric Environ 10:1451–1456
Song L, Ding W, Shen J et al (2008) Nitric oxide mediates abscisic acid induced thermotolerance in the calluses from two ecotypes of reed under heat stress. Plant Sci 175:826–832
Szepesi A, Csiszár J, Gémes K et al (2009) Salicylic acid improves acclimation to salt stress by stimulating abscisic aldehyde oxidase activity and abscisic acid accumulation, and increases Na^+ content in leaves without toxicity symptoms in *Solanum lycopersicum* L. J Plant Physiol 166:914–925
Tian X, Lei Y (2006) Nitric oxide treatment alleviates drought stress in wheat seedlings. Biol Plant 50:775–778
Tomashow MF (2010) Molecular basis of plant cold acclimation: insights gained from studying the CBF cold response pathway. Plant Physiol 154:571–577
Vogel JT, Zarka DG, Van Buskirk HA et al (2005) Roles of the CBF2 and ZAT12 transcription factors in configuring the low temperature transcriptome of *Arabidopsis*. Plant J 41:195–211
Xiao-Ping S, Xi-Gui S (2006) Cytokinin-and auxin-induced stomatal opening is related to the change of nitric oxide levels in guard cells in broad bean. Physiol Plant 128:569–579
Xing H, Tan L, An L et al (2004) Evidence for the involvement of nitric oxide and reactive oxygen species in osmotic stress tolerance of wheat seedlings: Inverse correlation between leaf abscisic acid accumulation and leaf water loss. Plant Growth Regul 42:61–68
Xu J, Wang W, Sun J et al (2011) Involvement of auxin and nitric oxide in plant Cd-stress responses. Plant Soil 346:107–119
Xuan Y, Zhou S, Wang L et al (2010) Nitric oxide functions as a signal and acts upstream of *AtCaM3* in thermotolerance in *Arabidopsis* seedlings. Plant Physiol 153:1895–1906
Zhao MG, Chen L, Zhang LL, Zhang WH (2009) Nitric reductase-dependent nitric oxide production is involved in cold acclimation and freezing tolerance in *Arabidopsis*. Plant Physiol 151:755–767
Zhao R, Xie H, Lv S et al (2013) LeMAPK4 participated in cold-induced ethylene production in tomato fruit. J Sci Food Agric 93:1003–1009
Zhu M, Dai S, Chen S (2012a) The stomata frontline of plant interaction with the environment-perspectives from hormone regulation. Front Biol 7:96–112
Zhu XF, Jiang T, Wang ZW et al (2012b) Gibberellic acid alleviates cadmium toxicity by reducing nitric oxide accumulation and expression of IRT1 in *Arabidopsis thaliana*. J Hazard Mater 239:302–307

Chapter 14
Tolerance of Plants to Abiotic Stress: A Role of Nitric Oxide and Calcium

M. Nasir Khan, Firoz Mohammad, M. Mobin and M. Ali Saqib

Abstract Plants are continuously exposed to changing environmental conditions such as temperature, drought, salinity, heavy metals, etc., which in their extreme limits pose serious threats and set the plants with impaired growth, physiological and biochemical activities that are witnessed by the losses in crop growth and yield. However, to cope with inimical stresses plants are equipped with a series of defense system that help them to perform normally even under stressful conditions. In order to activate the defense system, signaling networks in plants trigger the molecular machinery against that particular stress condition. Calcium (Ca^{2+}) has been proven as one of the important second messengers in eliciting responses to diverse biotic and abiotic stress signals. These stress signals elevate the cytosolic Ca^{2+} concentration which is sensed by Ca^{2+}-binding proteins such as calmodulin, calcium-dependent protein kinases, and calcineurin B-like proteins that initiate downstream events leading to changes in gene expression and plant adaptation to stress tolerance. Nitric oxide (NO) is a molecule with multifaceted roles in plant growth, development, and in the tolerance of plants to biotic and abiotic stresses. Besides, NO is involved in the elevation of cytosolic Ca^{2+} in response to biotic and abiotic stresses. Elevated level of Ca^{2+} concentration not only elicits specific physiological responses to a given signal but also serve to elevate and/or maintain NO generation. The present chapter is focused on the synergistic role of NO and calcium in eliciting responses to abiotic stress signals.

Keywords Calcium mobilization · Calcium-binding proteins · Nitric oxide · Protein kinase · Stress tolerance

---

M. N. Khan (✉) · M. Mobin · M. Ali Saqib
Faculty of Science, Department of Biology, University of Tabuk,
Tabuk 71491, Saudi Arabia
e-mail: khanmn01@gmail.com

F. Mohammad
Plant Physiology Section, Department of Botany, Aligarh Muslim University,
Aligarh 202002, India

M. N. Khan et al. (eds.), *Nitric Oxide in Plants: Metabolism and Role in Stress Physiology*, DOI: 10.1007/978-3-319-06710-0_14,

14.1 Introduction

Nitric oxide (NO) has emerged as an important biologically active molecule with a variety of functions in plants. Although the source of NO synthesis in plants is elusive, the accumulation of NO has been reported in several plant species. In the preceding section, a significant number of evidences have been presented on enzymatic and nonenzymatic sources of NO synthesis in plants. Involvement of NO in plant growth and stress tolerance has been extensively studied. NO is known to induce de-etiolation and an increase in chlorophyll in potato, lettuce, and *Arabidopsis* (Beligni and Lamattina 2000), and in the guard cells of pea leaves (Leshem et al. 1997). NO has been shown to delay senescence (Leshem and Pinchasov 2000) and induces adventitious root formation (Lanteri et al. 2006). NO has been shown to confer resistance against abiotic stresses such as drought, salt, heavy metal, chilling, and ultraviolet-B radiation and acts as an antioxidant and reduces the generation of superoxide formation and lipid peroxidation (Khan et al. 2012). Moreover, NO has also been reported to play a key role as signaling molecule in biotic and abiotic signal transduction pathways in plants.

Calcium (Ca^{2+}) is recognized as an essential nutrient for growth and development of plants and considered as one of the important signaling carriers in all the living organisms. Accumulation of Ca^{2+} in the cell wall helps in the adhesion of cells and plays a pivotal role in fertilization (Digonnet et al. 1998), pollen tube elongation (Pierson et al. 1996), circadian rhythms (Johnson et al. 1995), oxidative stress (Price et al. 1994), and pathogen infection (Xu and Heath 1998). Ca^{2+} increases the plant tissues' resistance to biotic and abiotic stresses and has been recognized as one of the most important ubiquitous second messengers in many signal transduction networks in plants. In plants, most of the cellular Ca^{2+} is sequestered in cytoplasmic organelles such as vacuole, endoplasmic reticulum (ER), or cell wall. However, the cytosolic Ca^{2+} concentration ($[Ca^{2+}]_{cyt}$) has been found to increase in response to various physiological stimuli. Elevation in $[Ca^{2+}]_{cyt}$ is sensed by specific sensors that initiate downstream events leading to changes in gene expression and plant adaptation to stress tolerance. In this way, the plant responds to stresses as individual cells and synergistically as a whole organism.

NO is known to be involved in the mobilization of intracellular Ca^{2+}, while the Ca^{2+} channels have been suggested as a potential NO target (Garcia-Mata and Lamattina 2007). It has been shown that NO contributes to $[Ca^{2+}]_{cyt}$ increases in plant cells exposed to biotic and abiotic stresses including hyperosmotic stresses and elicitors of defense responses (Lamotte et al. 2004, 2006; Vandelle et al. 2006). Moreover, several studies suggested the requirement of Ca^{2+} during NO synthesis in plants (Corpas et al. 2004, 2006; del Río et al. 2004). Holding in perspective the importance of Ca^{2+} and NO, the present chapter is an effort to present an overview of the synergistic role of nitrous oxide and calcium in the tolerance of plants to abiotic stresses.

14.2 Cross Talk Between NO and Calcium

It has been well established that NO acts as a key signaling molecule involved in plant growth, development, stress responses, and programmed cell death. Simple structure, small dimensions, and high diffusivity of NO facilitate its direct effect on second messengers. It has been shown that NO causes elevation in $[Ca^{2+}]_{cyt}$ in plant cells exposed to biotic and abiotic stresses (Lamotte et al. 2006). The movement of Ca^{2+} from intracellular stores to cytosol is energy-independent which is facilitated by calcium channels. On thc other hand, removal of accumulated Ca^{2+} from cytosol to intracellular stores or apoplast is energy-dependent catalyzed by Ca^{2+} pumps such as Ca^{2+}-ATPases and Ca^{2+}/H^{+} antiporters. Moreover, elevated level of $[Ca^{2+}]_{cyt}$ induces NO synthesis in plants via calcium-binding proteins. Thus, synergistic action of NO and calcium ultimately contributes to the tolerance to biotic and abiotic stresses by activating the defense mechanism of plants (Fig. 14.1).

14.2.1 Stress-Induced Ca^{2+} Mobilization by NO

Plants posses a highly conserved signal transduction network, in which Ca^{2+} contributes as one of the most significant second messengers. The Ca^{2+} signaling pathway comprised of three main steps, i.e., generation of Ca^{2+} changes, recognition of these changes and transduction. In plants under the resting state, $[Ca^{2+}]_{cyt}$ is maintained typically at ~200 nM (Bush 1995). Most of the cellular Ca^{2+} is stored in a number of vesicular compartments such as vacuole (100 mM), endoplasmic reticulum (ER) (1 mM), or the cell wall (1 mM). The Ca^{2+} sequestered in these cytoplasmic compartments is released into the cytoplasm in response to a wide range of environmental, developmental, and growth stimuli (Lecourieux et al. 2006; Dodd et al. 2010). A large difference in Ca^{2+} concentration between cytosol and vesicular compartments generates a massive electrochemical gradient across the plasma membrane, ER, and tonoplast that facilitate the energy-independent movement of Ca^{2+} into the cytosol. Movement of Ca^{2+} in and out of cell and organelles is controlled by specific Ca^{2+} channels and pores. The alterations in $[Ca^{2+}]_{cyt}$ in response to a particular stimulus are sensed by Ca^{2+} sensors such as calmodulins (CaMs), CDPKs (Ca^{2+}-dependent protein kinases), or annexins (Berridge et al. 2003), which serve as Ca^{2+}-decoding elements (Fig. 14.1).

The stress-induced changes in the $[Ca^{2+}]_{cyt}$ have been reported as early as in 1997 by Takano et al. and by Gong et al. (1998) in response to water and heat stresses in pea and tobacco, respectively. Durner et al. (1998) and Klessig et al. (2000) were first to report that Ca^{2+} might participate downstream of NO in plant signal transduction pathways. Later on, NO has been reported to act as Ca^{2+}-mobilizing agent and raises $[Ca^{2+}]_{cyt}$ in ABA-induced stomatal closure in guard cells (Garcia-Mata and Lamattina 2007) and in hyperosmotic-stressed and fungal

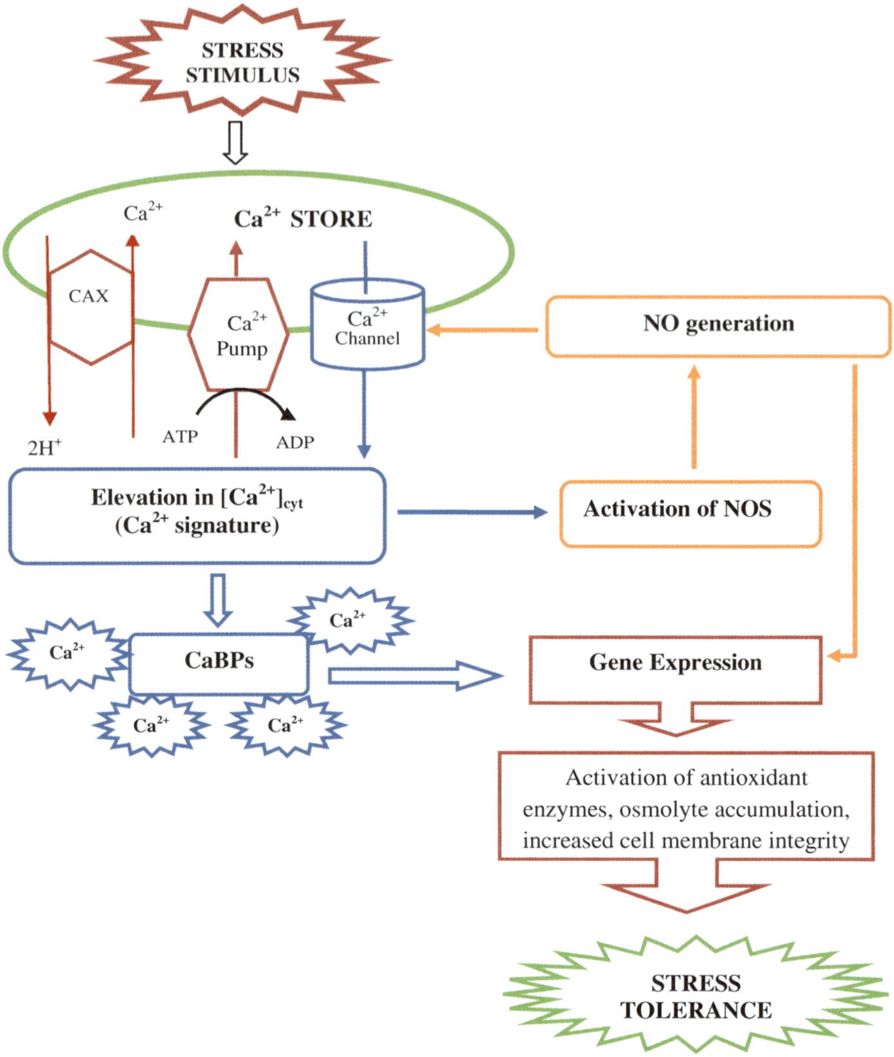

Fig. 14.1 Simplified illustration of synergistic role of nitric oxide (NO) and calcium (Ca^{2+}) in stress tolerance of plants. Perception of stress stimulus causes energy-independent Ca^{2+} mobilization from Ca^{2+} stores to cytosol through Ca^{2+} channels leading to elevation in cytosolic Ca^{2+} concentration ($[Ca^{2+}]_{cyt}$), which is sensed by Ca^{2+}-binding proteins (CaBPs) that initiate downstream events leading to changes in gene expression and plant adaptation to stress tolerance. Elevation in $[Ca^{2+}]_{cyt}$ also induces NO generation through nitric oxide synthase (NOS)-like activity. NO targets Ca^{2+} channels for mobilization of Ca^{2+} from Ca^{2+} stores as well as changes in gene expression leading to adaptation to stress tolerance. Normal functioning of cellular machinery and rapid signal-specific changes in cellular Ca^{2+} in response to stimuli require replenishment of Ca^{2+} stores and a return to resting Ca^{2+} levels, which is accomplished by the Ca^{2+} pump using energy of ATP and through cation exchange (CAX) proteins using the energy of proton gradient

elicitor cryptogein-treated tobacco cells (Gould et al. 2003; Lamotte et al. 2004). Similar results were reported in grapevine cells elicited by *Botrytis cinerea* endopolygalactouronase 1 (Vandelle et al. 2006). Recently, González et al. (2012) observed that rise in $[Ca^{2+}]_{cyt}$ was NO-specific and not associated with decomposition products of NO donor sodium nitroprusside (SNP), as the NO-scavenger cPTIO significantly inhibited the observed NO-mediated elevation in $[Ca^{2+}]_{cyt}$. Thus, based on the available evidences it can be postulated that NO plays an important role in the regulation of Ca^{2+} homeostasis in plant cells (Fig. 14.1).

14.2.2 Mechanism of NO-induced Changes in $[Ca^{2+}]_{cyt}$

Transport of Ca^{2+} in and out of cell and organelles is controlled by specific Ca^{2+} channels. These Ca^{2+} channels have been characterized in the plasma membrane, ER, tonoplast, nuclear, and plastid membranes. NO affects Ca^{2+} channels in two ways: (i) directly, which involves S-nitrosylation—the reversible formation of a covalent bond between a cysteine residue and an NO group and (ii) indirectly, involving cGMP and/or cyclic ADP ribose (cADPR).

Cyclic ADP ribose (cADPR), the Ca^{2+}-mobilizing second messenger synthesized from NAD^+ by ADP-ribosyl cyclase, is considered as the promoter of Ca^{2+} release from intracellular Ca^{2+} stores in a wide variety of animal and plant cells via activation of the Ca^{2+}-permeable channel ryanodine receptors (RYRs) (Allen et al. 1995). In animal system, pharmacological studies suggested RYR-like channels as the main targets of NO action and cADPR as a key intracellular messenger that mediates NO signals (Clementi 1998; Willmott et al. 1996). Klessig et al. (2000) emphasized the involvement of cADPR in the mediation of NO action on defense gene expression in tobacco leaves. They reported that 8-bromo-cADPR, a selective antagonist of cADPR, delays and reduces NO-induced accumulation of pathogenesis-related (*PR*)-1. However, Durner et al. (1998) observed that cADPR inhibitors induce, and RYR inhibitors suppress, the expression of *PR-1*.

NO-induced release from intracellular stores to raise $[Ca^{2+}]_{cyt}$ via cADPR together with cGMP-dependent cascade was also reported in plants. The cGMP is synthesized by the covalent bonding of NO with the heme group of soluble guanylate cyclase (sGC). Garcia-Mata et al. (2003) reported that the rise in $[Ca^{2+}]_{cyt}$ was blocked by the inhibitors of sGC and RYR, which confirms cGMP as a putative mediator for NO-induced activation of cADPR-dependent endomembrane Ca^{2+} channels. It has been observed that Ca^{2+}- and NO-mediated signaling pathways are implicated in the observed ABA inhibition of light-induced stomatal opening in *Vicia faba* guard cells (Garcia-Mata and Lamattina 2007). Moreover, pharmacological experiments suggested that NO is active upstream of $[Ca^{2+}]_{cyt}$ transients during the processes of ABA-induced stomatal closure and auxin-induced adventitious root formation (Lanteri et al. 2006). Lamotte et al. (2004) reported that increase in $[Ca^{2+}]_{cyt}$ of *N. plumbaginifolia* cells was sensitive to ruthenium red (RR), an inhibitor of mammalian RYRs when treated with the NO donor DEA/NO. Lamotte et al. (2006) showed that NO was able to activate both plasma membrane (PM) and intracellular Ca^{2+}-permeable channels via signaling cascades, involving PM depolarization, cADPR, and protein kinases in *N. plumbaginifolia* cells expressing the Ca^{2+} reporter apoaequorin subjected to hyperosmotic stress. On the other hand, Lecourieux et al. (2005) concluded that NO released by DEA/NO could not trigger any change in nuclear-free Ca^{2+} concentration in *N. plumbaginifolia* cells that express the Ca^{2+} reporter apoaequorin in the nucleus. Thus, this finding suggests that the effects of NO on Ca^{2+} homeostasis might be restricted to specific cellular compartments. cGMP has also been shown to activate cyclic nucleotide-gated ion channels (CNGCs), which are permeable to both monovalent and divalent cations and are directly activated by cGMP and/or cAMP leading to the elevation of $[Ca^{2+}]_{cyt}$ (Lemtiri-Chlieh et al. 2004; Bridges et al. 2005).

Another pathway in which Ca^{2+} is released from intracellular stores is through inositol 1,4,5-trisphosphate (IP3)-regulated channels. In animals, it has been reported that NO could activate the enzyme phospholipase C (PLC), which catalyzes the formation of IP3. Thus, PLC activity has been proposed to be part of the NO-dependent pathway that controls $[Ca^{2+}]_{cyt}$ via IP3(Clementi et al. 1995). Lanteri et al. (2006) observed that inhibitors of IP3-regulated Ca^{2+} channels promote a significant reduction in adventitious root formation in NO- and IAA-treated cucumber explants. Although several reports suggest the existence of animal like cADPR/ryanodine and IP3-regulated Ca^{2+} channels in plants, sequence analysis found no homologous proteins in *Arabidopsis thaliana* and rice (Nagata et al. 2004). However, both types of channels have been implicated in different plant processes by biochemical, electrophysiological, and pharmacological studies.

The other major mechanism by which NO affects Ca^{2+} channels is the reversible protein phosphorylation including protein kinases (PKs) and phosphatases. It is well established that in animals, NO modulates the activity of distinct classes of protein kinases, but in plants NO-induced modulation of protein kinase activities has been poorly investigated and most of the studies are based on the artificially synthesized NO. For instance, NO has been shown to induce PKs in *A. thaliana* (Capone et al. 2004), cucumber explants (Lanteri et al. 2006), and tobacco leaves (Klessig et al. 2000). However, none of these PKs has been identified except SIPK (salicylic acid-induced protein kinase), in tobacco (Klessig et al. 2000). Moreover, the involvement of protein kinases in mediating NO-induced changes in $[Ca^{2+}]_{cyt}$ was further testified by using inhibitors of protein kinases, such as K252a and staurosporine, these inhibitors caused reduction in $[Ca^{2+}]_{cyt}$ rise triggered by NO in *Vicia faba* guard cells and tobacco cell suspensions, indicating that protein kinases might be downstream effectors of NO action on $[Ca^{2+}]_{cyt}$ (Sokolovski et al. 2005; Lamotte et al. 2006).

14.3 The Ca^{2+} Signature

The spatial and temporal alteration in cellular Ca^{2+} by the synchronized activity of channels, pumps, and transporters, in response to a stimulus is termed as "Ca^{2+} signature." This Ca^{2+} signature initiates downstream events leading to changes in gene expression and plant adaptation to stress tolerance. In this way, the plant responds to stresses as individual cells and synergistically as a whole organism. The magnitude Ca^{2+} signature to a particular stimulus is specific and it varies with the variation in cell types, and the generation mechanisms for Ca^{2+} signature may be different to different stimuli. The Ca^{2+} signature elicited by a combination of stresses are likely to be different from those evoked by individual stresses.

A signal perception may elevate $[Ca^{2+}]_{cyt}$ levels up to 3 μM by the synchronized action of channels, pumps, and transporters (Scrase-Field and Knight 2003). The unambiguous Ca^{2+} signals for a specific stimulus are defined by spatial and kinetic features such as the magnitude and duration of Ca^{2+} elevation, and whether a single Ca^{2+} transient or multiple spikes occur, duration of spikes, and the lag time between the spikes (Tracy et al. 2008; McAinsh and Pittman 2009). These spatial and temporal patterns of cellular Ca^{2+} changes are expressed as Ca^{2+} signatures that are characteristic for a particular stimulus. Knight et al. (1991) observed the cold- and touch-induced similar calcium kinetics from different calcium sources and locations. Price et al. (1994) described that seedlings can respond to other stimuli, during the refractory periods, which further confirm that distinct signals mobilize calcium from different stores. It has been shown that Ca^{2+} signatures are cell type and organ specific in response to various abiotic stresses. To distinguish tissue-specific differences in Ca^{2+} signature, Kiegle (2000) used different transgenic plants transformed with a gene-encoding aequorin (a reporter gene for Ca^{2+}) targeted to the cytoplasm of the epidermis, endodermis, or pericycle of *Arabidopsis* roots. They noticed an extended oscillation in aequorin luminescence in the endodermis and pericycle that were distinct from the epidermis. It was accomplished that the same stimulus transduced differently depending on the cell type. Moreover, Pauly et al. (2000) and Logan and Knight (2003) deduced that Ca^{2+} signatures in nucleus and mitochondria, respectively, are independent of the cytosolic one.

14.4 Ca^{2+} Sensing and Signaling

The changes in Ca^{2+} concentration (Ca^{2+} signature) are detected by various Ca^{2+} sensor proteins also called as Ca^{2+}-binding proteins (CaBPs), which induce specific responses to a particular stimulus. The specificity of Ca^{2+} signaling is equally maintained by the interplay between Ca^{2+} signatures and Ca^{2+} sensors. The CaBPs have Ca^{2+}-binding sites with high affinity for Ca^{2+}. Upon binding Ca^{2+}, CaBPs undergo conformational and/or enzymatic changes and their subsequent interactions with target proteins can alter enzymatic activities, cytoskeletal orientation, protein phosphorylation cascades, and gene expression leading to stress tolerance or a developmental switch (Tuteja and Mahajan 2007). Most of the CaBPs bind Ca^{2+} using EF (elongation factor)-hand, responsible for the high affinity of CaBPs with Ca^{2+}. The EF-hand is a highly conserved 29 amino acid motif consisting of an α helix E, a loop which binds the Ca^{2+} ion and a second α-helix F (residue 19–29) (Moews and Kretsinger 1975).

14.4.1 Calcium-Binding Proteins (CaBPs)

In plants, two classes of calcium-binding proteins (CaBPs), differing in their Ca^{2+}-induced activation, are present (i) Ca^{2+} sensor relays which include calmodulin (CaM), CaM-like proteins (CMLs), calcineurin B-like proteins (CBLs) and (ii) Ca^{2+} sensor responders such as Ca^{2+}-dependent protein kinases (CDPKs), Ca^{2+}- and Ca^{2+}/CaM-dependent protein kinases (CCaMKs). These sensors are characterized by the presence of EF-hand motif to bind Ca^{2+}. Plants also contain some other CABPs, which lack EF-hand motifs. These include phospholipase D (PLD), annexins, calreticulin, and pistil-expressed Ca^{2+}-binding proteins, which have been found to be involved in Ca^{2+}-dependent signaling pathways. Activated CaBPs can directly bind to the promoters of specific genes and induce or repress their expression.

14.4.1.1 Ca^{2+} Sensor Relays

CaM and CaM-like proteins (*CMLs*): The Ca^{2+} sensor CaM is a small (17 k Da) acidic protein located in the cytosol, nucleus (Van Der Luit et al. 1999), peroxisome (Yang and Poovaiah 2002), and extracellular matrix (Ma et al. 1999). CaM is composed of 4 Ca^{2+}-binding EF-hands (Luan et al. 2002). CaM regulates the activity of target proteins in a calcium-dependent or calcium-independent manner. CaM on binding with Ca^{2+} interacts to other proteins and relays the signal resulting in activation or inactivation of interacting proteins (Zielinski 1998). CaM has been shown to play significant role in biotic and abioitc stress-induced physiological processes. Elevation in CaM level has been observed under high temperature and CaM antagonists have been shown to reduce heat stress tolerance (Liu et al. 2003). Sun et al. (2000) in maize reported that HSP (heat shock protein)-70 binds CaM in a Ca^{2+}-dependent manner. Heat stress-induced increase in $[Ca^{2+}]_{cyt}$ level has been shown to activate *CAM* gene expression and the synthesis of HSP in wheat (Liu et al. 2003). Liu et al. (2007) showed that AtPP7, the first plant protein Ser/Thr phosphatase ever described, interacts with both CaM and the heat shock transcription factor (HSF), which regulates the expression of *HSP* genes. Catalase is the major H_2O_2-scavenging enzyme, which is involved in the degradation of H_2O_2 into water and oxygen. Ca^{2+}/CaM binds to and activates plant catalases, but does not have any effect on catalases from other sources (Yang and Poovaiah 2002).

Changes in expression of CMLs have also been recorded in response to various abiotic stress stimuli. Xu et al. (2011) in Arabidopsis reported an induction in *OsMSR2*, a CML from rice, by multiple abiotic stress stimuli and overexpression of *OsMSR2* in enhanced tolerance to drought and salinity and increased sensitivity to exogenous ABA. Enhanced expression of *CML24* was observed in response to high or low temperature, oxidative stress, and exposure to ABA (Delk et al. 2005). However, repression of *CML24* expression by RNA silencing resulted in decreased sensitivity to ABA as well as enhanced resistance to various metal stresses (Delk et al. 2005). Therefore, it confirms that the CBPs regulate the gene expression either through activation or through repression.

Calcineurin B-like (CBL) proteins and CBL-interacting protein kinases (CIPKs): These are Ca^{2+}/calmodulin-dependent serine–threonine phosphatases which were discovered in *Arabidopsis* (Lu and Zhu 1998; Kudla et al. 1999). Like CaMs, CBL belongs to the family of EF-hand Ca^{2+}-binding proteins. The existence of CBL and CBL-interacting protein kinases (CIPKs) network has been shown to participate in the signal transduction under biotic and abiotic stress conditions (Gu et al. 2008; Yang et al. 2008; Huang et al. 2011). Not long ago, 10 CBLs and 26 CIPKs have been reported in Arabidopsis (Batistic and Kudla 2004). Furthermore, localization of specific CBLs to a specific compartment of the plant cell plays an important role in decoding the spatially distinct Ca^{2+} signatures. CBL4/SOS3 is myristoylated in vitro and associated with microsomal membranes (Ishitani et al. 2000), while CBL1 and CBL9 are targeted to the plasma membrane (Cheong et al. 2007). The specific pattern of localization of CBLs and their interacting kinases facilitates specific decoding of Ca^{2+} signatures, which are differentiated spatially within a given cell (Mahajan et al. 2008).

Batistic and Kudla (2004) observed differential regulation of *CBL* gene expression on exposure to cold, drought, salinity, and ABA. Quan et al. (2007) observed the involvement of CBL10 in association with SOS3 in salt tolerance (Quan et al. 2007). Under salt stress, intramolecular interaction causes inactivation of SOS2/CIPK24, which upon binding to SOS3/CBL4 is released that senses salinity-induced calcium increase. Subsequently, SOS3/CBL4 targets the active kinase to the plasma membrane where it phosphorylates and activates SOS1, leading to Na^{+} extrusion (Gong et al. 2004). Overexpression of *CBL5* gene in *Arabidopsis* induces drought tolerance and *CBL8* and *CIPK15* enhances tolerance to salt stress in rice (Cheong et al. 2010; Xiang et al. 2007). CBL1 regulates plant responses to salt, drought, and cold stress (Cheong et al. 2003). On the other hand, CBL9 acts as a negative regulator of ABA signaling, during germination and early development (Pandey et al. 2004).

14.4.1.2 Ca^{2+} Sensor Responders

Calcium-Dependent Protein Kinase (CDPK): Among the calcium-binding proteins, CDPKs are the most extensively studied calcium signaling kinases. CDPKs have been recognized to play significant role in the stress signaling and hormone-regulated developmental processes in plants. They are positive regulators of abiotic stress responses, and the overexpression of the respective kinase resulted in enhanced stress tolerance in plants (Boudsocq and Sheen 2013). CDPKs were found to be activated within 1–2 min after stress exposure (Boudsocq et al. 2010). They are serine/threonine protein kinases with a C-terminal calmodulin-like domain with four EF-hand motifs responsible to bind directly with Ca^{2+} and N-terminal myristoylation motif for potential association with membranes (Martín and Busconi 2000). Some workers have determined the kinase activity of CDPKs in the absence or presence of varying Ca^{2+} concentration (Dixit and Jayabaskaran 2012) or 14-3-3 proteins (Lachaud et al. 2013). The involvement of CDPKs in stress-induced gene transcription was demonstrated using a maize leaf protoplast transient expression system (Sheen 1996). In *Arabidopsis thaliana,* many of the CDPKs have been shown to act as a positive regulators of ABA-mediated signal transduction pathway under abiotic stress conditions. Mori et al. (2006) reported that AtCPK32 regulates ABA-mediated seed germination, while AtCPK3 and AtCPK6 were reported to be controlling ABA-mediated closure of stomata (Zhu et al. 2007). AtCPK21 and AtCPK23 proteins function as positive regulators in ABA-induced stomatal closure as well as in the regulation of long-term adaptive processes. The mutant lines *cpk21* and *cpk23* display an accumulation of stress-related metabolite and marker genes leading to an increase in stress tolerance (Ma and Wu 2007; Franz et al. 2011). Zou et al. (2010), in Arabidopsis leaves, observed a significant reduction in influx of K^+ with a concomitant rise in CDPK10 concentration within 30 min after drought stress which ultimately contributed to stomatal closure. Recently, Kong et al. (2013) in maize identified 40 CDPK genes as components of maize development and multiple transduction pathways. They observed that 12 CDPK genes were responsive to various stimuli, including salt, drought, and cold, ABA, and H_2O_2.

Calmodulin-dependent protein kinases (CaMKs) and Ca^{2+}/CaM-dependent protein kinases (CCaMKs): Calmodulin-dependent protein kinases (CaMKs) in plants are not as well elucidated as in animals. However, several plant homologs of CaMKs have been cloned from *Arabidopsis* (Liu et al. 2007) apple, maize, rice (Zhang et al. 2003), and wheat (Liu et al. 2003). CaMKs exhibit positive as well as negative effects on transcription factor. Liu et al. (2003, 2008) observed that CaMKs were involved in signal transduction during heat shock response in wheat and Arabidopsis and increases the DNA binding of heat shock transcription factors (Li et al. 2004). On the other hand, Choi et al. (2005) observed that CaMK inhibits CaM-binding transcription factor (CBT)-mediated transcriptional activation in rice.

Ca^{2+}/CaM-dependent protein kinases (CCaMKs): CCaMks, in addition to Ca^{2+}, require CaM for their activity. CCaMKs possess an N-terminal kinase domain followed by two regulatory domains: a CaM-binding domain which overlaps with an auto-inhibitory region and a C-terminal visinin-like domain containing three Ca^{2+}-binding EF-hands. CCaMK upon binding with Ca^{2+} get phosphorylated. This phosphorylated CCaMK stimulates its association with Ca^{2+}/CaM complex which

is required for substrate phosphorylation (Sathyanarayanan and Poovaiah 2004). An up-regulation of *PsCCaMK* (Pandey et al. 2002) and *NtCaMK1* (Zhang et al. 2003) in pea after cold and salt stress was noticed. Similarly, Hua et al. (2004) reported up-regulation of *NtCBK2*, a CaM-binding protein kinase by salt stress and gibberellins. However, very little has been elucidated about involvement of CCaMKs in stress responses. Therefore, further studies are required to explore the function of these proteins in signaling pathways under abiotic stresses.

14.4.2 Other Ca^{2+}-Binding Proteins

The Ca^{2+}-binding proteins discussed in the preceding section are characterized by the presence of Ca^{2+}-binding EF-hand motifs. However, there exist some other Ca^{2+}-binding proteins without EF-hand motifs but play considerable role in stress responses. These proteins have C2 domain and annexin fold.

Phospholipase D (PLD), Ca^{2+}-binding protein which contains a C2 domain, is involved in Ca^{2+}-dependent phospholipid binding (Reddy and Reddy 2004). PLD are involved in ethylene and ABA signaling, α amylase synthesis in aleurone cells, stomatal closure, pathogen responses, leaf senescence, and drought tolerance (Wang 2001). PLD activity is implicated through the generation of phosphatidic acid, which acts as an important second messenger in plant stress responses or by inducing membrane remodeling (Bargmann and Munnik 2006).

Annexins are a family of small proteins with four to eight repeats of approximately 70 amino acids (Clark and Roux 1995) that bind phospholipids in Ca^{2+}-dependent manner (Sathyanarayanan and Poovaiah 2004). Although exact function of annexins is not known, they may regulate target proteins at the plasma membrane to promote stress tolerance. Cantero et al. (2006) reported eight annexin genes *(AnnAt)* in Arabidopsis genome that display differential induction by salinity, dehydration, cold and heat shock. Salt stress enhanced elevation in protein levels of AnnAt1 and its association with plasma membrane, and knockout mutants of AnnAt1 and AnnAt4 exhibit hypersensitive response to osmotic stress and ABA (Lee et al. 2004).

Besides, there also exist several other Ca^{2+}-binding proteins such as calreticulin, ferisomes, pistil-expressed Ca^{2+}-binding proteins (PCP), and calnexins which also play essential role in Ca^{2+} binding during signaling cascade.

14.5 Elevated Levels of $[Ca^{2+}]_{cyt}$ and NO Synthesis

Elevated levels of $[Ca^{2+}]_{cyt}$ not only elicits specific and appropriate physiological responses to a given signal but also serve to elevate and/or maintain NO generation.

In plants, several potential sources of NO exist, including nitrate reductase (NR), nitric oxide synthase (NOS)-like enzymes and also nonenzymatic sources. In animals, NOS catalyzes the conversion of L-arginine to L-citrulline and NO. Although no NOS in plants similar to the mammalian one has been identified so far, however, during the last few years several studies claim the presence of NOS-like activity in plants. Growing bodies of evidence suggest the dependence of plant-NOS activity on Ca^{2+} and CaM (Corpas et al. 2004, 2006; del Rio et al. 2004). The involvement of Ca^{2+} in NOS activity was also confirmed when elicitor-induced NO synthesis in tobacco and grapevine cells was suppressed by the inhibitors of mammalian NOS activities and increases in $[Ca^{2+}]_{cyt}$ (Lamotte et al. 2004; Vandelle et al. 2006). Sang et al. (2008) showed that Ca^{2+}-CaM is required for ABA- and H_2O_2-induced NO production in leaves of maize plants. They also reported that Ca^{2+}- or ABA- or H_2O_2-induced increase in the activity of NOS in maize leaves was blocked by pretreatment with the Ca^{2+} chelator, Ca^{2+} channel blockers, and CaM antagonists. Lamotte et al. (2004) showed that addition of inhibitors of plasma membrane Ca^{2+}-permeable channels in the mid-course of cryptogein-induced NO synthesis in tobacco cell suspensions suppressed NO production within minutes. The findings of Ali et al. (2007) forged ahead our current understanding on the involvement of Ca^{2+} in NOS activity, they showed lipopolysaccharide-induced NO synthesis, which was controlled by an upstream Ca^{2+} influx mediated by the plasma membrane Ca^{2+}-permeable channel CNGC2. NO synthase-dependent NO production was strongly depressed by cadmium and treatment with Ca^{2+} prevented this effect (Rodríguez-Serrano et al. 2009). Thus, based on these reports, it can be concluded that $[Ca^{2+}]_{cyt}$ does not participate directly but changes in $[Ca^{2+}]_{cyt}$ might be involved in mediating NO synthesis in plant cells.

14.6 Ca^{2+} Homeostasis

Signaling-induced elevation in $[Ca^{2+}]_{cyt}$ can chelate negatively charged molecules in the cell and hence can cause cytotoxicity. Therefore, replenishment of intracellular and extracellular Ca^{2+} stores and a return to resting Ca^{2+} levels is necessary for normal functioning of cellular machinery, and to bring about rapid signal-specific changes in cellular Ca^{2+} in response to stimuli. Accomplishment of low $[Ca^{2+}]_{cyt}$ in plant cells requires active (energy-dependent) transport of Ca^{2+} from the cytosol, against electrochemical gradient, to apoplast or intracellular organelles. For this task, plants are equipped with antiporter proteins in plasma membrane and endomembranes which facilitate the removal of Ca^{2+} from cytosol. There are two major groups of antiporter proteins (i). Ca^{2+}-ATPase, uses the energy of ATP to pump Ca^{2+} out of the cytoplasm into organelles such as the vacuole and ER and (ii) calcium exchanger (CAX) proteins or Ca^{2+}/H^+ antiporter, found on the tonoplast membrane. Ca^{2+}/H^+ antiporter exchanges two protons

(H^+) for one Ca^{2+}, using the energy of the proton gradient to sequester cytoplasmic Ca^{2+} in the vacuole (Bose et al. 2011).

Ca^{2+}-ATPases are estimated to represent only <0.1 % of the membrane protein and are thus 30–100-fold less abundant than H^+-ATPases in the plasma membrane (3 %) and the endomembranes (5–10 %) (Tuteja et al. 2007). Ca^{2+}-ATPases are suggested to have high affinity (K_m = 1–10 m M) but low capacity for Ca^{2+} transport, are responsible for maintaining $[Ca^{2+}]_{cyt}$ homeostasis in the resting cells, whereas the Ca^{2+}/H^+ antiporters, which have lower affinities (K_m = 10–15 m M) but high capacities for Ca^{2+} transport, are likely to remove Ca^{2+} from the cytosol during $[Ca^{2+}]_{cyt}$ signals and thereby modulate $[Ca^{2+}]_{cyt}$ (Hirschi 2001). This hypothesis is corroborated by the fact that the Ca^{2+}/H^+ antiporter, but not the vacuolar Ca^{2+}-ATPase, resets $[Ca^{2+}]_{cyt}$ in yeast following hypertonic shock (Denis and Cyert 2002).

Removal of Ca^{2+} from cytosol, by antiporter proteins, is accompanied with several other important functions such as maintenance of low or resting $[Ca^{2+}]_{cyt}$ for cytoplasmic metabolism and to bring about rapid signal-specific changes in cellular Ca^{2+} in response to stimuli, replenishment of intracellular and extracellular Ca^{2+} stores for subsequent $[Ca^{2+}]_{cyt}$ signals, availability of Ca^{2+} in the ER for normal functioning of secretory system and removal of some other divalent cations from cytosol to prevent cytotoxicity (Hirschi 2001).

In order to combat salt stress, plants are fitted with salt overly sensitive (SOS) pathway, constituted by various ion pumps which work on ionic homeostasis strategy (Mahajan et al. 2008). Salt stress-induced elevation in $[Ca^{2+}]_{cyt}$ is sensed by a Ca^{2+} sensor SOS3 which upon interacting with protein kinase SOS2 transduces the signal downstream. This SOS3–SOS2 protein kinase complex phosphorylates, SOS1, a plasma membrane Na^+/H^+ antiporter, resulting in efflux of Na^+ ions (Mahajan et al. 2008). CAX1 (Ca^{2+}/H^+ antiporter) has been identified as an additional target of SOS2 activity reinstating cytosolic Ca^{2+} homeostasis under salt stress (Cheng et al. 2004).

14.7 Conclusion

Perception of stress stimuli triggers a complex series of cellular events leading to tolerance of plant to the perceived stress. NO has been established as an important signaling molecule involved in growth, development, and stress responses in plants. The physical and chemical properties of NO facilitate its direct effect on second messengers that mediates responses to developmental and stress stimuli in plants. Ca is considered as one of the important second messengers in eliciting responses to biotic and abiotic stress signals. Elevation in $[Ca^{2+}]_{cyt}$ in response to stresses acts as the transmission of stress signals to the cellular machinery leading to alterations in gene expression and plant adaptation to stress tolerance. Nitric oxide has been shown to play important role in the elevation of $[Ca^{2+}]_{cyt}$ by mobilizing Ca from intracellular stores. Moreover, elevated levels of Ca not only participate in signaling cascade in response to stress but also play significant role in the synthesis of NO. Although various studies have been suggested the role of NO in Ca mobilization in response to stress and involvement of Ca in NO synthesis, the precise mechanism involved in this complex domain is still baffling. Therefore, further study is needed to unravel underlying molecular events involved in the orchestration of interplay between NO and Ca in plants.

References

Ali R, Ma W, Lemtiri-Chlieh F et al (2007) Death don't have no mercy and neither does calcium: Arabidopsis CYCLIC NUCLEOTIDE GATED CHANNEL2 and innate immunity. Plant Cell 19:1081–1095

Allen GJ, Muir SR, Sanders D (1995) Release of Ca^{2+} from individual plant vacuoles by both InsP3 and cyclic ADP-ribose. Science 268:735–737

Bargmann BO, Munnik T (2006) The role of phospholipase D in plant stress responses. Curr Opin Plant Biol 9:515–522

Batistic O, Kudla J (2004) Integration and channeling of calcium signaling through the CBL calcium sensor/CIPK protein kinase network. Planta 219:915–924

Beligni MV, Lamattina L (2000) Nitric oxide stimulates seed germination and de-etiolation, and inhibits hypocotyl elongation, three light-inducible responses in plants. Planta 210:215–221

Berridge MJ, Bootman MD, Roderick HL (2003) Calcium signalling: Dynamics, homeostasis and remodelling. Nat Rev Mol Cell Biol 4:517–529

Bose J, Pottosin II, Shabala SS et al (2011) Calcium efflux systems in stress signaling and adaptation in plants. Front Plant Sci 2:85

Boudsocq M, Sheen J (2013) CDPKs in immune and stress signaling. Trends Plant Sci 18:30–40

Boudsocq M, Willmann MR, McCormack M et al (2010) Differential innate immune signalling via Ca^{2+} sensor protein kinases. Nature 464:418–422

Bridges D, Fraser ME, Moorhead GBG (2005) Cyclic nucleotide binding proteins in the Arabidopsis thaliana and *Oryza sativa* genomes. BMC Bioinform 6:6

Bush DS (1995) Calcium regulation in plant cells and its role in signaling. Annu Rev Plant Physiol Plant Mol Biol 46:95–122

Cantero A, Barthakur S, Bushart TJ et al (2006) Expression profiling of the *Arabidopsis* annexin gene family during germination, de-etiolation and abiotic stress. Plant Physiol Biochem 44:13–24

Capone R, Tiwari BS, Levine A (2004) Rapid transmission of oxidative and nitrosative stress signals from roots to shoots in Arabidopsis. Plant Physiol Biochem 42:425–428

Cheng NH, Pittman JK, Zhu JK, Hirschi KD (2004) The protein kinase SOS2 activates the *Arabidopsis* H^{+}/Ca^{2+} antiporter CAX1 to integrate calcium transport and salt tolerance. J Biol Chem 279:2922–2926

Cheong YH, Kim KN, Pandey GK et al (2003) CBL1, a calcium sensor that differentially regulates salt, drought, and cold responses in *Arabidopsis*. Plant Cell 15:1833–1845

Cheong YH, Pandey GK, Grant JJ et al (2007) Two calcineurin B-like calcium sensors, interacting with protein kinase CIPK23, regulate leaf transpiration and root potassium uptake in Arabidopsis. Plant J 52:223–239

Cheong YH, Sung SJ, Kim BG et al (2010) Constitutive overexpression of the calcium sensor CBL5 confers osmotic or drought stress tolerance in *Arabidopsis*. Mol Cells 29:159–165

Choi MS, Kim MC, Yoo JH et al (2005) Isolation of a calmodulin-binding transcription factor from rice (*Oryza sativa* L.). J Biol Chem 280:40820–40831

Clark GB, Roux SJ (1995) Annexins of plant cells. Plant Physiol 109:1133–1139

Clementi E (1998) Role of nitric oxide and its intracellular signaling pathways in the control of Ca^{2+} homeostasis. Biochem Pharmacol 55:713–718

Clementi E, Sciorati C, Nistico G (1995) Growth factor-induced Ca^{2+} responses are differentially modulated by nitric oxide via activation of a cyclic GMP-dependent pathway. Mol Pharmacol 48:1068–1077

Corpas FJ, Barroso JB, Carreras A et al (2004) Cellular and subcellular localization of endogenous nitric oxide in young and senescent pea plants. Plant Physiol 136:2722–2733

Corpas FJ, Barroso JB, Carreras A et al (2006) Constitutive arginine-dependent nitric oxide synthase activity in different organs of pea seedlings during plant development. Planta 224:246–254

Rio LA del, Corpas FJ, Barroso JB (2004) Nitric oxide and nitric oxide synthase activity in plants. Phytochem 65: 783–792

Delk NA, Johnson KA, Chowdhury NI, Braam J (2005) CML24, regulated in expression by diverse stimuli, encodes a potential Ca2 + sensor that functions in response to abscisic acid, daylength, and ion stress. Plant Physiol 139:240–253

Denis V, Cyert MS (2002) Internal Ca^{2+} release in yeast is triggered by hypertonic shock and mediated by a TRP channel homologue. J Cell Biol 156:29–34

Digonnet C, Aldon D, Leduc N et al (1998) First evidence of a calcium transient in flowering plants at fertilization. Development 124:2867–2874

Dixit AK, Jayabaskaran C (2012) Phospholipid mediated activation of calcium dependent protein kinase 1 (CaCDPK1) from chickpea: a new paradigm of regulation. PLoS ONE 7:e51591

Dodd AN, Kudla J, Sanders D (2010) The language of calcium signaling. Annu Rev Plant Biol 61:593–620

Durner J, Wendehemme D, Klessig DF (1998) Defense gene induction in tobacco by nitric oxide, cyclic GMP and cyclic ADP-ribose. Proc Natl Acad Sci USA 95:10328–10333

Franz S, Ehlert B, Liese A et al (2011) Calcium dependent protein kinase CPK21 functions in abiotic stress response in *Arabidopsis thaliana*. Mol Plant 4:83–96

Garcia-Mata C, Lamattina L (2007) Abscisic acid (ABA) inhibits light-induced stomatal opening through calcium- and nitric oxide-mediated signalling pathways. Nitric Oxide 17:143–151

Garcia-Mata C, Gay R, Sokolovski S et al (2003) Nitric oxide regulates K^+ and Cl^- channels in guard cells through a subset of abscisic acid-evoked signalling pathways. Proc Natl Acad Sci USA 100:11116–11121

Gong M, Van del Luit AH et al (1998) Heat shock induced change in intracellular calcium level in tobacco seedlings in relation to thermo tolerance. Plant Physiol 116:429–437

Gong D, Guo Y, Schumaker KS, Zhu JK (2004) The SOS3 family of calcium sensors and SOS2 family of protein kinases in *Arabidopsis*. Plant Physiol 134:919–926

González A, de los Ángeles Cabrera M, Josefa Henríquez M et al (2012) Cross talk among calcium, hydrogen peroxide, and nitric oxide and activation of gene expression involving calmodulins and calcium-dependent protein kinases in ulva compressa exposed to copper excess. Plant Physiol 158:1451–1462

Gould KS, Lamotte O, Klinguer A et al (2003) Nitric oxide production in tobacco leaf cells: a generalized stress response? Plant, Cell Environ 26:1851–1862

Gu Z, Ma B, Jiang Y et al (2008) Expression analysis of the calcineurin B-like gene family in rice (*Oryza sativa* L.) under environmental stresses. Gene 415:1–12

Hirschi K (2001) Vacuolar H^+/Ca^{2+} transport: who's directing the traffic? Trends Plant Sci 6:100–104

Hua W, Li RJ, Wang L, Lu YT (2004) A tobacco calmodulin binding protein kinase (NtCBK2) induced by high-salt/GA treatment and its expression during floral development and embryogenesis. Plant Sci 166:1253–1259

Huang C, Ding S, Zhang H et al (2011) CIPK7 is involved in cold response by interacting with CBL1 in *Arabidopsis thaliana*. Plant Sci 181:57–64

Ishitani M, Liu J, Halfter U et al (2000) SOS3 function in plant salt tolerance requires N-myristoylation and calcium binding. Plant Cell 12:1667–1678

Johnson CH, Knight MR, Kondo T et al (1995) Circadian oscillations of cytosolic and chloroplastic free calcium in plants. Science 269:1863–1865
Khan MN, Siddiqui MH, Mohammad F, Naeem M (2012) Interactive role of nitric oxide and calcium chloride in enhancing tolerance to salt stress. Nitric Oxide 27:210–218
Kiegle E, Moore CA, Haseloff J et al (2000) Cell-type-specific calcium responses to drought, salt and cold in the *Arabidopsis* root. Plant J 23:267–278
Klessig DF, Durner J, Noad R et al (2000) Nitric oxide and salicylic acid signaling in plant defense. Proc Natl Acad Sci USA 97:8849–8855
Knight MR, Campbell AK, Smith SM, Trewavas AJ (1991) Transgenic plant aequorin reports the effects of touch and cold-shock and elicitors on cytoplasmic calcium. Nature 352:524–526
Kong X, Lv W, Jiang S et al (2013) Genome-wide identification and expression analysis of calcium-dependent protein kinase in maize. BMC Genom 14:433
Kudla J, Xu Q, Harter K et al (1999) Genes for calcineurin B-like proteins in Arabidopsis are differentially regulated by stress signals. Proc Natl Acad Sci USA 96:4718–4723
Lachaud C, Prigent E, Thuleau P et al (2013) 14-3-3-regulated Ca2 + -dependent protein kinase CPK3 is required for sphingolipid-induced cell death in Arabidopsis. Cell Death Differ 20:209–217
Lamotte O, Gould K, Lecourieux D et al (2004) Analysis of nitric oxide signaling functions in tobacco cells challenged by the elicitor cryptogein. Plant Physiol 135:516–529
Lamotte OC, Courtois G, Dobrowolska A et al (2006) Mechanism of nitric-oxide-induced increase of free cytosolic Ca^{2+} concentration in *Nicotiana plumbaginifolia* cells. Free Radic Biol Med 40:1369–1376
Lanteri M, Pagnussat GC, Lamattina L (2006) Calcium and calcium-dependent protein kinases are involved in nitric oxide and auxin-induced adventitious root formation in cucumber. J Exp Bot 57:1341–1351
Lecourieux D, Lamotte O, Bourque S et al (2005) Elicitors induce specific changes in nuclear free calcium in tobacco cell suspensions. Cell Calcium 38:527–538
Lecourieux D, Ranjeva R, Pugin A (2006) Calcium in plant defence-signalling pathways. New Phytol 171:249–269
Lee S, Lee EJ, Yang EJ et al (2004) Proteomic identification of annexins, calcium-dependent membrane binding proteins that mediate osmotic stress and abscisic acid signal transduction in Arabidopsis. Plant Cell 16:1378–1391
Lemtiri-Chlieh F, Berkowitz GA (2004) Cyclic adenosine monophosphate regulates calcium channels in the plasma membrane of Arabidopsis leaf guard and mesophyll cells. J Biol Chem 279:35306–35312
Leshem YY, Pinchasov Y (2000) Non-invasive photoacoustic spectroscopic determination of relative endogenous nitric oxide and ethylene content stoichiometry during the ripening of strawberries *Fragaria anannasa* (Duch.) and avocados *Persea americana* (Mill.). J Exp Bot 51:1471–1473
Leshem YY, Haramaty E, Iluz D et al (1997) Effect of stress nitric oxide (NO): interaction between chlorophyll fluorescence, galactolipid fluidity and lipoxygenase activity. Plant Physiol Biochem 35:573–579
Li B, Liu HT, Sun DY, Zhou RG (2004) Ca^{2+} and calmodulin modulate DNA-binding activity of maize heat shock transcription factor in vitro. Plant Cell Physiol 45:627–634
Liu J, Zhu JK (1998) A calcium sensor homolog required for plant salt tolerance. Science 280:1943–1945
Liu HT, Li B, Shang Z-L et al (2003) Calmodulin is involved in heat shock signal transduction in Wheat. Plant Physiol 132:1186–1195
Liu HT, Li GL, Chang H et al (2007) Calmodulin-binding protein phosphatase PP7 is involved in thermotolerance in Arabidopsis. Plant, Cell Environ 30:156–164
Liu HT, Gao F, Li G-L et al (2008) The calmodulin-binding protein kinase 3 is part of heat shock signal transduction in *Arabidopsis thaliana*. Plant J 55:760–773
Logan DC, Knight MR (2003) Mitochondrial and cytosolic calcium dynamics are differentially regulated in plants. Plant Physiol 133:21–24

Luan S, Kudla J, Rodriguez Concepcion M et al (2002) Calmodulins and calcineurin B-like proteins: calcium sensors for specific signal response coupling in plants. Plant Cell 14(Suppl. S):S389–S400

Ma SY, Wu WH (2007) AtCPK23 functions in Arabidopsis responses to drought and salt stresses. Plant Mol Biol 65:511–518

Ma L, Xu X, Cui S, Sun D (1999) The presence of a heterotrimeric g protein and its role in signal transduction of extracellular calmodulin in pollen germination and tube growth. Plant Cell 11:1351–1364

Mahajan S, Pandey GK, Tuteja N (2008) Calcium- and salt-stress signaling in plants: Shedding light on SOS pathway. Arch Biochem Biophys 471:146–158

Martín ML, Busconi L (2000) Membrane localization of a rice calcium dependent protein kinase (CDPK) is mediated by myristoylation and palmitoylation. Plant J 24:429–435

McAinsh MR, Pittman JK (2009) Shaping the calcium signature New Phytol 181:275–294

Moews PC, Kretsinger RH (1975) Terbium replacement of calcium in carp muscle calcium-binding parvalbumin: an x-ray crystallographic study. J Mol Biol 9:229–232

Mori IC, Murata Y, Yang Y et al (2006) CDPKs CPK6 and CPK3 function in ABA regulation of guard cell S-type anion- and Ca (2 +)-permeable channels and stomatal closure. PLoS Biol 4:1749–1762

Nagata T, Iizumi S, Satoh K et al (2004) Comparative analysis of plant and animal calcium signal transduction element using plant full-length cDNA data. Mol Biol and Evol 21:1855–1870

Pandey GK, Reddy MK, Sopory SK, Pareek SLS (2002) Calcium homeostasis in plants: Role of calcium binding proteins in abiotic stress tolerance Indian. J Biotech 1:135–137

Pandey GK, Cheong YH, Kim KN et al (2004) The calcium sensor calcineurin B-like 9 modulates abscisic acid sensitivity and biosynthesis in *Arabidopsis*. Plant Cell 16:1912–1924

Pauly N, Knight MR, Thuleau P et al (2000) Control of free calcium in plant cell nuclei. Nature 405:754–755

Pierson ES, Miller DD, Callaham DA et al (1996) Tiplocalized calcium entry fluctuates during pollen tube growth. Developmental Biol 174:160–173

Price AH, Taylor A, Ripley SJ et al (1994) Oxidative signals in tobacco increase cytosolic calcium. Plant Cell 6:1301–1310

Quan R, Lin H, Mendoza I et al (2007) SCABP8/CBL10, a putative calcium sensor, interacts with the protein kinase SOS2 to protect *Arabidopsis* shoots from salt stress. Plant Cell 19:1415–1431

Reddy VS, Reddy ASN (2004) Proteomics of calcium-signaling components in plants. Phytochem 65:1745–1776

Rodríguez-Serrano M, Romero-Puertas MC, Pazmiño DM et al (2009) Cellular response of pea plants to cadmium toxicity: cross talk between reactive oxygen species, nitric oxide, and calcium. Plant Physiol 150:229–243

Sang J, Zhang A, Lin F (2008) Cross-talk between calcium-calmodulin and nitric oxide in abscisic acid signaling in leaves of maize plants. Cell Res 18:577–588

Sathyanarayanan PV, Poovaiah BW (2004) Decoding Ca^{2+} signals in plants. Crit Rev Plant Sci 23:1–11

Scrase-Field SA, Knight MR (2003) Calcium: just a chemical switch? Curr Opin Plant Biol 6:500–506

Sheen J (1996) Ca^{2+}-dependent protein kinases and stress signal transduction in plants. Science 274:1900–1902

Sokolovski S, Hills A, Gay R et al (2005) Protein phosphorylation is a prerequisite for intracellular Ca^{2+} release and ion channel control by nitric oxide and abscisic acid in guard cells. Plant J 43:520–529

Sun X-T, Li B, Zhou G-M et al (2000) Binding of the maize cytosolic Hsp70 to calmodulin, and identification of calmodulin-binding site in Hsp70. Plant Cell Physiol 41:804–810

Takano M, Takahashi H, Suge H (1997) Calcium requirement for the induction of hydrogen of hypotropism abd enhancement of calcium induced curvature by water stress in primary roots of pea *Pisum sativum* L. Plant Cell Physiol 38:385–391

Tracy FE, Gilliham M, Dodd AN et al (2008) NaCl-induced changes in cytosolic free Ca^{2+} in *Arabidopsis thaliana* are heterogeneous and modified by external ionic composition. Plant, Cell Environ 31:1063–1073

Tuteja N, Mahajan S (2007) Calcium signaling network in plants. Plant Sig Behav 2:79–85

Van der Luit AH, Olivari C, Haley A et al (1999) Distinct calcium signaling pathways regulate calmodulin gene expression in tobacco. Plant Physiol 121:705–714

Vandelle E, Poinssot B, Wendehenne D et al (2006) Integrated signalling network involving calcium, nitric oxide, and active oxygen species but not mitogen-activated protein kinases in BcPG1-elicited grapevine defences. Mol Plant Microbe Interact 19:429–440

Wang X (2001) Plant phospholipases. Annu Rev Plant Physiol Plant Mol Biol 52:211–231

Willmott N, Sethi JK, Walseth TF et al (1996) Nitric oxide induced mobilization of intracellular calcium via the cyclic ADP-ribose signalling pathway. J Biol Chem 271:3699–3705

Xiang Y, Huang Y, Xiong L (2007) Characterization of stress responsive CIPK genes in rice for stress tolerance improvement. Plant Physiol 144:1416–1428

Xu H, Heath MC (1998) Role of calcium in signal transduction during the hypersensitive response caused by basidiospore-derived infection of the cowpea rust fungus. Plant Cell 10:585–597

Xu G-Y, Rocha P, Wang M-L et al (2011) A novel rice calmodulin-like gene, *OsMSR2*, enhances drought and salt tolerance and increases ABA sensitivity in Arabidopsis. Planta 234:47–59

Yang T, Poovaiah BW (2002) Hydrogen peroxide homeostasis: activation of plant catalase by calcium/calmodulin. Proc Natl Acad Sci USA 99:4097–4102

Yang W, Kong Z, Omo-Ikerodah E et al (2008) Calcineurin B-like interacting protein kinase OsCIPK23 functions in pollination and drought stress responses in rice (Oryza sativa L.). J Genetics Genomics 35:531.S1–543.S2

Zhang L, Lu YT (2003) Calmodulin-binding protein kinases in plants. Trends Plant Sci 8:123–127

Zhu SY, Yu XC, Wang XJ et al (2007) Two calcium dependent protein kinases, CPK4 and CPK11, regulate abscisic acid signal transduction in Arabidopsis. Plant Cell 19:3019–3036

Zielinski RE (1998) Calmodulin and calmodulin-binding proteins in plants. Ann Rev Plant Physiol Plant Mol Biol 49:697–725

Zou JJ, Wei FJ, Wang C et al (2010) Arabidopsis calcium-dependent protein kinase CPK10 functions in abscisic acid- and Ca^{2+}-mediated stomatal regulation in response to drought stress. Plant Physiol 154:1232–12431

Chapter 15
Abiotic Stress Tolerance in Plants: Exploring the Role of Nitric Oxide and Humic Substances

V. Mora, M. Olaetxea, E. Bacaicoa, R. Baigorri, M. Fuentes, A. M. Zamarreño and J. M. Garcia-Mina

Abstract A number of studies have demonstrated the key role of nitric oxide in the regulation of many fundamental physiological processes that includes plant responses to abiotic and biotic stresses. On the other hand, beneficial action of humic substances on plant growth has been well corroborated, particularly when plants are subjected to abiotic stresses. Furthermore, several recent works have reported the functional links between the plant growth promoting action of humic substances and nitric oxide production and function in plants. In this article, we try to briefly review and discuss the main results showing the relationships between nitric oxide function and humic substances action on plants, also stressing the nitric oxide-dependent involvement of other plant growth regulators, such as auxin, ethylene, abscisic acid, and cytokinins.

Keywords Abiotic stress · Humic acid · Humic substances · Nitric oxide phytohormones · Plant growth

15.1 Introduction

As potential crop yields are closely related to the content of soil organic matter (SOM), in all soil types (MacCarthy et al. 1990; Magdoff and Weil 2004). It seems to be linked to the presence of a specific fraction of SOM, ordinarily known as humus. The relationships among soil fertility, crop yield, and humus were observed

V. Mora · E. Bacaicoa · R. Baigorri · M. Fuentes · A. M. Zamarreño ·
J. M. Garcia-Mina (✉)
CIPAV-Roullier Group, Poligono Arazuri-Orcoyen C/C, 31160 Orcoyen, Spain
e-mail: jgmina@timacagro.es

M. Olaetxea · J. M. Garcia-Mina
Faculty of Sciences, Department of Chemistry and Soil Chemistry, University of Navarra, Pamplona, Spain

M. N. Khan et al. (eds.), *Nitric Oxide in Plants: Metabolism and Role in Stress Physiology*, DOI: 10.1007/978-3-319-06710-0_15,

and quantified in studies from the beginning of modern era (MacCarthy et al. 1990). Since then, many studies have also reported the beneficial action of water-extracted-, and/or alkali-extracted soil humus on the growth and mineral nutrition of diverse plant species, cultivated in soil, inert substrates, and hydroponics (Nardi et al. 2002; MacCarthy et al. 1990; Magdoff and Weil 2004). The present chapter is focused on the key-regulatory role of root-nitric oxide (NO) in the humic substances (HS)-mediated promoting action on plant development.

15.2 Humic Substances

Regarding the chemical nature of humus and its production in soils and other natural environments, it is generally accepted that humic materials are produced in soil by the transformation of fresh organic matter, coming from plants or/and animals, by both randomized chemical and biochemical reactions normally involving soil microbiota, soil enzymes, and some physico-chemical environmental conditions (oxygen availability, soil water content, soil temperature) (Stevenson 1994; Hayes 2009; Huang and Hardie 2009). This transformation-degradation of fresh organic matter is normally known as humification process and can evolve for a long time, even thousands of years (Stevenson 1994). There exist diverse hypothesis about the main steps involved in humification (Huang and Hardie 2009). It is quite clear, however, that the chemical-biological pathways evolving during humification are influenced by the environment, mainly plant species, animal wastes, soil physico-chemical features (pH, texture), soil aeration, soil water content, soil microbiota, and so on (Huang and Hardie 2009). By analogy, composting processes—a specific technique generally used to induce the chemical and biological transformation of fresh organic matter under aerobic conditions—is considered as a kind of accelerated humification process that produces partially humified HS (Haug 1993). From a practical viewpoint, however, the main HS classification, which is normally used in the literature, is based on their water-solubility as a function of pH. Thus, humic acids are that HS fraction soluble at alkaline pH but insoluble at acid pH, while fulvic acids is soluble at all pH values, and humine is insoluble at all pH values (Stevenson 1994).

Two general conceptual physico-chemical models have been proposed in order to explain the main physico-chemical features of HS:

The classical view considers HS as natural polyelectrolytes with a polymer-based structure, which form polydisperse and heterogenous molecular systems with a major macromolecular behavior in solution as a function of pH, ionic strength, or elemental composition (Swift 1989; Clapp and Hayes 1999).

The modern view considers HS as supramolecular structures formed by relatively simple molecules held together through relatively weak,—no covalent-binding forces, such as hydrogen bonds, hydrophobic effect, or van der Waals forces (Piccolo 2001). In this context some studies also stressed the micellar character and surfactant properties of HS in solution (Clapp and Hayes 1999; Wershaw 1999).

More recently, several studies have shown that, in fact, both general views are probably simultaneously present in humic systems (Baigorri et al. 2007a, b), with their relative relevance being a function of the HS solubility features as a function of pH and ionic strength. Thus, humic acids that are insoluble at alkaline-neutral pH and high ionic strength presented a macromolecular behavior in solution, while fulvic acids had a clear supramolecular conformational behavior and humic acids that are soluble at both alkaline-neutral pH and high ionic strength shared both molecular models (Baigorri et al. 2007a, b). However, the fact that HS are operationally defined through a specific methodology of extraction makes potentially possible to obtain "Humic Acid (HA)", "Fulvic Acid (FA)" and "humine" fractions from whatever natural or modified (oxidized, reduced, polymerized) organic material, independently of its real humification degree (wood, biochar, carbohydrates, natural polymers). In this scenario, it would be very useful and convenient to introduce a new nomenclature related to the process origin of "HS". We proposed the following types of HS.

15.2.1 Types of Humic Substances

Organic substances extracted by IHSS-method from organic materials modified or transformed by using diverse alternative or complementary process different from composting, such as controlled pyrolysis (char, biochar) (Schulz et al. 2013), chemical oxidation, and polymerization-coal derived oxy-humates (Jooné et al. 2003), carbohydrate derived-fulvates (Sherry et al. 2013) and phenol-derived humates (Fuentes et al. 2013). These HS may be named Artificial HS (AHS), and their fractions: Artificial humic acids (AHA) and artificial fulvic acids (AFA).

Organic substances extracted by IHSS-method from intact, no-biologically or chemically modified, fresh (living) organic materials, such as plant or animal fresh residues (leaves, whole shoot, root, animal or fish flour, wood, seaweed). These HS may be named Fresh HS (FHS), and their fractions: Fresh humic acids (FHA) and fresh fulvic acids (FFA).

Organic substances extracted by IHSS-method from composted organic materials (Haug 1993). These HS may be named Compost HS (CHS), and their fractions: Compost humic acids (CHA) and compost fulvic acids (CFA).

Organic substances extracted by IHSS-method from naturally humified organic matter with sedimentary origin present in terrestrial (soils, coal, leonardite, peats) and aquatic (lakes, rivers, sea) environments (Stevenson 1994). These HS may be named as sedimentary HS (SHS) and their fractions, sedimentary humic acids (SHA) and sedimentary fulvic acids (SFA).

In principle, only SHS and, by analogy CHS, should properly be considered as real HS according to the diverse definitions of HS, briefly described above. Thus, all of these defined HS as the product of some reaction type (either biological-chemical degradation or simple aggregation) in natural environments.

15.3 Beneficial Effects of HS on Plant Growth and Mineral Nutrition

It is generally accepted that the presence of HS in soils or the application of HS to plants cultivated in soils, inert substrates and hydroponics affect the growth of both root and shoot as well as mineral nutrition. These HS actions are also normally reflected in plant processes related to physiology and yield (Nardi et al. 2002, 2009; Chen and Aviad 1990; Zandonadi et al. 2013). The HS actions on plant growth and mineral nutrition are expressed through diverse, but potentially complementary, effects. These effects are normally classified in two main types: indirect and direct effects.

15.3.1 Indirect Effects of HS

Actions of HS are mainly linked to their size-functional group distribution relationships and structural features, which involve the presence of both aromatic-aliphatic molecular regions and oxygen related functional groups in aliphatic and aromatic domains, with the ability to interact with organic and inorganic molecules and ions present in soil-substrate phases (Chen et al. 2004; Tipping 2002). One of the main HS actions, the effect on nutrient bioavailability, is directly associated with their ability to form complexes with metal cations (Stevenson 1994), which affects the bioavailability of micronutrients (iron, cooper, zinc or manganese); and macronutriens (phosphorus), especially under soil conditions favoring nutrient deficiency (Chen and Aviad 1990; Chen et al. 2004).

15.3.2 Direct Effects of HS

Direct effects of HS action can be explained by both unspecific and/or specific local-effects of HS at plant cell membranes that can trigger molecular and biochemical processes at transcriptional and post-transcriptional levels, in roots and/or shoots. In principle, the specific local-HS effect involves the uptake of HS into the plant. Studies using ^{14}C-labelled HS showed that a small fraction of HS could penetrate into root apoplast area, mainly those with lower apparent molecular weight (MW). This fact can facilitate the action of HS on nutrient uptake molecular systems and signaling pathways in root cell membranes (Vaughan and Malcom 1985; Vaughan 1986; Nardi et al. 2002, 2009). However, the real significance of this type of direct effects remains partially unknown. In this context, it is also possible that the presence of an unspecific action of HS on root or leaf surface might also affect molecular and biochemical processes at transcriptional and post-transcriptional levels, in both root and shoot (Asli and Neumann 2010; Aguirre et al. 2009; Calderín García et al. 2012, 2014; Canellas et al. 2002; Cordeiro et al. 2011; Muscolo et al. 2007; Olaetxea et al. 2012; Quaggiotti et al. 2004; Trevisan et al. 2010).

15.4 Factors Affecting Action of HS on Plant Growth

HS action on plant growth, metabolism and mineral nutrition, also depends on several factors that may be considered as extrinsic or intrinsic.

15.4.1 Extrinsic Factors and HS Action

Main characteristics of the substrate. In general, plants growing on soil (or substrate) conditions involving potential stress are more sensitive to the beneficial action of HS. This is the case for soils or substrates with low organic matter content and low nutrient, principally micronutrient bioavailability (Chen and Aviad 1990; Chen et al. 2004).

Crop type and management. Seed germination or seedling-young plant plantation, tillage or non-tillage, weed control-herbicide application, pest control-pesticide application: In general several studies have shown that HS are more efficient to improve plant growth when they were applied at the first step of plant cycle, mainly at seedling state (Chen and Aviad 1990; Chen et al. 2004; Olk et al. 2013).

Environmental conditions. Although controlled studies involving HS in crops subjected to abiotic stress are scarce, field experiments clearly show that their beneficial effects on yield and quality are more significant in crops cultivated under abiotic stress conditions mainly, limitation in water-availability, drought, salinity, and temperature (Olk et al. 2013; Chen and Aviad 1990; Chen et al. 2004).

Plant species. Several studies have shown that HS-effects are also dependent on the plant species (Tan 2003). However, the studies discussed below show significant effects of HS in both graminaceous and non-graminaceous plants (Olk et al. 2013; Chen and Aviad 1990; Chen et al. 2004).

Type and time of application. Type of application also affects the action of HS (Chen and Aviad 1990; Chen et al. 2004; Olk et al. 2013). At the beginning of plant cycle (seedlings and young plants) or later on, at flowering, fruit setting, or ripening. Results show that HS early application is normally associated with more consistent beneficial effects on plant development. In this sense, HS-effects could be transient with time, showing strong effects in crops suffering stress at the beginning of plant cycle (Chen and Aviad 1990).

15.4.2 Intrinsic Factors and HS Action

Physico-chemical features of HS, mainly structural (aromatic and aliphatic domains), functional (chemical functional groups such as, carboxylic, phenolic, amines, amides), and conformational (molecular size, molecular aggregation, molecular charge, and electrostatic potential at molecular surface) related features (Fuentes et al. 2013; Nardi et al. 2000, 2009; Stevenson 1994).

As for the relative relevance of indirect and direct effects in the plant growth promoting action of HS, some authors proposed that the beneficial action of HS on plant growth is closely related to their ability to improve the plant root uptake of mineral nutrients, principally iron, and zinc (Chen et al. 2004), while others sustain that besides the indirect effects, HS also have the ability to directly affect plant physiology and metabolism (Nardi et al. 2009, 2002; Zandonadi et al. 2013). In any case, it becomes clear that the relative role of indirect and direct effects in HS plant stimulation may depend on many factors mainly coupled with plant-soil (substrate)-HS system experimental-environmental conditions. In fact, some studies reported that in soils with very low organic matter and therefore very low HS in soil solution, like calcareous alkaline soils, the positive effects on the growth of diverse plant species (alfalfa and wheat) were mainly related to improvements in the bioavailability of deficient micronutrients such as iron, copper, zinc, or manganese (García-Mina et al. 2004). This was also the case for plants cultivated in organic acid soils deficient in copper and zinc (García-Mina et al. 2004). In both soil types the direct effects of HS was not significant. However, in the same plant-soil systems the increase in the concentration of applied HS (in this experiment HS concentration was triplicate) could not be only explained by improvements in micronutrient nutrition, thus suggesting the presence of some type of HS direct effect on the root in the rhizosphere (García-Mina 1999). This additional, direct, action of HS on plant growth could be relevant in ecosystems with high dissolved organic-humic matter in soil solution. Likewise, Chen and coworkers also demonstrated the very predominant role of indirect-nutritional HS-effects linked to HS complexing features, in plants cultivated in hydroponics (Chen et al. 2004). However, in soil-plant systems involving the external applications of adequate concentration of HS the involvement of some type of direct effect of HS on plant root, although artificial or induced by agronomical practices, cannot be easily questioned (Olk et al. 2013).

In conclusions, these results show that the relative importance of direct and indirect effects of HS on plant development will be closely related to plant nutritional status (plant nutrient deficiency), nutrient bioavailability in soil, SOM content, external applied HS concentration (for instance in drip irrigation, foliar treatments) and, probably, plant root features. In this context the concentration of HS needed to act on plant growth through indirect (nutritional) effects is rather low, since it is conditioned by the micronutrient level that is critical to allow plant growth. Conversely, HS direct effects are conditioned by the HS concentration needed to mediate plant growth enhancement, which seem to be significantly higher than that needed to indirect (nutritional) effects. So, in general, indirect effects have a clear constitutive role in the dynamics of nutrients and soil-plant systems in natural ecosystems, while the direct effects may have a constitutive role in plant growth regulation in some, specific, natural ecosystems very rich in organic matter, and an induced-role in some crop-production systems, such us drip irrigation.

All these inter-connected levels involved in HS action on plant development are tentatively summarized in Fig. 15.1.

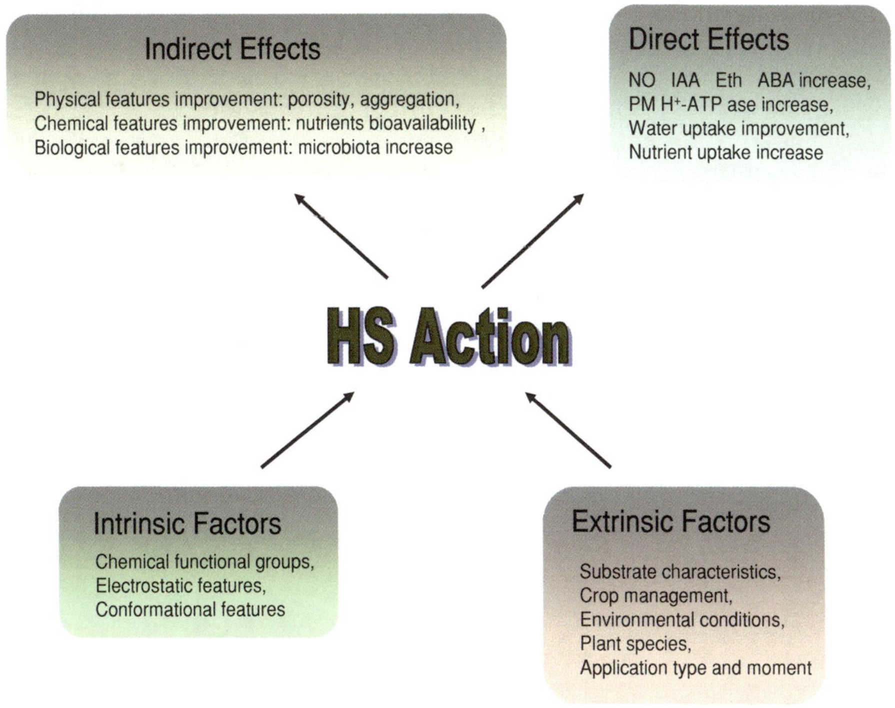

Fig. 15.1 Relationships between HS-effects and structural-chemical features

15.5 Interactive Role of NO, Other Phytohormones and HS in Plant Root- and Shoot-Growth, and Mineral Nutrition

Nitric oxide (NO) is an important bioactive free radical that regulates diverse physiological functions in plant growth and development interacting with plant growth regulators (Lamattina et al. 2003; Palanav-Unsal and Arisan 2009; Mur et al. 2013; Freschi 2013). Recently it was demonstrated that NO signaling is involved in the humic acids (HA) effects on root and shoot development (Mora et al. 2010, 2012, 2013; Zandonadi et al. 2010).

Although the existence, especially under specific experimental or environmental conditions, of HS direct effects on plant physiology and development is generally accepted, the mode of action as well as the mechanisms involved in this HS action are only partially elucidated (Nardi et al. 2002; Trevisan et al. 2010). As will be discussed below, the available results show that a large group of these direct effects are related to improvements in the molecular systems involved in both root uptake and plant use efficiency of several mineral nutrients.

Numerous studies have reported the ability of AHS, CHS, and SHS to affect the transcriptional regulation of gene-networks, mainly transporters, and transcription factors involved in the root uptake and further metabolisms of Fe (Aguirre et al. 2009; Pinton et al. 1999; Billard et al. 2013), Cu (Billard et al. 2013), metal- (Billard et al. 2013), nitrogen-nitrate (Quaggiotti et al. 2004; Pinton et al. 1999; Mora et al. 2010; Jannin et al. 2012), and sulfur (Jannin et al. 2012). These nutritional-linked effects were often reflected in positive effects at molecular regulation of genes involved in diverse aspects of plant metabolism (Aguirre et al. 2009; Trevisan et al. 2010; Jannin et al. 2012) and physiological level (photosynthesis, chloroplast preservation, anti-senescence action, shoot growth) (Merlo et al. 1991; Liu et al. 1998; Azcona et al. 2011; Jannin et al. 2012; Billard et al. 2013). Some studies also showed that the above-mentioned HS-mediated effects were associated with concomitant effects on molecular events regulated by changes in the root- and shoot- concentrations of several phytoregulators (Mora et al. 2010, 2012, 2013).

15.5.1 Interactive Role of NO, Other Phytohormones and HS in Plant Root

NO interacts with different phytohormones, such as auxins (IAA), cytokinins (CK), abscisic acid (ABA), ethylene and others (Freschi 2013). Numerous studies have shown the relevant relationships between the HS ability to affect both root features and biological activity and that of auxins (Quaggiotti et al. 2004; Muscolo et al. 2007; Zandonadi et al. 2007; Canellas et al. 2011; Trevisan et al. 2009); as well as ethylene, another phytohormone closely linked to auxin activity (Mora et al. 2009, 2012, 2013).

Studies carried out by different research groups, involving CHS and SHS as well as diverse plant species (for instance, *Arabidopsis*, tomato, cucumber, maize, and rice) have reported that the application of these HS on the root area affected many molecular and/or biochemical pathways also regulated by auxin activity (Mora et al. 2012, 2013; Zandonadi et al. 2007; Dobbss et al. 2007; Canellas et al. 2008; Calderín García et al. 2012; Trevisan et al. 2009). Thus, a number of studies carried out by Serenella Nardi's group have observed the ability of HS mainly CHS, to affect the expression of auxin-dependent genes in diverse plant species (Trevisan et al. 2010, 2011). These effects mediated by HS were significantly blocked by the presence of inhibitors of auxin transport and functionality (Nardi et al. 1994). These authors also showed that HS-effects on root morphology mimicked those produced by IAA (Trevisan et al. 2010; Muscolo et al. 2013). In line with these results, several studies carried out by Façanha-Canellas's group showed the close relationships between the auxin-like effects of some CHS and CHS-mediated root morphological changes, mainly those related to lateral root proliferation (Dobbss et al. 2007, 2010; Zandonadi et al. 2007; Canellas et al. 2002, 2011). These authors linked HS auxin-like action with the ability of HS to increase root plasma membrane (PM) H^{+}-ATPase activity. In fact the ability of HS to enhance proton pumps has been described by different authors working on diverse plant species and cell organelles (Mora et al. 2010; Zandonadi et al. 2007). Some of these studies also showed the time-course correlation of HS-mediated increase in root PM H^{+}-ATPase and the increase in the root uptake of several mineral nutrients such as nitrogen (as nitrate) (Mora et al. 2010), sulfur (as sulfate) and Fe (Billard et al. 2013; Jannin et al. 2012). These effects were also accompanied by an up-regulation of genes involved in the root transport of nitrate (Jannin et al. 2012), sulfate (Jannin et al. 2012) and Fe (Aguirre et al. 2009; Billard et al. 2013). In this context, some authors suggested that root growth promoting effects of HS could derive from the IAA-mediated increase in root PM H^{+}-ATPase activity in line with the acid growth theory (Hager 2003; Rayle and Cleland 1992).

In line with the above-mentioned results, studies using SHS extracted from peat or leonardite showed that their application to the root of cucumber plants caused a significant, and dose-dependent, increase in the root concentration of IAA (Mora et al. 2009, 2012). Furthermore, SHA root application also produced a significant concomitant increase in the root production of ethylene (Mora et al. 2009, 2012) (Fig. 15.2). However, these studies also showed that, although some SHS-mediated effects on root morphology (mainly lateral root proliferation) are probably IAA- and ethylene-dependents, other effects (such as root dry weight, secondary root number, and primary-secondary root thickness) were not significantly affected by the use of inhibitors of both IAA-functionality (PCIB) and ethylene synthesis (cobalt (II)) and action (STS) (Mora et al. 2012). These findings were well corresponded with Schmidt et al. (2007), who carried out experiments on postembryonic Arabidopsis roots, including water-extractable sphagnum peat humic acids (WSHA), another type of SHA. They demonstrated that WSHA application induced an ensemble of changes in the expression of genes directly involved in cell fate differentiation, which, however, could not be explained by a WSHA-action mediated by auxin- and, probably, ethylene-dependent pathways.

More recently, other studies have shown that all these hormonal-related events observed in the roots treated with several HS types, both CHS and SHS, could be triggered by a HS-mediated enhancement of root NO production. Zandonadi et al. (2010) showed the functional relationships between the HS (a CHA)-induced increase in PM H^{+}-ATPase activity and its ability to both increase the root production of NO and affect IAA-dependent pathways. Additionally, Mora et al. (2009, 2012, 2013) showed the potential functional links between the SH (a SHA)-mediated increase in both IAA and ethylene root concentrations and a previous, transient, increase in NO root production induced by SHA root application in cucumber.

Mora et al. (2009, 2013) observed that the SHA-mediated increase in both IAA and an ethylene root concentration was preceded by a transient increase in NO concentration in diverse root segments (Fig. 15.3). These effects on NO (after 4 h), IAA and ethylene root (after 24 h) concentrations were associated with a significant increase in root dry weight, secondary root density, and root thickness. However, further works involving specific inhibitors of IAA-transport (TIBA) and function (PCIB), ethylene synthesis (cobalt (II)) and function (STS), and a NO-

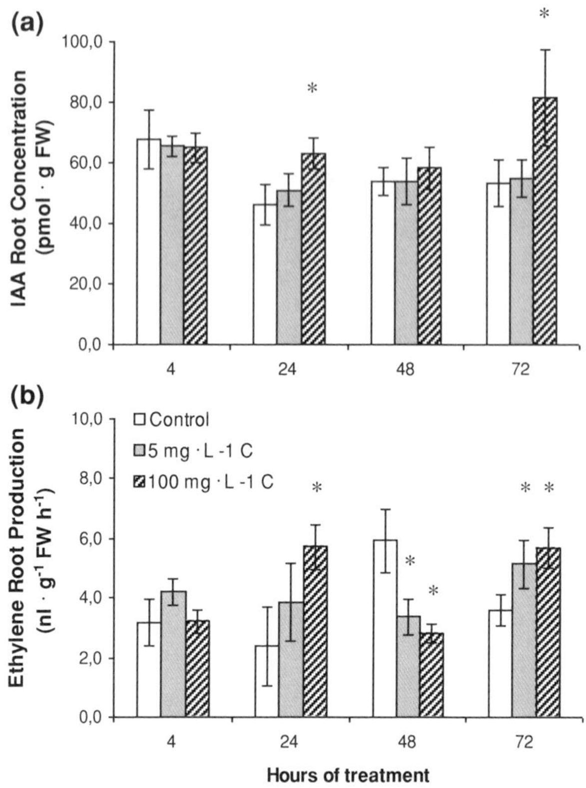

Fig. 15.2 Production of Indole-acetic acid (IAA) (**a**), and ethylene (**b**) in the root of humic acid-treated and control plants. Data are mean ±SE ($n = 5$). Differences between mean values indicated as * are significant ($p \leq 0.05$) according to Tukey-b test. (Mora et al. Unpublished results)

scavenger (PTIO) (Fig. 15.4); showed that SHA-mediated increases in both root dry weight and secondary root density as well as primary and secondary root thickness, in cucumber, were not explained by SHA effects on NO, IAA, and ethylene root concentrations (Mora et al. 2012). Recently, complementary studies indicated that SHA-mediated increase in ethylene root production is IAA-dependent, while the increase caused by SHA in IAA root concentration was expressed through both dependent- and nondependent-NO pathways (Mora et al. 2013). This study also showed that SHA application caused an increase in ABA-root concentration that was IAA and ethylene-dependent (Mora et al. 2013). This effect on ABA concentration in roots caused by SHA might also be involved in the SHS-mediated effects on root development and PM H^+-ATPase activity.

Zandonadi et al. (2010) working with a CHA extracted and purified from a vermicompost of cattle manure, reported that the CHA effects on both root development and PM H^+-ATPase activity were mediated by a NO-dependent pathway in maize seedling roots. These authors studied the effect of CHA, IAA, and NO-precursors (SNP and GSNO) on PM H^+-ATPase activity and some parameters involved in root development (primary root length as well as lateral root emergence rates and density) in maize seedlings. In order, to discriminate

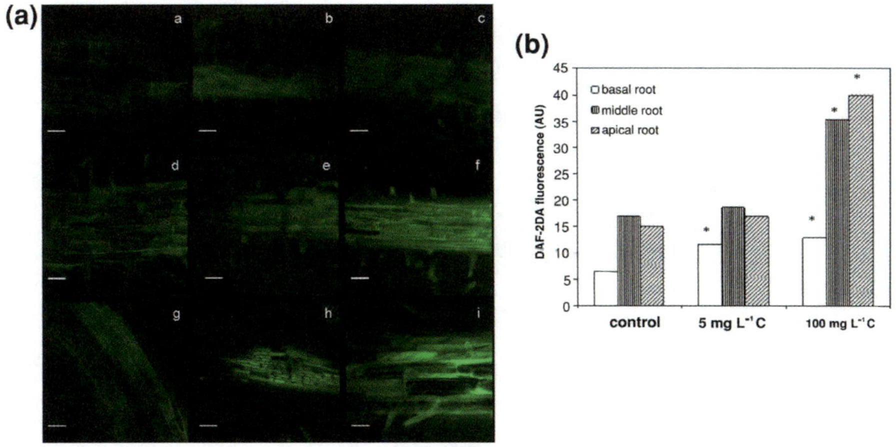

Fig. 15.3 **a** Representative images illustrating the confocal laser scanning microscopy (CLSM) detection of endogenous nitric oxide in cucumber plants roots using 10 μM DAF-2 DA as fluorescent probes at 4 h. Basal root sections of control, 5 and 100 mg L^{-1} C of HA, respectively (*a–c*), middle root sections of control, 5 and 100 mg L^{-1} C of HA, respectively (*d–f*), apical root sections of control, 5 and 100 mg L^{-1} C of HA, respectively (*g–i*). Bar = 50 μm. **b** Nitric oxide in cucumber plants root detected by DAF-2 DA. Photographs were analyzed with ImageJ (NIH) software and fluorescence intensity was estimated by measuring the average pixel intensity. Data are expressed as arbitrary units (AU). (Mora et al. Unpublished results)

among the different effectors and their causal inter-relationships, separated studies were also carried out in the presence of inhibitors of IAA-transport and functionality (TIBA and PCIB), and a NO-scavenger (PTIO). The results showed that CHA-mediated activation of PM H^+-ATPase was both IAA- and NO-dependents. Likewise, the IAA-mediated activation of PM H^+-ATPase was inhibited by IAA-inhibitors and PTIO, thus indicating that this action of IAA is NO-dependent. The authors did not show the effect of IAA-inhibitors on NO-mediated PM H^+-ATPase activation. Besides, IAA-inhibitors and PTIO did not affect primary root length but significantly reduced the CHA-mediated increase in lateral root emergence and density. However, the results of Zandonadi et al. (2010) showed that CHA-treated plants growing in the presence of IAA-inhibitors or the NO-scavenger PTIO have higher lateral root emergence rates and density than control plants growing in the presence of IAA-inhibitors or PTIO. This fact suggests that these specific effects of CHA on root development could also involve other pathways different from those regulated by IAA- and/or NO. These results were in line with the above-mentioned results in cucumber (Mora et al. 2012).

As for the potential crosstalk pathways between NO and the other phytohormones affected by HS, in the case of NO-IAA interactions three main functional pathways have been described (Freschi 2013; Xu et al. 2010; Terrile et al. 2012). A number of studies have shown that IAA can promote NO production in the root of diverse plant species and experimental conditions, such as the NO-dependent

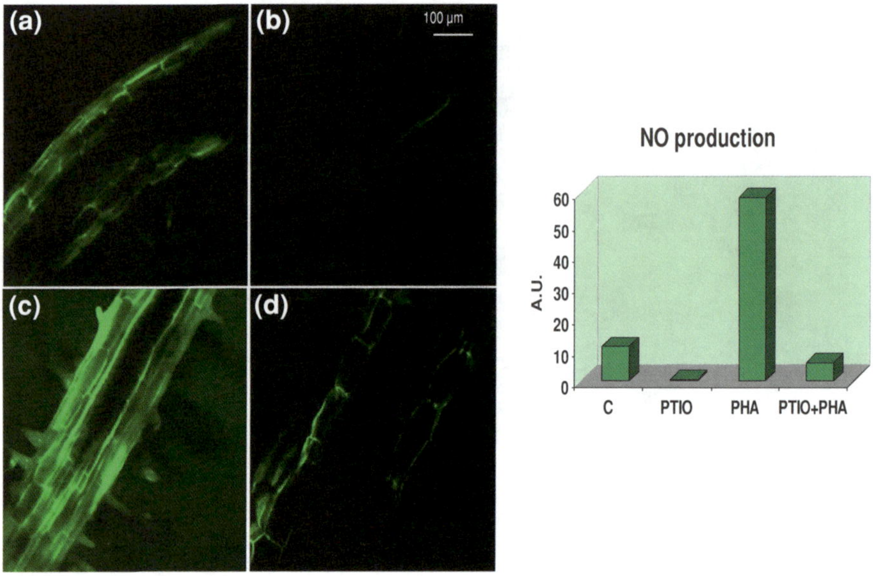

Fig. 15.4 Confocal laser scanning microscopic (CLSM) image showing nitric oxide production in roots of cucumber plants. Root section of control plant (11.06 AU) (**a**), root section of a plant treated with 200 μM PTIO (0.44 AU) (**b**), root section of a plant treated with 100 mg L^{-1} C of PHA (58.43 AU) (**c**), root section of plant treated with PTIO plus PHA (5.73 AU) (**d**). Photographs were analyzed with ImageJ (NIH) software and fluorescence intensity was estimated by measuring the average pixel intensity. Data are expressed as arbitrary units (AU). Scale bar = 100 μm. (Mora et al. Unpublished results)

molecular regulation of Fe (III) chelate reductase by IAA (Chen et al. 2010; Freschi 2013). Other studies reported that NO enhances IAA functional activity through the *S*-nitrosylation of specific moieties (cys-180 and cys-480) of TIR1-region of auxin receptor (Terrile et al. 2012). Finally, NO can also favour IAA accumulation by repressing IAA-oxidase activity (Xu et al. 2010). In the case of the NO-IAA interactions involved in HS action on root development and functionality, the results observed by Mora et al. (2013) are compatible with an effect of SHA-promoted root NO on root IAA concentration by inhibiting IAA degradation via IAA-oxidase. Nevertheless, this mechanism is also compatible with a prior-effect of SHA on root IAA-functionality (IAA root distribution or IAA-receptor sensitivity), and a further effect on IAA-function (TIR1/S- nitrosylation). The results of Mora et al. (2013) also indicated the involvement of a NO-independent pathway in SHA-mediated IAA root increase. In the same way, the results of Zandonadi et al. (2010) showing that the CHA-mediated increase in PM H^+-ATPase activity is IAA-dependent and NO-dependent, and that the IAA-pathway involved in IAA-mediated PM H^+-ATPase activation is also NO-dependent, are also compatible with NO-IAA interactions involving the three above-mentioned mechanisms.

As for NO-ethylene interaction, although the antagonistic action of both molecules has been extensively studied and documented in the case of fruit ripening and abscission as well as leaf or flower senescence (Zhu and Zhou 2007; Freschi 2013), other studies showed that NO can promote ethylene concentration in non-senescent leaf tissues and roots (Freschi 2013; Mur et al. 2013; Fischer 2012). In some specific cases, as the regulation of Fe-deficiency root responses, ethylene could also induce NO production (Garcia et al. 2011). In the case of the HS-mediated increase in root ethylene production described by Mora et al. (2009, 2012, 2013) this effect was IAA-dependent but its causal relationship with NO could not be assessed due to the increase in ethylene caused by the NO-scavenger used in the experiments (PTIO). In any case, it is likely that NO may affect ethylene by an IAA-dependent pathway.

Regarding NO-ABA interactions, several studies have shown that NO is downstream step in ABA signaling pathway in the leaves of plant under osmotic stress (Freschi 2013; Hancock et al. 2011). In these cases, the ABA-mediated NO production seems to be linked to ROS production, mainly H_2O_2 (Freschi 2013; Bright et al. 2006; Zhang et al. 2011). However, in the regulation of specific developmental events, such as seed dormancy breaking, NO and ABA seem to play antagonistic roles (Freschi 2013). However, most of experiments showed the ABA-NO crosstalk in the leaves but not in the root. In the case of the SHA-mediated increase in ABA-root concentration, this process was clearly IAA- and ethylene-dependent. Moreover, its link with the SHA-induced NO root generation is not clear since PTIO application caused a significant increase in ABA concentration of roots of non-SHA treated (control) plants.

In summary, taken together, the above-discussed results indicate that the ability of HA isolated from different naturally (sedimentary) or induced (composted) humified materials, to enhance root growth and modify root architecture involve, among others, the following transcriptional and post-transcriptional events in the root (Table 15.1).

15.5.2 *Interactive Role of NO, Other Phytohormones and HS in Plant Shoot*

Although the studies about the mechanisms responsible for the effects of HS on root development are numerous, those concerning the mechanisms involved in the shoot promoting HS-action are scarce (Mora et al. 2010).

The key role of active cytokinins (CKs) root to shoot translocation in the signal action of nitrate promoting shoot growth has been well established (Sakakibara et al. 2006; Garnica et al. 2010). In this context, Mora et al. (2010) explored the potential relationships between the known effects of HS on root PM H^+-ATPase activity and nitrate root uptake with CKs root-shoot translocation and HS-promoted shoot growth in cucumber. The results showed that the application of a SHA to cucumber roots caused an increase in root PM H^+-ATPase activity and

Table 15.1 Effect of humic acids on transcriptional and post-transcriptional events in root

Hormonal	References	Transcriptional	References	Post-transcriptional	References
An increase in endogenous NO root production	Mora et al. (2012), Zandonadi et al. (2010)	Regulation of genes involved in epidermal cell fate specification	Schmidt et al. (2007)	An increase in root PM H^+-ATPases activities	Zandonadi et al. (2007), Mora et al. (2010)
A NO-dependent and independent increase in IAA root concentration	Mora et al. (2013)	Up-regulation of auxin regulated genes	Trevisan et al. (2009)	An increase in the root Fe(III)-chelate reductase activity	Aguirre et al. (2009)
An IAA-dependent increase in ethylene root production	Mora et al. (2012) Mora et al. (2013)	Up-regulation of genes codifying PM H^+-ATPase protein synthesis	Quaggiotti et al. (2004)	A time-dependent regulation of nitrate reductase activity	Mora et al. (2012)
An IAA- and ethylene-dependent increase in ABA root concentration	Mora et al. (2013)	Up-regulation of genes codifying root Fe(III)-chelate reductase activity	Aguirre et al. (2009)	An increase in nutrient root uptake, mainly nitrate, sulfate, iron, copper, magnesium, manganese	Mora et al. (2010), Billard et al. (2013), Jannin et al. (2012)
		Up-regulation of genes codifying nitrate-, sulfate- iron-, copper-, and metal-transporter protein synthesis	Aguirre et al. (2009), Jannin et al. (2012), Billard et al. (2013)	An increase in root development that is not only NO-, IAA- and ethylene- dependent	Mora et al. (2012), Billard et al. (2013), Jannin et al. (2012), Schmidt et al. (2007)
		Regulation of genes involved in nitrogen, sulfur and carbon metabolism, cell and phytohormone metabolisms, and photosynthesis senescence metabolism	Jannin et al. (2012)		

nitrate root-shoot translocation that was associated with a concomitant increase in CKs root-shoot translocation and CKs leaf concentration. The SHA-mediated action on CKs plant distribution was also linked to changes in polyamine shoots and root concentrations (Mora et al. 2010). These results showed that SHA shoot promoting action are likely linked to those events responsible for the SHA-mediated increase in both root PM H^+-ATPase activity and nitrate root uptake. These effects of SHA were also linked to significant improvements in the root to shoot translocation of the main nutrients, an effect that is consistent with the sink action of CKs in leaves. Further studies carried out in rapeseed showed that the root application of a SHA affected the expression of genes involved in CKs signaling pathways in the shoot, and improved chloroplast functionality delaying senescence (Jannin et al. 2012). These effects were associated with a significant improvement in net photosynthetic rates (Jannin et al. 2012). In fact, the close relationships between HS action in shoot and CKs signaling and function was also supported by the significant similarity between the physiological effects and gene regulation of two SHA application and a seaweed extract very rich in CKs (Billard et al. 2013). HS-mediated effects on shoot development are summarized in Table 15.2.

As discussed above, HS-mediated root PM H^+-ATPase activation is probably regulated by IAA and NO through pathways probably involving ethylene (Mora et al. 2013; Waters et al. 2007; Lucena et al. 2006; Garcia et al. 2011). It was, therefore, possible that these plant growth regulators are also involved in the HS-mediated increase in plant shoot. This hypothesis was explored by Mora et al. (2013). The results showed that the ability of a SHA to enhance shoot growth was removed in the presence of an IAA-inhibitor (PCIB) or a NO-scavenger (PTIO) but was not affected by an inhibitor of ethylene biosynthesis (cobalt (II)). These results clearly indicated that the SHA-mediated action increasing shoot growth was dependent on the effect of SHA in the root promoting NO generation and IAA concentration. These results are consistent with the effects of SHA in the root on PM H^+-ATPase activity and nitrate root-shoot translocation since the HS-mediated actions were also NO- and IAA- dependent (Mora et al. 2010, 2012).

Regarding the potential crosstalk between NO and CKs, experimental evidence show that both phytohormones can act with each other, and work either in a synergetic or antagonistic way, depending on the process studied, the plant physiological state and experimental conditions (Freschi 2013). Anyway, most of these studies reported CKs-NO interactions in leaves but not involving root to shoot cross talk. In this context, although the possibility that the HS-mediated NO generation in the root might be extended to the shoot was not investigated, the causal links between the HS-mediated action on NO and IAA in the root, CKs in the shoot, and the HS shoot growth promoting ability, remains unclear. Anyway, current evidence suggests that these HS-mediated events could be inter-connected by the HS-mediated root PM H^+-ATPase activation.

In summary, these HS-effects on both root and shoot, which evidently may be extended to other actions associated with HS application in plant roots as revealed in many in vitro and in vivo studies (Vaughan and Malcom 1985) are tentatively integrated in Fig. 15.5.

Table 15.2 Effect of humic acids on transcriptional and post-transcriptional events in shoot

Hormonal	References	Transcriptional	References	Post-transcriptional	References
An increase in the root to shoot translocation of active CKs and polyamines	Mora et al. (2010)	Regulation of genes involved in nitrogen, sulfur and carbon metabolism, cell and phytohormone metabolisms, stress control, nutrient transport, photosynthesis senescence metabolism	Jannin et al. (2012)	An increase in nutrient root to shoot translocation and efficiency	Aguirre et al. (2009)
		Regulation of genes involved in CKs signaling pathways	Billard et al. (2013)	An increase in chloroplast number and functionality	Jannin et al. (2012)
				An increase in net photosynthetic rates	Jannin et al. (2012)
				A delay in senescence	Jannin et al. (2012)
				A time-dependent regulation of nitrate reductase activity	Mora et al. (2010)

Fig. 15.5 Integrated effects of HS on plant development. NO is primarily activated by HS on root. This NO activation is followed by a complex auxin/phytohormones pathway concluding in NO_3^- (as signal) and CKs root-to-shoot transport that increases chloroplast number. Finally this process is reflected in better plant development and shoot growth

15.6 Concluding Remarks and Future Perspectives

Current evidence strongly indicates that both NO and ROS play a key role in the expression of HS-beneficial effects on plant growth. These two molecules are directly involved in plant responses under stress conditions. Although controlled experiments on the effects of HS on plants subjected to abiotic stress different from nutrient deficiency are scarce, general experience indicate the beneficial action of HS is more significant and relevant under abiotic stress conditions.

It becomes clear that all above-mentioned events caused by HS in plant root and, thereby, in plant shoot, as well as their main physico-chemical action in root surface have to be related to both the tridimensional conformation (supramolecular and/or macromolecular aggregates) and functional group distribution in HS. However, although many studies have dealt with this topic, consistent knowledge is scarce. This fact is probably due to the structural heterogeneity and complexity of HS. In this sense, further studies are needed to elucidate more in depth the intrinsic and extrinsic mechanisms responsible for the HS-beneficial action on plant-soil systems.

References

Aguirre E, Lemenager D, Bacaicoa E et al (2009) The root application of a purified leonardite humic acid modifies the transcriptional regulation of the main physiological root responses to Fe deficiency in Fe-sufficient cucumber plants. Plant Physiol Biochem 47:215–223

Asli S, Neumann PM (2010) Rhizosphere humic acid interacts with root cell walls to reduce hydraulic conductivity and plant development. Plant Soil 336:313–322

Azcona I, Pascual I, Aguirreolea J, Fuentes M, García-Mina JM, Sánchez-Díaz M (2011) Growth and develpment of pepper are affected by humic subtsances derived from composted sludge. J Plant Nutr Soil Sci 174:916–924

Baigorri R, Fuentes M, Gonzalez-Gaitano G, García-Mina JM (2007a) Simultaneous presence of diverse molecular patterns in humic substances in solution. J Phys Chem B 111:10577–10582

Baigorri R, Fuentes M, González-Gaitano G, García-Mina JM (2007b) Analysis of molecular aggregation in humic substances in solution. Colloid Surface A 302:301–306

Billard V, Etienne P, Jannin L, Garnica M, Cruz F, García-Mina JM, Yvin JC, Ourry A (2013) Two biostimulants derived from algae or humic acid induce similar responses in the mineral content and gene expression of winter oilseed rape (*Brassica napus* L.). J Plant Growth. doi:10.1007/s00344-013-9372-2

Bright J, Desikan R, Hancock JT et al (2006) ABA-induced NO generation and stomatal closure in Arabidopsis are dependent on H_2O_2 synthesis. Plant J 45:113–122

Calderín García A, Azevedo Santos L, Guridi Izquierdo F et al (2012) Vermicompost humic acids as an ecological pathway to protect rice plant against oxidative stress. Ecol Eng 47:203–208

Calderín García A, Azevedo Santos L, Guridi Izquierdo F et al (2014) Potentialities of vermicompost humic acids to alleviate water stress in rice plants (*Oryza sativa* L.). J Geochem Explor 136:48–54

Canellas LP, Olivares FL, Okorokova-Facanha AL, Façanha AR (2002) Humic acids isolated from earthworm compost enhance root elongation, lateral root emergence, and plasma membrane H^+-ATPase activity in maize roots. Plant Physiol 130:1951–1957

Canellas LP, Teixeira Junior LRL, Dobbss LB et al (2008) Humic acids crossinteractions with root and organic acids. Ann Appl Biol 153:157–166
Canellas LP, Dantas DJ, Aguiar NO et al (2011) Probing the hormonal activity of fractionated molecular humic components in tomato auxin mutants. Ann Appl Biol 159:202–211
Chen Y, Aviad T (1990) Effects of humic substances on plant growth. In: MacCarthy P, Malcom RE, Clapp CE, Bloom PR (eds) Humic substances in soil and crop science: selected readings. Americal Society of Agronomy and Soil Science Society of America Madison, USA, pp 161–187
Chen Y, De Nobili M, Aviad T (2004) Stimulatory effects of humic substances on plant growth. In: Magdoff F, Weil R (eds) Soil organic matter in sustainable agriculture. CRC Press, Boca Ratona, pp 103–129
Chen WW, Yang JL, Qin C et al (2010) Nitric oxide acts downstream of auxin to trigger root ferric-chelate reductase activity in response to iron deficiency in *Arabidopsis*. Plant Physiol 154:810–819
Clapp CE, Hayes MHB (1999) Sizes and shapes of humic substances: micelles, subunits, and the mediterranean sun. Soil Sci 164:777–789
Cordeiro FC, Santa-Catarina C, Silveira V, de Souza SR (2011) Humic acid effect on catalase activity and the generation of reactive oxygen species in corn (*Zea mays* L.). Biosci Biotechnol Biochem 75:70–74
Dobbss LB, Medici LO, Peres LEP et al (2007) Changes in root development of *Arabidopsis* promoted by organic matter from oxisols. Ann Appl Biol 151:199–211
Dobbss LB, Canellas LP, Lopes Olivares F et al (2010) Bioactivity of chemically transformed humic matter from vermicompost on plant root growth. J Agric Food Chem 58:3681–3688
Fischer AM (2012) The complex regulation of senescence. Crit Rev Biotechn 31:124–147
Freschi L (2013) Nitric oxide and phytohormone interactions: current status and perspectives. Front Plant Sci 4:398
Fuentes M, Olaetxea M, Baigorri R et al (2013) Main binding sites involved in Fe(III) and Cu(II) complexation in humic-based structures. J Geochem Explor 129:14–17
Garcia MJ, Suarez V, Romera FJ et al (2011) A new model involving ethylene, nitric oxide and Fe to explain the regulation of Fe-acquisition genes in Strategy I plants. Plant Physiol Biochem 49:537–544
García-Mina JM (1999) Efectos de las sustancias húmicas sobre el desarrollo general de las plantas II. Mecanismos básicos de actuación. Agrícola Vergel 18:512–528
García-Mina JM, Antolín MC, Sanchez-Diaz M (2004) Metal-humic complexes and plant micronutrient uptake: a study based on different plant species cultivated in diverse soil types. Plant Soil 258:57–68
Garnica M, Houdusse F, Zamarreno AM, García-Mina JM (2010) The signal effect of nitrate supply enhances active forms of cytokinins and indole acetic content and reduces abscisic acid in wheat plants grown with ammonium. J Plant Physiol 167:1264–1272
Hager A (2003) Role of the plasma membrane H^+-ATPase in auxin-induced elongation growth: historical and new aspects. J Plant Res 116:483–505
Hancock JT, Neill SJ, Wilson ID (2011) Nitric oxide and ABA in the control of plant function. Plant Sci 181:555–559
Haug RT (1993) The practical handbook of compost engineering. Lewis Publishers, USA
Hayes MHB (2009) Evolution of concepts on environmental natural nonliving organic matter. In: Senesi N, Xing B, Huang PM (eds) Biophysico-chemical processes involving natural nonliving organic matter in environmental systems. Wiley, New Jersey, pp 1–39
Huang PM, Hardie AG (2009) Formation mechanisms of humic substances in the environment. In: Senesi N, Xing B, Huang AT (eds) Biophysico-chemical processes involving natural nonliving organic matter in environmental systems. Wiley, New Jersey, pp 41–109
Jannin L, Arkoun M, Ourry A et al (2012) Microarray analysis of humic acid effects on *Brassica napus* growth: Involvement of N, C and S metabolisms. Plant Soil 359:297–319

Jooné GK, Dekker J, van Rensburg CE (2003) Investigation of the immunostimulatory properties of oxihumate. Z Naturforsch C 58:263–267

Lamattina L, Garcia-Mata C, Graziano M, Pagnussat GC (2003) Nitric oxide: the versatility of an extensive signal molecule. Annu Rev Plant Biol 54:109–136

Liu C, Cooper RJ, Bowman DC (1998) Humic acid application affects photosynthesis, root development, and nutrient content of creeping bentgrass. Hort Sci 33:1023–1025

Lucena C, Waters BM, Romera FJ et al (2006) Ethylene could influence ferric reductase, iron transporter, and H^+-ATPase gene expression by affecting FER (or FER-like) gene activity. J Exp Bot 57:4145–4154

MacCarthy P, Clapp CE, Malcom RE, Bloom PR (1990) Humic substances in soil and crop sciences: selected readings. American Society of Agronomy and Soil Science Society of America, Madison

Magdoff F, Weil R (2004) Soil organic matter in sustainable agriculture. CRC Press LLC, USA

Merlo L, Ghisi R, Rascio N, Passera C (1991) Effects of humic substances on carbohydrate metabolism of maize leaves. Can J Plant Sci 71:419–425

Mora V, Bacaicoa E, Zamarreño AM et al (2009) Relationships among the growth promoting action of a purified humic acid and its effects on the nutritional status and hormonal balance of cucumber plants cultivated in hydroponics. In: XVIII Reunión de la Sociedad Española de Fisiología Vegetal—XI Congreso Hispano-Luso de Fisiología Vegetal, Zaragoza, España, 8–11 September 2009

Mora V, Bacaicoa E, Zamarreno AM et al (2010) Action of humic acid on promotion of cucumber shoot growth involves nitrate-related changes associated with the root-to-shoot distribution of cytokinins, polyamines and mineral nutrients. J Plant Physiol 167:633–642

Mora V, Baigorri R, Bacaicoa E et al (2012) The humic acid-induced changes in the root concentration of nitric oxide, IAA and ethylene do not explain the changes in root architecture caused by humic acid in cucumber. Environ Exp Bot 76:24–32

Mora V, Bacaicoa E, Baigorri R et al (2013) NO and IAA key regulators in the shoot growth promoting action of humic acid in *Cucumis sativus* L. J Plant Growth Regul. doi:10.1007/s00344-013-9394-9

Mur LA, Mandon J, Persijn S et al (2013) Nitric oxide in plants: an assessment of the current state of knowledge. AoB Plants 5: pls052

Muscolo A, Sidari M, Francioso O et al (2007) The auxin-like activity of humic substances is related to membrane interactions in carrot cell cultures. J Chem Ecol 33:115–129

Muscolo A, Sidari M, Nardi S (2013) Humic substance: Relationship between structure and activity. Deeper information suggest univocal findings. J Geochem Explor 129:57–63

Nardi S, Panuccio MR, Abenavoli MR, Muscolo A (1994) Auxin-like effect of humic substances extracted from faeces of *Allolobophora caliginosa* and *Allolobophora rosea*. Soil Biol Biochem 23:833–836

Nardi S, Pizzeghello D, Reniero F, Rascio N (2000) Chemical and biochemical properties of humic substances isolated from forest soils and plant growth. Soil Sci Soc Am 64:639–645

Nardi S, Pizzeghello D, Muscolo A, Vianello A (2002) Physiological effects of humic substances on higher plants. Soil Biol Biochem 34:1527–1536

Nardi S, Carletti P, Pizzeghello D, Muscolo A (2009) Biological activities of humic substances. In: Senesi N, Xing B, Huang PM (eds) Biophysico-chemical processes involving natural nonliving organic matter in environmental systems. Wiley, New Jersey

Olaetxea M, Mora V, Baigorri R et al (2012) Plant growth and water uptake effects promoted by humic acid. In: The 16th meeting of the international humic substances society: functions of natural Organic matter in changing environment, Hangzhou, China, 9–14 September 2012

Olk DC, Dana DL, Callaway C, Raske M (2013) On-farm evaluation of a humic product in lowa (U.S.) maize production. In: Xu J, Wu J, He W (eds) Functions of natural organic matter in a changing enviroment. Springer, Dordrecht, pp 1047–1050

Palanav-Unsal N, Arisan D (2009) Nitric oxide signalling in plants. Bot Rev 75:203–229

Piccolo A (2001) The supramolecular structure of humic substances. Soil Sci 166:810–832

Pinton R, Cesco S, Iacolettig G et al (1999) Modulation of NO_3^- uptake by water-extractable humic substances: involvement of root plasma membrane H^+-ATPase. Plant Soil 215:155–161
Quaggiotti S, Ruperti B, Pizzeghello D et al (2004) Effect of low molecular size humic substances on nitrate uptake and expression of genes involved in nitrate transport in maize (*Zea mays* L.). J Exp Bot 55:803–813
Rayle DL, Cleland RE (1992) The Acid Growth Theory of auxin-induced cell elongation is alive and well. Plant Physiol 99:1271–1274
Sakakibara H, Takei K, Hirose N (2006) Interactions between nitrogen and cytokinin in the regulation of metabolism and development. Trends Plant Sci 11:440–448
Schmidt W, Santi S, Pinton R, Varanini Z (2007) Water-extractable humic substances alter root development and epidermal cell pattern in *Arabidopsis*. Plant Soil 300:259–267
Schulz H, Dunst G, Glaser B (2013) Positive effects of composted biochar on plant growth and soil fertility. Agron Sustain Dev 33:817–827
Sherry L, Millhouse E, Lappin DF et al (2013) Investigating the biological properties of carbohydrate derived fulvic acid (CHD-FA) as a potential novel therapy for the management of oral biofilm infections. BMC Oral Health 13:47
Stevenson FJ (1994) Humus chemistry: genesis, composition, reactions. Wiley, USA
Swift RS (1989) Molecular weight, size, shape, and charge characteristics of humic substances: some basic considerations. In: Hayes MHB, MacCarthy P, Malcom RE, Swift RS (eds) Humic substances: II. Wiley, New York, pp 449–465
Tan KH (2003) Humic matter in soil and the environment. Principles and controversies. Marcel Dekker Inc, New York
Terrile MC, Paris R, Calderon-Villalobos LI et al (2012) Nitric oxide influences auxin signaling through S-nitrosylation of the Arabidopsis TRANSPORT INHIBITOR RESPONSE 1 auxin receptor. Plant J 70:492–500
Tipping E (2002) Cation binding by humic substances. Cambridge University Press, Cambridge
Trevisan S, Pizzeghello D, Ruperti B et al (2009) Humic substances induce lateral root formation and expression of the early auxin-responsive IAA19 gene and DR5 synthetic element in Arabidopsis. Plant Biol 12:604–614
Trevisan S, Francioso O, Quaggiotti S, Nardi S (2010) Humic substances biological activity at the plant-soil interface: from environmental aspects to molecular factors. Plant Signal Behav 5:635–643
Trevisan S, Botton A, Vaccaro S et al (2011) Humic substances affect *Arabidopsis* physiology by altering the expression of genes involved in primary metabolism, growth and development. Environ Exp Bot 74:45–55
Vaughan D (1986) Effetto delle sostanze umiche sui processi metabolici delle piante. In: Burns RG, Dell'Agnola G, Miele S et al (eds) Sostanze Umiche effetti sul terreno e sulle piante. Ramo Editoriale degli Agricoltori, Roma, pp 59–81
Vaughan D, Malcom RE (1985) Influence of humic substances on growth and physiological processes. In: Vaughan D, Malcom RE (eds) Soil organic matter and biological activity. Martinus Nijhoff/Junk W Dordrecht, The Netherlands, pp 37–76
Waters BM, Lucena C, Romera FJ et al (2007) Ethylene involvement in the regulation of the $H(^+)$-ATPase CsHA1 gene and of the new isolated ferric reductase CsFRO1 and iron transporter CsIRT1 genes in cucumber plants. Plant Physiol Biochem 45:293–301
Wershaw RL (1999) Molecular aggregation of humic substances. Soil Sci 164:803–813
Xu J, Wang WY, Yin HX et al (2010) Exogenous nitric oxide imporves antioxidative capacity and reduces auxin degradation in roots of *Medicago truncatula* seedlings under cadmiun stress. Plant Soil 326:321–330
Zandonadi DB, Canellas LP, Façanha AR (2007) Indolacetic and humic acids induce lateral root development through a concerted plasmalemma and tonoplast H^+ pumps activation. Planta 225:1583–1595
Zandonadi DB, Santos MP, Dobbss LB et al (2010) Nitric oxide mediates humic acids-induced root development and plasma membrane H^+-ATPase activation. Planta 231:1025–1036

Zandonadi DB, Santos MP, Busato JG et al (2013) Plant physiology as affected by humified organic matter. Theor Exp Plant Physiol 25:12–25

Zhang A, Zhang J, Ye N et al (2011) Nitric oxide mediates brassinosteroid-induced ABA biosynthesis involved in oxidative stress tolerance in maize leaves. Plant Cell Physiol 52:181–192

Zhu S, Zhou J (2007) Effect of nitric oxide on ethylene production in strawberry fruit during storage. Food Chem 100:1517–1522

Chapter 16
Nitric Oxide in Relation to Plant Signaling and Defense Responses

Mui-Yun Wong, Mansour Salati and Yee-Min Kwan

Abstract Nitric oxide (NO) is an important signaling molecule in a biological system as the molecule is involved in various aspects of plant growth and responses to both abiotic and biotic stresses. Abiotic stress is caused by drought, high or low temperature, high salinity, heavy metals and oxidative stress while biotic stress is caused by pathogens and herbivores. This chapter will focus on the role of NO in plant defense responses against pathogen attack including the role of chitosan as an inducer of the signaling pathway leading to these defense responses. The crosstalk between abiotic and biotic stress responses is also discussed.

Keywords Chitosan · Crosstalk · Elicitors · Hypersensitive response · PR proteins · Transcription factors

16.1 Introduction

Plant defense responses include preformed and inducible. Preformed defense encompasses the presence of preformed peptides, e.g., defensins (Broekaert et al. 1995); protease inhibitors, e.g., cysteine protease inhibitor (Joshi et al. 1999). Preformed antimicrobial compounds are used by plants to deter pathogen infection (Filippone et al. 1999). Upon pathogen attack, both host and nonhost plants are able to detect elicitors produced by the pathogens or degraded plant cell wall components, thus, elicit inducible defense responses. Such defense responses

M.-Y. Wong (✉) · Y.-M. Kwan
Faculty of Agriculture, Department of Plant Protection, Institute of Tropical Agriculture, Universiti Putra Malaysia, 43400 Serdang, Selangor, Malaysia
e-mail: muiyun@upm.edu.my

M. Salati
Department of Plant Protection, Golestan Agricultural and Natural Resources Research Centre, Golestan, Iran

M. N. Khan et al. (eds.), *Nitric Oxide in Plants: Metabolism and Role in Stress Physiology*, DOI: 10.1007/978-3-319-06710-0_16,

include cell wall fortification with hydroxyproline-rich glycoprotein (HRGP) accumulation (Smallwood et al. 1994), formation of papillae (Bayles et al. 1990), callose deposition (Parker et al. 1993), extensins accumulation (Showalter 1993), elevation of lignin content (Tilburzy and Reisener 1990; Moerschbacher et al. 1990), activation of systemic acquired resistance (SAR), accumulation of salicylic acid (SA), activation of pathogenesis-related (PR) proteins, accumulation of antimicrobial compounds such as phytoalexins (Abel 2003; Modolo et al. 2002) and cysteine-rich thionins (Epple et al. 1997), and hypersensitive cell death response (HR) (D'Silva et al. 1998; del Pozo and Lam 1998). Some elicitors that had been identified in fungi and bacteria are summarized in Table 16.1.

16.2 Induction of Nitric Oxide Signaling Pathway by Chitosan

Chitosan is a biopolymer consists of repeated units of β-1,4 glucosamines with unique properties such as biodegradable, non-toxic and biocompatible which could be used as health food, feed additive, antimicrobial agent, plant growth stimulator, and elicitor of plant immune system (Yin et al. 2010). For effective biological applications, chitosan needs to be made water-soluble by hydrolyzing the biopolymer into oligochitosan. The main source of chitosan is the shell of marine crustaceans such as shrimps, crabs, and crayfishes (Kurita 2006).

Many researchers had reported on the antimicrobial activity of chitosan against pathogenic fungi including *Alternaria alternate*, *Botryitis cinerea*, *Colletotrichum gloeosporioides*, *Rhizopus stolonifer*, *Penicillium digitatum* (El Ghaouth et al. 1992, Bautista-Banos et al. 2006), *Sclerotinia sclerotiorum* (Cheah et al. 1997), *Fusarium oxysporum* f. sp. *radices-lycopersici* (Benhamou 1992), *Fusarium oxysporum* f. sp. *cubense* (Al-Hetar et al. 2011); against pathogenic bacteria including food borne Gram negative bacteria (Helander et al. 2001); and against viruses and viroids including *alfalfa mosaic virus* (Pospiezny et al. 1991) and *potato spindle tuber viroid* (Pospiezny 1997).

Chitosan acts as an elicitor of plant immune system and was validated as a potent plant immunomodulator in 24 plant species against various groups of plant pathogens (Yin et al. 2010). Chitosan elicits defense signaling pathway involving Ca^{2+}, reactive oxygen species (ROS), nitric oxide (NO), jasmonic acid (JA), salicylic acid (SA), and abscisic acid (ABA) (Kluesener et al. 2002; Li et al. 2009; Rakwal et al. 2002; Obara et al. 2002; Iriti and Faoro 2008). A number of chitosan responsive genes were identified including a novel Ser/Thr protein kinase (*oipk*) (Feng et al. 2006); *MAPK* (Yin et al. 2010); *SKP1* (Zhang et al. 2007); *OPR1* (Jang et al. 2009); *WRKY* (Hofmann et al. 2008); and genes associated with JA/ethylene (ET) pathway such as *OPR3* (JA synthase gene), *MPK4* and *EREBP* (Yin et al. 2006). On the other hand, chitosan responsive proteins were also identified, mainly are pathogenesis related (PR) proteins such as glucanase, chitinase, and peroxidase (Yafei et al. 2009;

Table 16.1 Summary of elicitors produced by fungi and bacteria

Producing species	Type of elicitor	Host	References
Fungi			
Acremonium strictum	AsES,	Strawberry	Chalfoun et al. (2013)
	Subtilisin		
Alternaria tenuissima	Hrip1	Tobacco	Kulye et al. (2012)
Botrytis cinerea	Cerato-platanin, BcSpl1	Tomato cv Moneymaker, Tobacco cv Havana, *Arabidopsis*	Frías et al. (2011)
	No description	*Elegans xylogenic*	Takeuchi et al. (2013)
Ceratocystis platani	CP	*Platanus acerifolia*	Lombardi et al. (2013)
Ceratocystis populicola	Pop1	*Platanus acerifolia*	Lombardi et al. (2013)
Fusarium sp.	No description	*Euphorbia pekinensis*	Gao et al. (2012)
Fusarium moniliforme var. subglutinans	No description	*Mangifera indica* L. cv Amrapali	Yadav et al. (2012)
Fusarium oxysporum	No description	*Cucumis melo* L. var *reticulatus*,	Sanchez-Estrada et al. (2009),
		Tomato	Troncoso-Rojas et al. (2013)
Magnaporthe grisea	Cerato-platanin, MgSM1,	*Arabidopsis*	Yang et al. (2009)
	MoHrip1	Rice	Chen et al (2012),
	Chitin oligosaccharides	Rice	Kishimoto et al. (2010),
		Arabidopsis	Libault et al. (2007)
Phomopsis sp.	No description	*Betula Platyphylla*	Fan et al. (2013)
Phytophthora sp.	Cryptogein	Tobacco	Bourque et al. (1999)
Phytophthora cryptogea	Cryptogein	Tobacco	Adam et al. (2012)
Phytophthora sojae	Oligopeptide	*Petroselinum crispum*	Eulgem et al. (1999)
Rhynchosporium secalis	Effector protein, NIP1	Barley	Van't et al. (2007)
Trichoderma asperellum	EplT4	Soybean	Wang et al. (2013)
Trichoderma virens	Small protein (Sm1), Epl1	Maize	Vargas et al. (2008)
Trichoderma viride	Xylanase	Rice	Hamada et al. (2012)
Bacteria			
Bacillus cereus	Dimethyl disulfide	Tobacco	Huang et al. (2012)
		Corn	
Bacillus subtilis	Mycosubtilin Surfactin	Wheat	Khong et al. (2012)
	Fengycin		
Eubacterial	Flagellin	Tomato	Felix et al. (1999)
Pseudomonas aeruginosa	Rhamnolipids	Grapevine	Varnier et al. (2009)
Pseudomonas syringae pv. Syringae	Harpins	Tobacco	Nurnberger (1999)

Ferri et al. 2009); phenylalanine ammonia lyase (PAL), catalase, polyphenoloxidase and superoxide dismutase (SOD) (Yafei et al. 2009).

In our study, we described the defense responses of chitosan-treated cucumber (*Cucumis sativus* L.) plants after inoculation with *Pseudoperonospora cubensis*, the causal agent of downy mildew in regards to the production of NO and enzyme activity of two PR proteins, i.e., chitinase and β-1,3-glucanase. To find out the relationship between chitosan and the source of NO production, we used NOS inhibitor, L-NAME (*N*-nitro-*L*-arginine methyl ester hydrochloride) and NO scavenger cPTIO [2-(carboxyphenyl)-4,4,5,5-tetramethylimidazoline-1-oxyl-3-oxide potassium salt]. To detect NO in plant cells, a fluorescent probe that labels NO, DAF-FM diacetate (4-amino-5-methylamino-2,7-difluorofluorescein diacetate) was used. This probe has been reported to be very sensitive in detecting NO in real-time imaging (Gould et al. 2003; Lamattina et al. 2003). DAF-FM diacetate is a new alternative to DAF-2 DA (4,5-diaminofluorescein diacetate) to overcome the limitation of detection only at neutral or basic pH but also at acidic pH (Zhang et al. 2003).

We also evaluated the effect of chitosan on disease suppression of the infected plants and the expression of a nitric oxide associated (NOA) gene isolated from cucumber in this study and was designated *CsNOA1*. *CsNOA1* is a homologue of *AtNOA1* (Accession number NM 180335.1), which has been identified as a putative regulator of nitric oxide synthase (NOS) activity in plants (Crawford 2006) and later characterized as a cGTPase (Moreau et al. 2008). The hypothesis was that increased level of NO would be produced in plants treated with chitosan and NO would play an important role in defence responses of cucumber.

The results showed that both chitinase and β-1,3-glucanase activities peaked at 12 h.a.i. (hour after pathogen inoculation) in chitosan treated plants, whereas plants treated with chitosan coupled with L-NAME and cPTIO, respectively, did not show increased activities of both enzymes compared to the baseline level (Fig. 16.1a, b). NO generation in chitosan-treated plants was also increased and peaked at 8 h.a.i. (Fig. 16.2), which was 4 h earlier than that of chitinase and β-1, 3-glucanase activities. NO level in plants, treated with L-NAME (NOS inhibitor), increased significantly and peaked at 10 h.a.i. compared to initial level but was significantly lower than that of chitosan treated plants. In contrast, NO level in plants treated with cPTIO showed no significant difference from the baseline after 12 h.a.i. The increased level of NO detected in plants treated with L-NAME implies that there are other sources that generate NO in the plant cells besides NOS-like enzyme.

Transcript level of cPTIO also increased similar to that of chitosan treatment compared to transcript level of the control and actin (internal normalizer), which showed constant level throughout the sampling period (Fig. 16.3). The increase of both NO generation and gene expression of NOA after chitosan + L-NAME treatment implies that NO was generated from sources other than NOS-like enzyme since L-NAME is a NOS inhibitor and indicated that NOA was involved in NO generation. On the other hand, NOA expression increased after chitosan + cPTIO

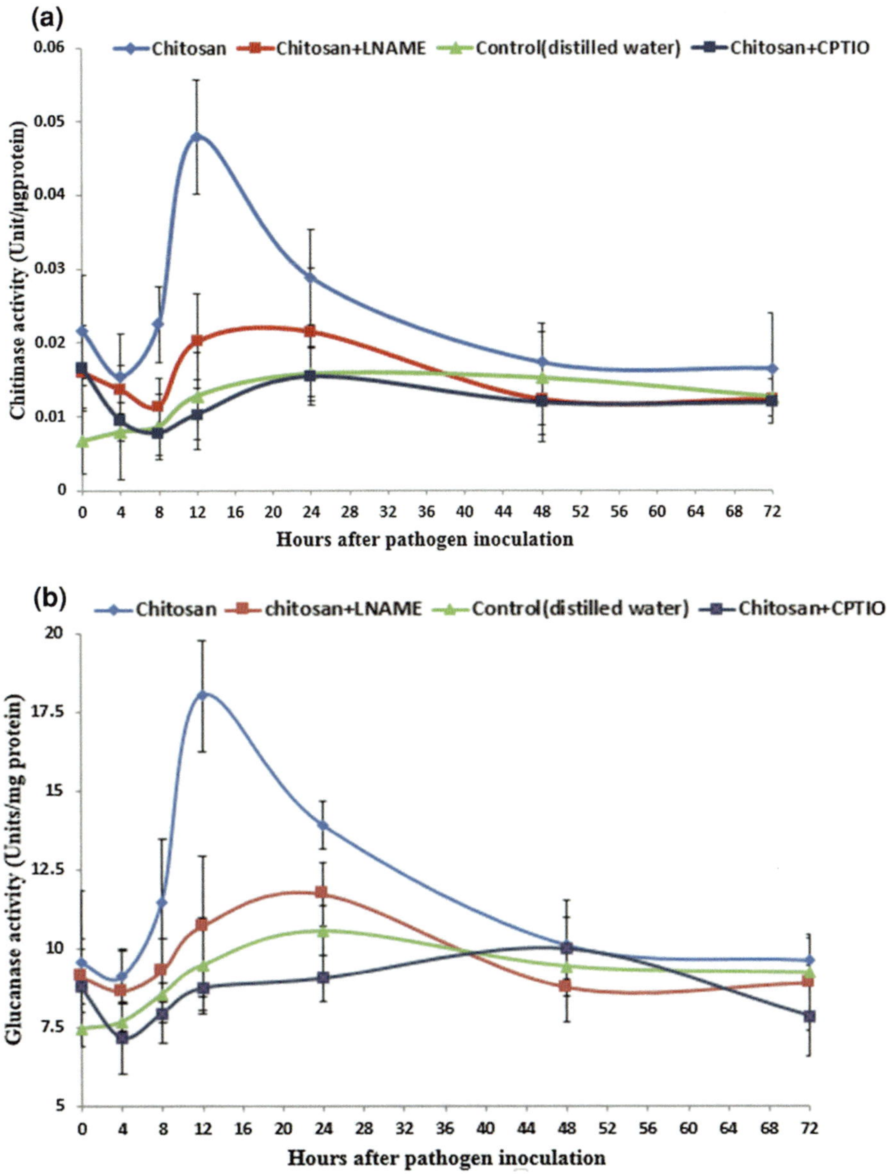

Fig. 16.1 **a** Chitinase and **b** β-1,3-glucanase activities in competitive study. Mean value ±SD ($n = 3$); L-NAME, N^G-nitro-L-arginine methyl ester; cPTIO, 2-phenyl-4,4,5,5-tetramethylimidazoline-1-oxyl-3-oxide

treatment while there was no further increase in NO generation (at baseline) which implies that the generated NO was scavenged by cPTIO even though there was an increase in NOA expression.

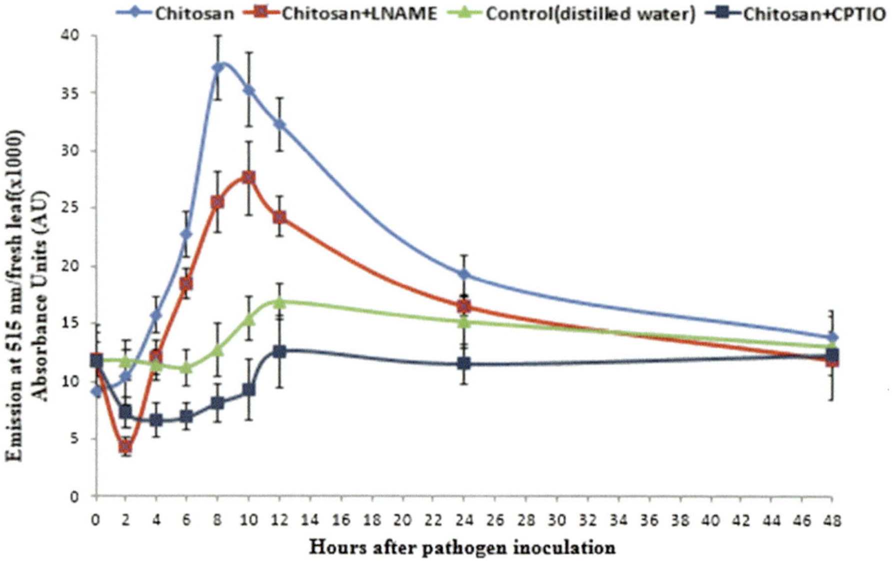

Fig. 16.2 NO generation in cucumber after pathogen inoculation; mean value ±SD ($n = 3$). L-NAME, N^G-nitro-L-arginine methyl ester; cPTIO, 2-phenyl-4,4,5,5-tetramethylimidazoline-1-oxyl-3-oxide

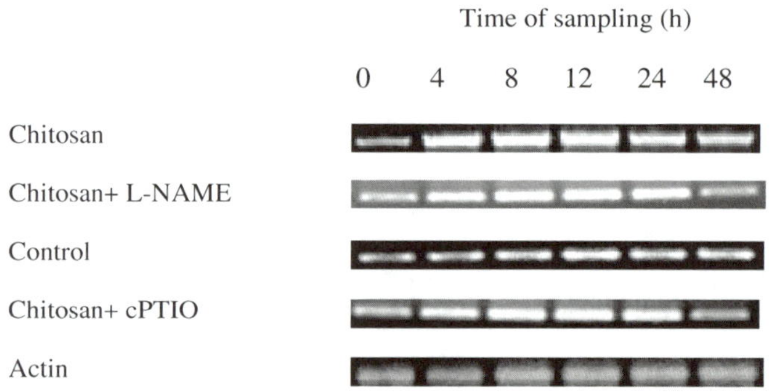

Fig. 16.3 Pattern of transcript accumulation of NOA gene of cucumber after various treatments. L-NAME, N^G-nitro-L-arginine methyl ester; cPTIO, 2-phenyl-4,4,5,5-tetramethylimidazoline-1-oxyl-3-oxide

The results also showed that the application of chitosan alone on cucumber plants prior *P. cubensis* inoculation suppressed downy mildew infection with only 42 % disease incidence compared to the control plants (97 %) (Fig. 16.4). In contrast, plants treated with chitosan coupled with L-NAME or cPTIO demonstrated higher disease incidence which were 61 and 71 %, respectively, though

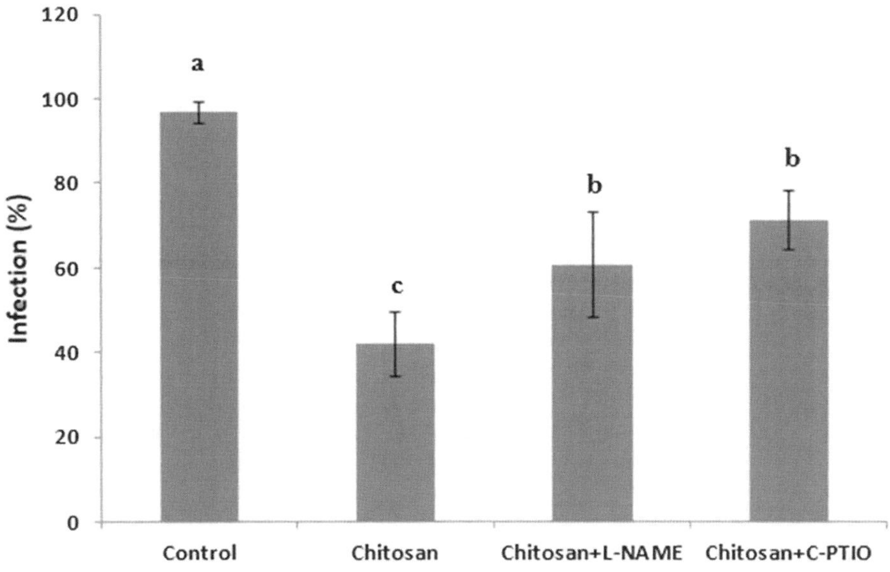

Fig. 16.4 Disease assessment of downy mildew in cucumber in the competitive study. Means with same letter are not significantly different (LSD test; $P = 0.05, n = 3$); *error bars* represent standard deviations. L-NAME, N^G-nitro-L-arginine methyl ester; cPTIO, 2-phenyl-4,4,5,5- tetramethylimidazoline-1-oxyl-3-oxide

there was no significant difference between disease incidence resulted from the two treatments.

These findings demonstrated that NO played a critical role in mediating the defense responses in cucumber against downy mildew. A disruption in NO generation conferred negative effects on chitinase and β-1, 3-glucanase activities and increased disease incidence significantly. Chitosan induced the generation of NO and defense related genes (chitinase and β-1, 3-glucanase) in cucumber to counter the attack of *P. cubensis*. The induced expression of NOA and increased NO generation in chitosan treated plants indicated the involvement of NOA in NO generation in cucumber. However, further study is needed to confirm these findings.

16.3 Nitric Oxide Signaling and Defense Responses

For plants to be able to elicit inducible defense responses caused by pathogen attack, they require signal molecules such as Ca^{2+}, reactive oxygen species (ROS) and NO at early events leading to signaling network via phytohormones SA, JA, ABA, and ET. Upon perception of an elicitor, early signaling events detected in elicitor-treated cells include activation of protein kinases (PK) which triggers Ca^{2+} influx and the production of ROS and NO (Allan and Fluhr 1997; Delledonne et al.

1998; Lecourieux-Ouaked et al. 2000), and anion effluxes (Wendehenne et al. 2002). For the biosynthesis of NO, refer to an earlier chapter written in this book. The major routes in which NO mediates its signaling mechanism are probably through interaction with ROS and thiol groups (Wilson et al. 2008). NO and H_2O_2 are required to activate SA-dependent signaling pathway leading to host cell death (Durner et al. 1998; Delledonne et al. 2001). H_2O_2 was demonstrated to induce NO generation via increasing NO-generating enzyme activity (Lum et al. 2002) but on the other hand, NO was shown to regulate elicitor-mediated HR independently of H_2O_2 (Lamotte et al. 2004).

In addition, apart from directly affecting NO synthesis and accumulation, H_2O_2 is able to compete for potential NO binding sites such as reacting to thiol groups (Wilson et al. 2008). Thus, it has been suggested that NO and H_2O_2 regulate the biosynthesis of each other (Garcia-Brugger et al. 2006). Anion effluxes involving Cl^- and NO^{3-} have been shown to play a critical role in mediating elicitor-induced events, and it occurs at the very early stage of the signaling pathway (Wendehenne et al. 2002). Voltage-dependent gated anion channels (VDACs) and voltage-dependent chloride channels (CLCs) have been identified in plants but their role as anion channels in plasma membrane of plants is still unclear (Barbier-Brygoo et al. 2000).

Downstream of NO in the signaling pathway, cyclic GMP (cGMP) and cyclic ADP-ribose (cADPR) are two key players. NO induces cGMP synthesis (McDonald and Murad 1995; Delledonne et al. 1998; Durner et al. 1998) but exactly how NO induces cGMP synthesis in plants requires further investigations. In animal system, NO binds to the haem domain of soluble cGMP-producing guanylate cyclases (sGC), which leads to increased cGMP levels (Stamler 1994) but such sGC is yet to be found in plants. The identification of a novel protein with guanylyl cyclase activity in *Arabidopsis* was demonstrated to be unaffected by NO (Ludidi and Gehring 2003). In animals, cGMP mediates cADPR signaling to increase the level of free Ca^{2+} in the cells via intracellular ryanodine receptor calcium channel (RYR) (Wendehenne et al. 2001). Similarly, in plants, the application of cADPR elevated transcripts of PAL and PR-1 in tobacco and the expression of these two genes was prevented when RYR inhibitors were applied (Durner et al. 1998; Klessig et al. 2000).

NO also activates mitogen-activated protein kinase (MAPK) cascade where MPK4 has been identified to function as a negative regulator of SA signaling while positively regulating JA signaling in *Arabidopsis* (Kumar and Klessig 2000). The accumulation of SA is induced by NO and was shown to be required for the expression of PAL, PR-1 and chalcone synthase (Durner et al. 1998; Wendehenne et al. 2001; Wilson et al. 2008). SA-dependent pathway involves several key factors such as SA receptor, NPR1, NPR3, and NPR4; NPR1-associated TGA transcription factors; and WRKY transcription factors (Wu et al. 2012; Fu et al. 2012; Fu and Dong 2013). The *Arabidopsis* WRKY70 was shown to act as a positive regulator of SA signaling and a negative regulator of JA signaling (Li et al. 2004) while WRKY62 negatively regulates JA signaling (Mao et al. 2007).

At the post-translational level, NO signaling via S-nitrosylation involving GSNO (S-nitrosylated glutathione) reductase has also been shown to play an important role in defense responses and abiotic stresses (Feechan et al. 2005; Barroso et al. 2006) but how NO regulates this event remains unclear. In *Arabidopsis* undergoing HR, 16 S-nitrosylated proteins were identified, mostly are enzymes serving intermediary metabolism, signaling and antioxidant defense (Romero-Puertas et al. 2008) which the authors suggested that NO can modulate the concentration of metabolites and the balance of energy status. This finding will pave the way for further study on the role of S-nitrosylation during plant defense.

16.4 Crosstalk Between Abiotic and Biotic Stress Responses

Plants have the ability to respond effectively to both abiotic and biotic stresses in their natural environments. This ability is regulated by phytohormones SA, JA, ET, and ABA where they play important roles in various signaling networks either synergistically or antagonistically, which is known as signaling crosstalk. Accumulating evidence have demonstrated the involvement and interactions between ROS, NO, SA, JA, ET, and ABA. Both ROS and NO are shown to mediate crosstalk between abiotic and biotic stress signaling pathways (Fujita et al. 2006, Arasimowicz and Floryszak-Wieczorek, 2007). Apart from mediating plant defense against pathogen attack, ROS and NO are required to enhance drought and salinity tolerance, and induce ABA synthesis in plants (Garcia-Mata and Lamattina 2001; Zhao et al. 2001; Valderrama et al. 2007). NO is also involved in protecting plants against UV-B radiation, heavy metals, herbicides and extreme temperatures (Hung et al. 2002; Kopyra and Gwozdz 2003; Neill et al. 2003; Zhao et al. 2004; Shi et al. 2005).

SA, JA, and ET are commonly involved in biotic stress signaling pathway upon pathogen attack (Ferrari et al. 2003; Guo and Stotz 2007; Zhu et al. 2011; Wang et al. 2012). In contrast, ABA is involved in abiotic stress signaling pathway and developmental processes such as seed development, dormancy, germination, and stomatal movement (Finkelstein et al. 2002; Tuteja 2007; Jia and Zhang 2008; Nakashima and Yamaguchi-Shinozaki, 2013). ABA is shown to act as a negative regulator of defense responses in plants against various biotrophic and necrotrophic pathogens by suppressing SAR induction (Adie et al. 2007; de Torres-Zabala et al. 2007; Yasuda et al. 2008).

Downstream of ROS/NO and phytohormones are MAPKs which play a crucial role in both abiotic and biotic signaling pathways. For example, MPK3, MPK4, and MPK6 are involved in cold and salt stress responses, ABA signaling during post-germination stage and ROS signaling (Ichimura et al. 2000; Yuasa et al. 2001; Droillard et al. 2002). Further down the crosstalk, some transcription factors that regulate both abiotic and biotic signaling leading to stress responses have been identified. AtMYC2 (the basic helix-loop-helix (bHLH) transcription factor) together with AtMYB2 (the *R2R3MYB*-type transcription factor which binds *cis*-elements in the dehydration-inducible *RD22* gene) were demonstrated to be involved in the ABA signaling pathway during drought and osmotic stress (Abe et al. 2003). RD26 (a dehydration-responsive NAC transcription factor) isolated from dehydrated *A. thaliana* was suggested to play a role in the crosstalk among defense, senescence and ABA-mediated stress-signaling pathways (Fujita et al. 2004). Zat12 (C_2H_2-type zinc-finger transcription factor) was also suggested to be a multiple-stress-responsive transcription factor involved in wounding, ROS and abiotic stress signaling (Davletova et al. 2005a) and is regulated by Heat Shock Factor (HSF) 21 (a redox-sensitive transcription factor) (Davletova et al. 2005b).

16.5 Conclusions and Future Prospects

NO is a crucial signal molecule involved in regulating defense responses in plants against pathogen attack. However, the source of NO production during defense responses is still unclear. NO-mediated signaling pathway apparently is a complex phenomenon and may be accomplished through directly modifying proteins or activating secondary messengers such as cGMP. Exactly how NO induces cGMP synthesis in plants requires further investigations. Two direct protein modifications involved in plant defense via NO signaling are S-nitrosylation and tyrosine nitration (Tada et al. 2008; Cecconi et al. 2009; Yun et al. 2011). Studies on the role of S-nitrosylation during plant defense are lacking and require intense investigation especially by drawing insights from similar studies in animal system. Furthermore, detailed knowledge regarding signal cascades upstream and downstream of NO is still lacking (Gaupels et al. 2011) and identification of NO-related second messengers will be required to understand how NO is perceived and activates defense responses in plants.

Acknowledgment The authors thank the Malaysian Ministry of Higher Education for providing a research grant for the chitosan project.

References

Abe H, Urao T, Ito T et al (2003) *Arabidopsis* AtMYC2 (bHLH) and AtMYB2 (MYB) function as transcriptional activators in absicisic acid signaling. Plant Cell 15:63–78

Abel AJ (2003) Role of reactive oxygen species in the response of barley to necrotrophic pathogens. Protoplasma 221:137–143

Adam T, Bouhidel K, Der C et al (2012) Constitutive expression of clathrin hub hinders elicitor-induced clathrin-mediated endocytosis and defense gene expression in plant cells. FEBS Lett 586:3293–3298

Adie BA, Perez-Perez J, Godoy M et al (2007) ABA is an essential signal for plant resistance to pathogens affecting JA biosynthesis and the activation of defenses in *Arabidopsis*. Plant Cell 19:1665–1681

Al-Hetar MY, Zainal Abidin MA, Sariah M, Wong MY (2011) Antifungal activity of chitosan against *Fusarium oxysporum* f. sp. *cubense*. J Appl Polym Sci 120:2434–2439

Allan AC, Fluhr R (1997) Two distinct sources of elicited reactive oxygen species in tobacco epidermal cells. Plant Cell 9:1557–1572

Arasimowicz M, Floryszak-Wieczorek J (2007) Nitric oxide is a bioactive signaling molecule in plant stress responses. Plant Sci 172:876–887

Barbier-Brygoo H, Vinauger M, Colcombet J et al (2000) Anion channels in higher plants: functional characterization, molecular structure and physiological role. Biochim Biophys Acta 1465:199–218

Barroso J, Corpas FJ, Carreras A et al (2006) Localization of S-nitrosoglutathione and expression of *S*-nitrosoglutathione reductase in pea plants under cadmium stress. J Exp Bot 57:1785–1793

Bautista-Baños S, Hernández-Lauzardo AN, Velázquez-del Valle MG et al (2006) Chitosan as a potential natural compound to control pre and postharvest diseases of horticultural commodities. Crop Prot 25:108–118

Bayles CJ, Ghemawat MS, Aist JR (1990) Inhibition by 2-deoxy-D-glucose of callose formation, papilla deposition, and resistance to powdery mildew in an *ml-o* barley mutant. Physiol Mol Plant P 36:63–72

Benhamou N (1992) Ultrastructural and cytochemical aspects of chitosan on *Fusarium oxysporum* f. sp. *radicis-lycopersici*, agent of tomato crown and root rot. Phytopathology 22:1185–1193

Bourque S, Binet M-N, Ponshet M et al (1999) Characterization of the cryptogein binding sites on plant plasma membranes. J Biol Chem 274:34699–34705

Broekaert WF, Terras FRG, Cammue BPA, Osborn RW (1995) Plant defensins: novel antimicrobial peptides as components of the host defense system. Plant Physiol 108:1353–1358

Chalfoun NR, Grellet-Bournonville CF, Martínez-Zamora MG et al (2013) Purification and characterization of AsES protein: a subtilisin secreted by *Acremonium strictum* is a novel plant defense elicitor. J Biol Chem 288:14098–14113

Cheah LH, Page BBC, Sheperd R (1997) Chitosan coating for inhibition of sclerotina carrots. New Zeal J Crop Hort 25:89–92

Chen M, Zeng H, Qiu D et al (2012) Purification and characterization of a novel hypersensitive response-inducing elicitor from *Magnaporthe oryzae* that triggers defense response in rice. PLoS ONE 7:e37654

Cecconi D, Orzetti S, Vandelle E et al (2009) Protein nitration during defense response in *Arabidopsis thaliana*. Electrophoresis 30:2460–2468

Crawford (2006) Mechanisms for nitric oxide synthesis in plants. J Exp Bot 57:471–478

Davletova S, Schlauch K, Coutu J, Mittler R (2005a) The zinc-finger protein Zat12 plays a central role in reactive oxygen and abiotic stress signaling in *Arabidopsis*. Plant Physiol 139:847–856

Davletova S, Rizhsky L, Liang H et al (2005b) CYTOSOLIC ASCORBATE PEROXIDASE 1 is a central component of the reactive oxygen gene network of *Arabidopsis*. Plant Cell 17:268–281

de Torres-Zabala M, Truman W, Bennett MH et al (2007) *Pseudomonas syringae* pv. tomato hijacks the *Arabidopsis* abscisic acid signalling pathway to cause disease. EMBO J 26:1434–1443

del Pozo O, Lam E (1998) Caspases and programmed cell death in the hypersensitive response of plants to pathogens. Curr Biol 8:1129–1132

Delledonne M, Xia Y, Dixon RA, Lamb C (1998) Nitric oxide functions as a signal in plant disease resistance. Nature 394:585–588

Delledonne M, Zeier J, Marocco A, Lamb C (2001) Signal interaction between nitric oxide and reactive oxygen intermediates in the plant hypersensitive disease resistance response. Proc Natl Acad Sci USA 98:13454–13459

Droillard M, Boudsocq M, Barbier-Brygoo H, Lauriere C (2002) Different protein kinase families are activated by osmotic stresses in *Arabidopsis thaliana* cell suspensions. Involvement of the MAP kinases AtMPK3 and AtMPK6. FEBS Lett 527:43–50

D'Silva I, Poirier GG, Heath MC (1998) Activation of cysteine proteases in cowpea plants during the hypersensitive response- a form of programmed cell death. Exp Cell Res 245:389–399

Durner J, Wendehenne D, Klessig DF (1998) Defense gene induction in tobacco by nitric oxide, cyclic GMP, and cyclic ADP-ribose. Proc Natl Acad Sci USA 95:10328–10333

El Ghaouth A, Arul J, Asselin A, Benhamou N (1992) Antifungal activity of chitosan on post-harvest pathogens: induction of morphological and cytological alterations in *Rhizopus stolonifer*. Mycol Res 96:769–779

Epple P, Apel K, Bohlmann H (1997) Overexpression of an endogenous thionin enhances resistance of Arabidopsis against *Fusarium oxysporum*. Plant Cell 9:509–520

Eulgem T, Rushton PJ, Schmelzer E et al (1999) Early nuclear events in plant defence signalling: rapid gene activation by WRKY transcription factors. EMBO J 18:4689–4699

Fan G, Zhai Q, Zhan Y (2013) Gene expression of lupeol synthase and biosynthesis of nitric oxide in cell suspension cultures of *Betula Platyphylla* in response to a *Phomopsis* elicitor. Plant Mol Biol Rep 31:296–302

Feechan A, Kwon E, Yun B et al (2005) A central role for *S*-nitrosothiols in plant disease resistance. Proc Natl Acad Sci USA 102:8054–8059

Felix G, Duran JD, Volko S, Boller T (1999) Plants have a sensitive perception system for the most conserved domain of bacterial flagellin. Plant J 18:265–267

Feng B, Chen Y, Zhao C et al (2006) Isolation of a novel Ser/Thr protein kinase gene from oligochitosan-induced tobacco and its role in resistance against tobacco mosaic virus. Plant Physiol Biochem 44:596–603

Ferrari S, Plotnikova JM, De Lorenzo G, Ausubel FM (2003) *Arabidopsis* local resistance to *Botrytis cinerea* involves salicylic acid and camalexin and requires *EDS4* and *PAD2*, but not *SID2*, *EDS5* or *PAD4*. Plant J 35:193–205

Ferri M, Tassoni A, Franceschetti M et al (2009) Chitosan treatment induces changes of protein expression profile and stilbene distribution in *Vitis vinifera* cell suspensions. Proteomics 9:610–624

Filippone MP, Ricci JD, de Marchese AM et al (1999) Isolation and purification of a 316 Da preformed compound from strawberry (*Fragaria ananassa*) leaves active against plant pathogens. FEBS Lett 459:115–118

Finkelstein RR, Gampala SSL, Rock CD (2002) Abscisic acid signaling in seeds and seedlings. Plant Cell 14(Suppl):s15–s45

Fujita M, Fujita Y, Maruyama K et al (2004) A dehydration-induced NAC protein, RD26, is involved in a novel ABA-dependent stress-signaling pathway. Plant J 39:863–876

Frías M, González C, Brito N (2011) BcSpl1, a cerato-platanin family protein, contributes to *Botrytis cinerea* virulence and elicits the hypersensitive response in the host. New Phytol 192:483–495

Fujita M, Fujita Y, Noutoshi Y et al (2006) Crosstalk between abiotic and biotic stress responses: a current view from the points of convergence in the stress signaling networks. Curr Opin Plant Biol 9:436–442

Fu ZQ, Dong X (2013) Systemic acquired resistance: turning local infection into global defense. Annu Rev Plant Biol 64:839–863

Fu ZQ, Yan S, Saleh A et al (2012) NPR3 and NPR4 are receptors for the immune signal salicylic acid in plants. Nature 486:228–232

Gao F, Ren C, Dai C (2012) Signaling effects of nitric oxide, salicylic acid, and reactive oxygen species on isoeuphpekinensin accumulation in *Euphorbia pekinensis* suspension cells induced by an endophytic fungal elicitor. J Plant Growth Regul 31:490–497

Garcia-Brugger A, Lamotte O, Vandelle E et al (2006) Early signaling events induced by elicitors of plant defenses. Mol Plant-Microbe Interact 19:711–724

Garcia-Mata C, Lamattina L (2001) Nitric oxide induces stomatal closure and enhances the adaptive plant responses against drought stress. Plant Physiol 126:1196–1204

Gaupels F, Kuruthukulangarakoola GT, Durner J (2011) Upstream and downstream signals of nitric oxide in pathogen defence. Curr Opin Plant Biol 14:707–714

Gould KS, Lamotte O, Klinguer A et al (2003) Nitric oxide production in tobacco leaf cells: a generalized stress response? Plant, Cell Environ 26:1851–1862

Guo X, Stotz HU (2007) Defense against *Sclerotinia sclerotiorum* in *Arabidopsis* is dependent on jasmonic acid, salicylic acid, and ethylene signaling. Mol Plant-Microbe Interact 20:1384–1395

Hamada H, Kurusu T, Okuma E et al (2012) Regulation of a proteinaceous elicitor-induced Ca^{2+} influx and production of phytoalexins by a putative voltage-gated cation channel, OsTPC1, in cultured rice cells. J Biol Chem 287:9931–9939

Helander IM, Nurmiaho-Lassila EL, Ahvenainen R et al (2001) Chitosan disrupts the barrier properties of the outer membrane of Gram-negative bacteria. Int J Food Microbiol 71:235–244

Hofmann MG, Sinha AK, Proels RK, Roitsch T (2008) Cloning and characterization of a novel LpWRKY1 transcription factor in tomato. Plant Physiol Biochem 46:533–540

Huang CJ, Tsay JF, Chang SY et al (2012) Dimethyl disulfide is an induced systemic resistance elicitor produced by *Bacillus cereus* C1L. Pest Manag Sci 68:1306–1310

Hung KT, Chang CJ, Kao CH (2002) Paraquat toxicity is reduced by nitric oxide in rice leaves. Plant Physiol 159:159–166

Ichimura K, Mizoguchi T, Yoshida R et al (2000) Various abiotic stresses rapidly activate *Arabidopsis* MAP kinases ATMPK4 and ATMPK6. Plant J 24:655–665

Iriti M, Faoro F (2008) Abscisic is involved in chitosan-induced resistance to tobacco necrosis virus (TNV). Plant Physiol Biochem 46:1106–1111

Jang S, Cho K, Shibato J et al (2009) Rice OsOPRs: transcriptional profiling responses to diverse environmental stimuli and biochemical analysis of OsOPR1. Plant Biol 52:229–243

Jia W, Zhang J (2008) Stomatal movements and long-distance signaling in plants. Plant Signal Behav 3:772–777

Joshi BN, Sainani MN, Bastawade B et al (1999) Pearl millet cysteine protease inhibitor. Evidence for the presence of two distinct sites responsible for anti-fungal and anti-feedent activities. Eur J Biochem 265:556–563

Khong NG, Randoux B, Tayeh C et al (2012) Induction of resistance in wheat against powdery mildew by bacterial cyclic lipopeptides. Commun Agric Appl Biol Sci 77:39–51

Kishimoto K, Kouzai Y, Kaku H et al (2010) Perception of the chitin oligosaccharides contributes to disease resistance to blast fungus *Magnaporthe oryzae* in rice. Plant J 64:343–354

Klessig DF, Durner J, Noad R et al (2000) Nitric oxide and salicylic acid signaling in plant defense. Proc Natl Acad Sci USA 97:8849–8855

Kluesener B, Young JJ, Murata Y et al (2002) Convergence of calcium signaling pathways of pathogenic elicitors and abscisic acid in *Arabidopsis* guard cells. Plant Physiol 130:2152–2163

Kopyra M, Gwozdz EA (2003) Nitric Oxide stimulates seed germination and counteracts the inhibitory effect of heavy metals and salinity on root growth of *Lupinus luteus*. Plant Physiol Biochem 41:1011–1017

Kumar D, Klessig DF (2000) Differential induction of tobacco MAP kinases by the defence signals nitric oxide, salicylic acid, ethylene and jasmonic acid. Mol Plant Microbe Interact 13:347–351

Kurita K (2006) Chitin and chitosan: functional biopolymers from marine crustaceans. Mar Biotechnol 8:203–226

Kulye M, Liu H, Zhang Y et al (2012) Hrip1, a novel protein elicitor from necrotrophic fungus, *Alternaria tenuissima*, elicits cell death, expression of defence-related genes and systemic acquired resistance in tobacco. Plant, Cell Environ 35:2104–2120

Lamattina L, Garcia-Mata C, Graziano M, Pagnussat G (2003) Nitric oxide: the versatility of an extensive signal molecule. Annu Rev Plant Biol 54:109–136

Lamotte O, Gould K, Lecourieux D et al (2004) Analysis of nitric oxide signaling functions in tobacco cells challenged by the elicitor cryptogein. Plant Physiol 135:516–529

Lecourieux-Ouaked F, Pugin A, Lebrun-Garcia A (2000) Phosphoproteins involved in the signal transduction of cryptogein, an elicitor of defense reactions in tobacco. Mol Plant Microbe Interact 13:821–829

Li J, Brader G, Palva ET (2004) The WRKY70 transcription factor: a node of convergence for jasmonate-mediated and salicylate-mediated signals in plant defence. Plant Cell 16:319–331

Li Y, Yin H, Wang Q et al (2009) Oligochitosan induced *Brassica napus L.* production of NO and H_2O_2 and their physiological function. Carbohydr Polym 75:612–617

Libault M, Wan J, Czechowski T et al (2007) Identification of 118 *Arabidopsis* transcription factor and 30 ubiquitin-ligase genes responding to chitin, a plant-defense elicitor. Mol Plant-Microbe Interact 20:900–911

Lombardi L, Faoro F, Luti S et al (2013) Differential timing of defense-related responses induced by cerato-platanin and cerato-populin, two non-catalytic fungal elicitors. Physiol Plant. doi:10.1111/ppl.12041

Ludidi N, Gehring C (2003) Identification of a novel protein with guanylyl cyclase activity in *Arabidopsis thaliana*. J Biol Chem 278:6490–6494

Lum HK, Butt YKC, Lo SCL (2002) Hydrogen peroxide induces a rapid production of nitric oxide in mung bean (*Phaseolus aureus*). Nitric Oxide Biol Chem 6:205–213

Mao P, Duan M, Wei C, Li Y (2007) WRKY62 transcription factor acts downstream of cytosolic NPR1 and negatively regulates jasmonate-reponsive gene expression. Plant Cell Physiol 48:833–842

McDonald LJ, Murad F (1995) Nitric oxide and cGMP signaling. Adv Pharmacol 34:263–275

Modolo LV, Cunha FQ, Braga MR, Salgado I (2002) Nitric oxide synthase mediated phytoalexin accumulation in soybean cotyledons in response to the *Diaporthe phaseolorum* f. sp. *meridionalis* elicitor. Plant Physiol 130:1288–1297

Moerschbacher B, Noll U, Gorrichon L, Reisener HJ (1990) Specific inhibition of lignification breaks hypersensitive resistance of wheat to stem rust. Plant Physiol 93:465–470

Moreau M, Gyu IL, Wang Y et al (2008) AtNOS/AtNOA1 is a functional *Arabidopsis thaliana* cGTPase and not a nitric oxide synthase. J Biol Chem 283:32957–32967

Nakashima K, Yamaguchi-Shinozaki K (2013) ABA signaling in stress-response and seed development. Plant Cell Rep 32:959–970

Neill SJ, Desikan R, Hancock JT (2003) Nitric signaling in plants. New Phytol 159:11–35

Nürnberger T (1999) Signal perception in plant pathogen defense. Cell Mol Life Sci 55:167–182

Obara N, Hasegawa M, Kodama O (2002) Induced volatiles in elicitor-treated and rice blast fungus-inoculated rice leaves. Biosci Biotechnol Biochem 66:2549–2559

Parker JE, Szabo V, Staskawicz B et al (1993) Phenotypic characterization and molecular mapping of the *Arabidopsis thaliana* locus *RPP5*, determining disease resistance to *Peronospora parasitica*. Plant J 4:821–831

Pospiezny H (1997) Antiviriod activity of chitosan. Crop Prot 16:105–106

Pospiezny H, Chirkov S, Atabekov J (1991) Induction of antiviral resistance in plants by chitosan. Plant Sci 79:63–68

Rakwal R, Tamogami S, Agrawal GK, Iwahashi H (2002) Octadecanoid signaling component "burst" in rice (*Oryza sativa* L.) seedling leaves upon wounding by cut and treatment with fungal elicitor chitosan. Biochem Biophys Res Commun 295:1041–1045

Romero-Puertas M, Campostrini N, Matte A et al (2008) Proteomic analysis of S-nitrosylated proteins in Arabidopsis thaliana undergoing hypersensitive response. Proteomics 8:1459–1469

Sánchez-Estrada A, Tiznado-Hernández M, Ojeda-Conteras AJ et al (2009) Induction of enzymes and phenolic compounds related to the natural defence response of netted melon fruit by a bio-elicitor. J Phytopathol 157:24–32

Shi S, Wang G, Wang Y et al (2005) Protective effect of nitric oxide against oxidative stress under ultraviolet-B radiation. Nitric oxide Biol Chem 13:1–9

Showalter AM (1993) Structure and function of plant cell wall proteins. Plant Cell 5:9–23

Smallwood M, Beven A, Donovan N et al (1994) Localization of cell wall proteins in relation to the developmental anatomy of the carrot root apex. Plant J 5:237–246

Stamler JS (1994) Redox signaling: nitrosylation and related target interactions of nitric oxide. Cell 78:931–936

Tada Y, Spoel SH, Pajerowska-Mukhtar K et al (2008) Plant immunity requires conformational changes of NPR1 via S-nitrosylation and thioredoxins. Science 321:952–956

Takeuchi C, Nagatani K, Sato Y (2013) Chitosan and a fungal elicitor inhibit tracheary element differentiation and promote accumulation of stress lignin-like substance in Zinnia *Elegans xylogenic* culture. J Plant Res. doi:10.1007/s10265-013-0568-0

Tilburzy R, Reisener HJ (1990) Resistance of wheat to *Puccinia graminis* f. sp. *tritici:* association of the hypersensitive reaction with the cellular accumulation of lignin-like material and callose. Physiol Mol Plant P 36:109–120

Troncoso-Rojas R, Sánchez-Estrada A, Carvallo T et al (2013) A fungal elicitor enhances the resistance of tomato fruit to *Fusarium oxysporum* infection by activating the phenylpropanoid metabolic pathway. Phytoparasitica 41:133–142

Tuteja N (2007) Abscisic acid and abiotic stress signaling. Plant Signal Behav 2:135–138

Valderrama R, Corpas FJ, Carreras A et al (2007) Nitrosative stress in plants. FEBS Lett 581:453–461

Van't SKAE, Gierlich A, Knogge W (2007) A single binding site mediates resistance- and disease-associated activities of the effector protein NIP1 from the barley pathogen *Rhynchosporium secalis*. Plant Physiol 144:1654–1666

Vargas WA, Djonović S, Sukno SA, Kenerly CM (2008) Dimerization controls the activity of fungal elicitors that trigger systemic resistance in plants. J Biol Chem 283:19804–19815

Varnier AL, Sanchez L, Vasta P et al (2009) Bacterial rhamnolipids are novel MAMPs conferring resistance to *Botrytis cinerea* in grapevine. Plant, Cell Environ 32:178–193

Wang Y, Song J, Wu Y et al (2013) Eplt4 proteinaceous elicitor produced in *Pichia pastoris* has a protective effect against *Cercosporidium sofinum* infections of soybean leaves. Appl Biochem Biotechnol 169:722–737

Wang Z, Tan X, Zhang Z et al (2012) Defense to *Sclerotinia sclerotiorum* in oilseed rape is associated with the sequential activations of salicylic acid signaling and jasmonic acid signaling. Plant Sci 184:75–82

Wendehenne D, Lamotte O, Frachisse J-M et al (2002) Nitrate efflux is an essential component of the cryptogein signaling pathway leading to defense responses and hypersensitive cell death in tobacco. Plant Cell 14:1937–1951

Wendehenne D, Pugin A, Klessig DF, Durner J (2001) Nitric oxide: comparative synthesis and signaling in animal and plant cells. Trends Plant Sci 6:177–183

Wilson ID, Neill SJ, Hancock JT (2008) Nitric oxide synthesis and signaling in plants. Plant, Cell Environ 31:622–631

Wu Y, Zhang D, Chu JY et al (2012) The *Arabidopsis* NPR1 protein is a receptor for the plant defense hormone salicylic acid. Cell Rep 1:639–647

Yadav SP, Chand G, Chakrabarti DK (2012) Defence-related biochemical changes by elicitor of malformation pathogen, *Fusarium mangiferae*, in mango. Arch Phytopathol Plant Protect 45:1356–1368

Yafei C, Yong Z, Xiaoming Z et al (2009) Functions of oligochitosan induced protein kinase in tobacco mosaic virus resistance and pathogenesis related proteins in tobacco. Plant Physiol Biochem 47:724–731

Yang Y, Zhang H, Li G et al (2009) Ectopic expression of MgSM1, a Cerato-platanin family protein from *Magnaporthe grisea*, confers broad-spectrum disease resistance in Arabidopsis. Plant Biotechnol J 7:763–777

Yasuda M, Ishikawan A, Jikumaru Y et al (2008) Antagonistic interaction between systemic acquired resistance and the abscisic acid–mediated abiotic stress response in *Arabidopsis*. Plant Cell 20:1678–1692

Yin H, Bai XF, Zhao XM, Du YG (2010) Molecular cloning and characterization of a *Brassica napus* L. MAP kinase involved in oligochitosan-induced defense signaling. Plant Mol Biol Rep 28:292–301

Yin H, Li S, Zhao X, Du Y, Ma X (2006) cDNA microarray analysis of gene expression in *Brassica napus* treated with oligochitosan elicitor. Plant Physiol Biochem 44:910–916

Yuasa T, Ichimura K, Mizoguchi T, Shinozaki K (2001) Oxidative stress activates ATMPK6, an *Arabidopsis* homologue of MAP kinase. Plant Cell Physiol 42:1012–1016

Yun B, Feechan A, Yin M et al (2011) S-nitrosylation of NADPH oxidase regulates cell death in plant immunity. Nat Lett 478:264–268

Zhang C, Czymmek KJ, Shapiro AD (2003) Nitric oxide does not trigger early programmed cell death events but may contribute to cell-to-cell signaling governing progression of the *Arabidopsis* hypersensitive response. Mol Plant-Microbe Interact 16:962–972

Zhang FY, Feng B, Li W et al (2007) Induction of tobacco genes in response to oligochitosan. Mol Biol Rep 34:35–40

Zhao Z, Chen G, Zhang C (2001) Interaction between reactive oxygen species and nitric oxide in drought-induced abscisic acid synthesis in root tips of wheat seedlings. Aust J Plant Physiol 28:1055–1061

Zhao L, Zhang F, Guo J et al (2004) Nitric oxide functions as a signal in salt resistance in the calluses from two ecotypes of reed. Plant Physiol 134:849–857

Zhu L, Li Y, Li L et al (2011) Ethylene is involved in leafy mustard systemic resistance to *Turnip mosaic virus* infection through the mitochondrial alternative oxidase pathway. Physiol Mol Plant Path 76:166–172

Chapter 17
The Role of Nitric Oxide in Programmed Cell Death in Higher Plants

Hu-Yi He, Ming-Hua Gu and Long-Fei He

Abstract Programmed cell death (PCD) is a genetically controlled biological process involved in defense, development, and stress response. Generally, the characters of plant PCD are similar to animal apoptosis, for instance cytoplasm shrinkage, chromatin condensation, and DNA fragmentation. An important signaling molecule, nitric oxide (NO) has been implicated in environmental-induced plant PCD, but its signaling and controlling network is still unknown. Whether NO promotes or suppresses PCD depends on NO sources and concentration in different plant species and environmental conditions. The effects of NO on developmental PCD were extensively studied. NO not only plays a crucial role in hypersensitive response (HR) during plant-pathogen interactions, but is also involved in abiotic stress-induced PCD including heat shock, salt, drought, cold, UV radiation, ozone, and heavy metals (mainly cadmium, aluminum). Previous studies showed the mitochondrion as a modulating center of PCD and also control NO level *in planta*. Vacuole processing enzyme (VPE) and caspase-like protein are involved in PCD. NO regulates the expression of PCD-associated genes via mitogen-activated protein kinase (MAPK) cascade, *S*-nitrosylation, and cGMP-dependent pathway. In addition, there are diverse interactions between NO and other signals such as hydrogen peroxide, calcium, ethylene, and salicylic acid (SA) during PCD. Based on understanding of related knowledge, NO signaling network in response to PCD in higher plants is presented in this chapter.

Keywords Higher plants · Nitric oxide · Programmed cell death · Signaling network · Stress

H.-Y. He · M.-H. Gu · L.-F. He (✉)
College of Agronomy, Guangxi University, Nanning 530004, People's Republic of China
e-mail: lfhe@gxu.edu.cn

M. N. Khan et al. (eds.), *Nitric Oxide in Plants: Metabolism and Role in Stress Physiology*, DOI: 10.1007/978-3-319-06710-0_17,

17.1 Introduction

Programmed cell death (PCD) occurs in various forms throughout the plant life cycle, probably with both common and specific aspects. PCD is a genetically controlled biological process which activates an intrinsic suicide program of cells. It not only controls the degradation of intracellular components, but facilitates removal of unwanted, incorrect, or damaged cells from multicellular organisms. It plays an important role in defense response, development, and environmental stress. Leaf senescence, hypersensitive response (HR), lysigenous aerenchyma formation, and aleurone degradation are all the forms of PCD in pants.

Generally, the characters of plant PCD are similar to animal apoptosis, such as cytoplasm shrinkage, chromatin condensation, membrane blebbing, DNA fragmentation, and selective cleavage of proteins. During HR development, plant also can form apoptotic bodies. Although individual processes differ in the triggering factors such as vacuole collapse, releasing sequestered hydrolases, may be the universal trigger of plant PCD. It is indicated that the molecular machinery underlying PCD is well conserved in eukaryotic organisms. The executive phases and typical hallmarks of PCD differ under different occasions. Cleavage of genomic DNA during apoptotic PCD is divided into two subsequent steps; an early cleavage into high molecular weight fragments, whose sizes coincidence with chromatin loop domains, and later an intense fragmentation, usually forming oligonucleosomal fragments (Brotner et al. 1995), that can be detected by DNA electrophoresis in the whole tissue or cell population, also visualized by TUNEL reaction (TdT-mediated dUTP nick-end labeling) in individual cells (Gavrieli et al. 1992; Zhan et al. 2013).

Nitric oxide (NO) a simple diatomic, diffusible, gaseous free radical, involved in many physiological processes such as PCD, seed germination, lateral root initiation, flowering, stomatal closure, and responses to stress in plants. Moreover, as an important signaling molecule, NO has been implicated in environmental-induced plant PCD, but its signaling network is still unknown. Whether NO promotes or suppresses PCD is dependent on sources and concentration of NO in different plant species.

17.2 Evolution of NO and Dual Function During Plant Programmed Cell Death

Owing to the essential function of NO in plant signaling network, its endogenous source is very important. There are two ways to generate NO in plants viz. L-arginine-dependent nitric oxide synthase (NOS) pathway and nitrite-dependent nitrate reductase (NR) pathway. Although NOS-like activity has been detected in plants, this enzyme remains enigmatic. No gene or protein with sequence homology to known mammalian type NOS has been found (Crawford 2006).

NR works as a major enzymatic source of NO production in plants. It can convert nitrite to NO in vitro and in vivo (Desikan et al. 2002). In *Arabidopsis*, NR is encoded by two genes, *NIA1* and *NIA2*, which contribute differently to the synthesis of NO in different tissues.

Another is nonenzymatic conversion of nitrite to NO in the apoplast. NO produced in plants at low concentration may rapidly eliminate lipid peroxyl radicals, alter the species and components of reactive oxygen species (ROS), and block the injuries from ROS, induce the expression of antioxidant genes and the activity of antioxidant enzymes (Lamattina et al. 2003).

17.3 Effects of NO on Developmental PCD

The effects of NO on developmental PCD have been extensively studied (Table 17.1). Gibberellin (GA)-induced PCD in barley aleurone layers is mediated by ROS, because GA greatly reduces the amount of CAT (catalase) and SOD (superoxide dismutase). NO donors, SNP (sodium nitroprusside) and SNAP (*S*-nitroso-*N*-acetylpenicillamine) delay the loss of two enzymes and PCD in barley aleurone layers treated with GA, but stimulate slightly the secretion of α-amylase. It is suggested that NO may be an endogenous modulator of PCD in barley aleurone layers (Beligni et al. 2002).

Leaf senescence is a highly coordinated process that involves PCD. Early stages of leaf senescence occurring during normal leaf ontogenesis, but not triggered by stress factors, are poorly known. Kolodziejek et al. (2007) found that both nDNA fragmentation and chromatin condensation occurred quite early during barley leaf senescence and always in the same order. NO was localized in vivo and in situ within the cytoplasm, mainly in mitochondria, in leaves at the same stage as those in which chromatin condensation was observed. The highest concentration of NO was found in the cytoplasm of mesophyll cells in the earliest stage of senescence, and lower concentrations were found during later stages that might suggest that NO plays an inductive role in PCD in leaf senescence.

During the seed development, the cells of the nucleus suffer a degenerative process early after fertilization as the cellular endosperm expands and accumulates reserves. Nuclear cell degeneration has been characterized as a form of developmental PCD. Lombardi et al. (2010) showed that nucleus PCD is accompanied by a considerable production of both NO and hydrogen peroxide (H_2O_2), and each of the two molecules is able to induce the production of the other and to cause PCD when applied to a living nucleus. Xylem cells have to be killed so as to facilitate the formation of rigid hollow tubes specialized for water transport. NO is also a key factor regulating PCD and lignification during xylem formation (Neill et al. 2005).

Table 17.1 Reports of nitric oxide-mediated PCD in plants

PCD types	Inducing factors	Species	NO alteration	Effect	Reference
Growth and development	Gibberellin	Barley aleurone layers	Decrease	Protect against oxidative damage	Beligni et al. (2002)
	Xylogenesis	*Zinnia elegans*			Neill et al. (2005)
	Early senescence	Barley leaves	Increase		Kolodziejek et al. (2007)
	Seed development	*Sechium edule nucellus*	Increase		Lombardi et al. (2010)
	Self-incompatibility	Papaver	Increase	Actin reorganization	Wilkins et al. (2011)
HR	*Pseudomonas syringae*	Tobacco suspension cells	Increase	cGMP pathway	Durner et al. (1998)
		Soybean suspension cells	Increase	NO/H_2O_2 cooperation	Delledonne et al. (2001)
		Arabidopsis	Increase	Intercellular signal	Zhang et al. (2003)
					Guo et al. (2004)
	Yeast elicitor	*Cupressus lusitanica*	Increase	Production of peroxynitrite	Zhao et al. (2007)
		Arabidopsis thaliana suspension cultures	Increase		Clarke et al. (2000)
		sycamore (*Acer pseudoplatanus*) cultured cells		Actin depolymerization	Malerba et al. (2008)
Abiotic stress	Cadmium	*Arabidopsis* suspension cells	Increase	Induce *SAG12* expression	De Michele et al. (2009)
		Tobacco BY-2 cells	Increase	Promote Cd^{2+} accumulation	Ma et al. (2010)
		Arabidopsis	Increase	MPK6-mediated caspase-3-like activation	Ye et al. (2012)
		Yellow lupine	Increase		Arasimowicz-Jelonek et al. (2012)

(continued)

Table 17.1 (continued)

PCD types	Inducing factors	Species	NO alteration	Effect	Reference
	High light	Tobacco leaf	Increase	Cross-talk between NO and H_2O_2	
		Rice leaf	Increase	Protein *S*-nitrosylation	Zago et al. (2006) Lin et al. (2011)
	Ozone	Tobacco BY-2 cells	Increase	Antioxidant systems	De Pinto et al. (2002)
		Arabidopsis leaf	Increase	SA signaling genes	Ahlfors et al. (2009a, b)
	NO donor	*Arabidopsis thaliana* and *Nicotiana tabacum* cells	Increase	Suppression of ROS-scavenging systems	Murgia et al. (2004)
		Tobacco protoplasts	Increase	Mitochondrial pathway regulated by Ca^{2+}	Wang et al. (2010a, b)

17.4 Role of NO in Hypersensitive Response

NO plays a crucial role in HR during plant-pathogen interactions. NO and H_2O_2 function in combination with each other all along HR cell death (Table 17.1).

Administration of NO donors or recombinant mammalian NOS to tobacco plants or tobacco suspension cells triggered expression of the defense-related genes encoding pathogenesis-related 1 protein and phenylalanine ammonia lyase (PAL). These genes were also induced by cyclic guanosine monophosphate (cGMP) and cyclic ADP-ribose, two molecules that can serve as secondary messengers for NO signaling in mammals. Consistent with cGMP acting as a secondary messenger in tobacco, NO treatment induced dramatic and transient increases in endogenous cGMP levels. Unregulated NO levels drive a diffusion limited reaction with O_2^- to generate peroxynitrite ($ONOO^-$), which is a mediator of cellular injury in many biological systems but not a mediator of HR. The HR is triggered only by balanced production of NO and reactive oxygen intermediates. Increasing the level of O_2^- reduces NO-mediated toxicity. HR is activated after interaction of NO not with O_2^- but with H_2O_2. During the HR, SOD accelerates O_2^- dismutation to H_2O_2 to minimize the loss of NO by reaction with O_2^- and to trigger HR through NO/H_2O_2 cooperation. The rates of production and dismutation of O_2^- generated during the oxidative burst play a crucial role in the modulation and integration of NO/H_2O_2 signaling in the HR (Delledonne et al. 2001). The researches on the kinetics of NO production and hypersensitive cell death showed that NO accumulation contributed to HR. NO was first seen as punctate foci at the cell surface. Subsequent NO accumulation patterns were consistent with NO being an intercellular signal that functions in cell-to-cell spread of the HR (Zhang et al. 2003).

Arabidopsis suspension cultures generate elevated levels of NO in response to challenge by avirulent bacteria, and NO are sufficient to induce cell death in *Arabidopsis* cells independently of ROS. NO-induced cell death is a form of PCD, requiring gene expression, and has a number of characteristics of PCD such as chromatin condensation and caspase-like activity in *Arabidopsis cells* (Clarke et al. 2000). Phytotoxin fusicoccin induces another form of cell death in sycamore (*Acer pseudoplatanus* L.) cultured cells, likely mediated by NO and independent of cytochrome c release, and they make it tempting to speculate that changes in actin cytoskeleton are involved in this form of PCD (Malerba et al. 2008).

17.5 Involvement of NO in Abiotic Stress-Induced PCD

NO is also involved in PCD induced by abiotic stress including heat shock, salt, drought, cold, UV radiation, ozone, and heavy metals (mainly cadmium, aluminum) (Table 17.1).

Arabidopsis thaliana cell suspension cultures underwent a PCD process when exposed to 100 and 150 mM $CdCl_2$. As suggested by the expression of the marker senescence-associated gene12 (*SAG12*), this process resembled an accelerated senescence. $CdCl_2$ treatment was accompanied by a rapid increase in NO and phytochelatin (PC) synthesis, which continued to be high as long as cells remained viable. NO is actually required for Cd^{2+}-induced cell death, because the inhibition of NO synthesis by NG-monomethylarginine monoacetate (L-NMMA) resulted in partial prevention of H_2O_2 increase, *SAG12* expression, and mortality. NO also modulated the extent of PC content and their function by *S*-nitrosylation (De Michele et al. 2009). Tobacco BY-2 cells exposed to 150 μM $CdCl_2$ underwent PCD with TUNEL-positive nuclei, significant chromatin condensation and the increasing expression of a PCD-related gene *Hsr203J*. Accompanied with the PCD, the production of NO increased significantly. NO played a positive role in $CdCl_2$-induced PCD by modulating Cd^{2+} uptake and thus promoting Cd^{2+} accumulation in BY-2 cells (Ma et al. 2010). The roots of 3-day-old yellow lupine seedlings exposed to 89 mM $CdCl_2$ resulted in PCD starting from 24 h of stress duration. Cd-induced PCD was preceded by a relatively early burst of NO localized mainly in the root tips. Above changes were accompanied by the NADPH-oxidase-dependent superoxide anion ($O_2^{\cdot-}$) production. NADPH-oxidase inhibitor and NO-scavenger significantly reduced $O_2^{\cdot-}$ and NO production, respectively, as well as diminished the pool of cells undergoing PCD (Arasimowicz-Jelonek et al. 2012).

Tobacco leaves, exposed to moderate high light, dramatically potentiated NO-mediated cell death in catalase-deficient (CAT1AS) but not in wild-type plants. The results consolidate significant crosstalk between NO and H_2O_2, and provide new insight into the early transcriptional response of plants to increased NO and H_2O_2 levels, and identify target genes of the combined action of NO and H_2O_2 during the induction of plant cell death (Zago et al. 2006). Lin et al. (2011) identified an NO accumulation mutant *noe1* (*nitric oxide excess 1*) in rice and analyzed its role in NO-mediated leaf cell death. The *NOE1*, encoded a rice catalase *OsCATC*, that increased the H_2O_2 in the leaves, which consequently promoted NO production via activation of NR. Removal of excess NO reduced cell death in both leaves and suspension cultures derived from *noe1* plants, implicating NO as an important endogenous mediator of H_2O_2-induced leaf cell death.

Ozone (O_3) induced a rapid accumulation of NO, which started from guard cells, spread to adjacent epidermal cells and eventually moved to mesophyll cells. NO production coincided with the formation of HR-like lesions. SNP and O_3 individually induced a large set of defense-related genes; however, in a combined treatment SNP attenuated the O_3 induction of salicylic acid (SA) biosynthesis and other defense-related genes. SNP treatment decreased O_3-induced SA accumulation. The O_3-sensitive mutant *rcd1* was found to be an NO overproducer; in contrast, *Atnoa1/rif1* (*Arabidopsis* nitric oxide associated 1/resistant to inhibition by FSM1), a mutant with decreased production of NO, was also O_3 sensitive. NO can modify signaling, hormone biosynthesis and gene expression in plants during O_3 exposure. NO is an important signaling molecule, which production is needed for a proper O_3 response (Ahlfors et al. 2009a, b).

The involvement of cellular antioxidant metabolism in the signal transduction triggered by these bioactive molecules has been investigated. NO and ROS levels were singularly or simultaneously increased in tobacco (*Nicotiana tabacum* cv

Bright-Yellow 2) cells by the addition of NO and/or ROS generators to the culture medium. The generation of NO did not cause an increase in PAL activity or induction of cellular death. It only induced minor changes in ascorbate (ASC) and glutathione (GSH) metabolisms. An increase in ROS induced oxidative stress in the cells, causing an oxidation of the ASC and GSH redox pairs; however, it had no effect on PAL activity and did not induce cell death at low concentrations. In contrast, the simultaneous increase of NO and ROS activated a process of death with the typical cytological and biochemical features of hypersensitive PCD and a remarkable rise in PAL activity. Under the simultaneous generation of NO and ROS, the cellular antioxidant capabilities were also suppressed (De Pinto et al. 2002). Treatment of tobacco protoplasts with SNP resulted in a rapid $[Ca^{2+}]_{cyt}$ accumulation and decrease in mitochondrial membrane potential (potential ($\Delta\Psi$m) before the appearance of PCD. NO-induced PCD could be largely prevented not only by cPTIO, but also by Ca^{2+} chelator, EGTA (*ethylene glycol tetraacetic acid*), Ca^{2+}-channel blocker $LaCl_3$ (Lanthanum chloride) or CsA (a specific mitochondrial permeability transition pore inhibitor, which also inhibit Ca^{2+} cycling by mitochondria). NO-induced PCD is mediated through mitochondrial pathway and regulated by Ca^{2+} (Wang et al. 2010a, b). The effects of different NO-donors releasing NO with either NO^+ (SNP) or NO^- (SNAP, GSNO, NOC-18) character have been compared in plant cells. SNP behaves differently than the other NO-donors tested; indeed, SNP induces accumulation of ferritin transcripts in *Arabidopsis*, whereas SNAP (*S*-nitroso-*N*-acetyl-penicillamine) inhibits its accumulation. Only SNP caused PCD and suppression of ROS-scavenging systems (Murgia et al. 2004). Artificial NO donors are widely used as tools to study the role of NO in plants. However, reliable and reproducible characterizations of metabolic responses induced by different NO donors are complicated by the variability of their NO release characteristics. NO release characteristics of the donors SNP, *S*-nitrosoglutathione (GSNO) and NOS, both in vitro and *in planta* (*Nicotiana tabacum* L. cv. BelW3) were evaluated and their effects on NO dependent processes such as the transcriptional regulation of the mitochondrial alternative oxidase (AOX) gene, accumulation of H_2O_2 and induction of cell death were assessed. Contrary to NOS and SNP, GSNO is not an efficient NO generator in leaf tissue. In spite of the different NO release signatures by SNP and NOS in tissue, the NO-dependent responses examined were similar, suggesting that there is a critical threshold for the NO response (Ederli et al. 2009).

17.6 Regulation of NO on PCD–Associated Genes Expression

Vacuole processing enzymes (VPEs) are a vacuole-localized cysteine protease, which exhibit caspase-1-like protein activity. It can mediate the activation of caspase-3-like protein to provoke PCD and is involved in virus-induced hypersensitive cell death in tobacco (Hatsugai et al. 2004). VPE activity is also required for aluminum (Al)-induced PCD in plants. Ced-9 inhibited both the Al-induced activity of caspase-like VPE and Al-induced PCD in tobacco (Wang et al. 2009). Senescence-associated gene 12 (*SAG12*) is considered the best molecular marker of senescence. The expression of *SAG12* increased at 2 and 3 d after 100 μM $CdCl_2$ treatment (De Michele et al. 2009). Al-induced PCD was promoted by *AhSAG*, a senescence-associated gene in *Arachis hypoganea* (Zhan et al. 2013). As one of the few endogenous cell death inhibitors in plants, bax inhibitor-1 (BI-1) is potentially a core regulator of PCD (Huckelhoven 2004). PpBI-1 can attenuate Al-induced PCD and enhance Al tolerance in transgenic yeast (Zheng et al. 2007). The programmed cell death 5 (*PDCD5*) gene encodes a protein that shares significant homology with the corresponding proteins of species ranging from yeast to mice (Liu et al. 1999). Overexpression of *OsPDCD5* gene induces PCD in rice (Attia et al. 2005).

In tobacco, mechanical wounding induced the rapid transcript accumulation and activation of wound-induced protein kinase (WIPK) (Seo et al. 1995). Transgenic tobacco plants ectopically expressing *AhMPK3* exhibited enhanced resistance to first and second instar larvae of *Spodoptera litura* (Kumar et al. 2009). The conditional overexpression of *AhMPK6* resulted in HR-like cell death in tobacco (Kumar and Kirti 2010). MPK kinase 6-mediated activation of VPE modulates heat shock-induced PCD in *Arabidopsis* (Li et al. 2012). NO promotes *MPK6*-mediated caspase-3-like activation in cadmium-induced *Arabidopsis thaliana* PCD (Ye et al. 2012). Over-expression of *OsGSNOR* reduced intracellular SNO levels, which regulates global levels of protein *S*-nitrosylation, alleviated leaf cell death in *noe1* plants (Lin et al. 2011).

Cytochorme c gets to the cytoplasm at least via two mechanisms. One is via formation of a transient mitochondrial permeability transition pore (MPTP), which is produced by the voltage-dependent anion channel (VDAC) on the outer membrane, the adenine nucleotide transporter (ANT) from the inner membrane and cyclophilin D in the matrix (Green and Reed 1998). Another is directly via the VDAC (Shimizu et al. 1999). Because the expression of AOX, the unique respiratory terminal oxidase in plants, can scavenge excess superoxide anion so that the balance of NO and H_2O_2 is destroyed, AOX plays protective roles in Al-induced *Arabidopsis* protoplast death (Li and Xing 2011). As a molecular chaperone, mitochondrial HSP70 may be involved in PCD initiation by reducing $\Delta\psi_m$ in mitochondrial outer membrane (Chen et al. 2009). Through NO/H_2O_2 cooperation, SOD accelerates O_2^- dismutation to H_2O_2 to minimize the loss of NO by reaction with O_2^- and to trigger hypersensitive cell death (Delledonne et al. 2001). Some genes associated with PCD are listed in Table 17.2.

17.7 Interaction Between NO and Other Signaling Molecules During Plant PCD

There are diverse interactions between NO and other signaling molecules such as H_2O_2, calcium, ethylene, and SA during PCD. The interaction between NO and H_2O_2 can be cytotoxic or protective. NO/H_2O_2 cooperation triggers hypersensitive

Table 17.2 Genes in relation to PCD

Genes	Expression	Signal molecule	Species	PCD type	References
VPE	+	ROS	Tobacco	HR, aluminum	Hatsugai et al. (2004), Wang et al. (2010a, b)
SAG	+		*Arachis hypoganea*	Aluminum	Zhan et al. (2013)
SAG12	+	NO	*Arabidopsis*	Cadmium	De Michele et al. (2009)
BI-1	+	Ca^{2+}	*Phyllostachys prdecox*	Aluminum	Zheng et al. (2007)
PDCD5	+		Rice	Development	Attia et al. (2005)
MPK3			Tobacco	Wounding	Seo et al. (1995)
MPK6	+	ROS/Ca^{2+}, NO	*Arabidopsis*	Heat shock, cadmium	Li et al. (2012), Ye et al. (2012)
MKK4	+	H_2O_2	*Arabidopsis*	HR	Ren et al. (2002)
MEKK1			*Arabidopsis*	Innate immunity	Asai et al. (2002)
WRKY22/WRKY 29			*Arabidopsis*	Innate immunity	Asai et al. (2002)
GSNOR	+	NO	Rice	High light	Lin et al. (2011)
ANT					Green and Reed (1998)
VDAC					Shimizu et al. (1999)
AOX	+	NO, ROS	Tobacco, *Arabidopsis*	SNP, aluminum	Ederli et al. (2009), Li and Xing (2011)
HSP70	+		Rice	Salt	Chen et al. (2009)
SOD	+	NO/H_2O_2	Soybean	HR	Delledonne et al. (2001)

+ (increase)

cell death in soybean cell suspensions (Delledonne et al. 2001). Boosted NO and O_2^- production is required for Cd-induced PCD in lupine roots. Moreover, the NO-dependent Cd-induced PCD in roots of 14-day-old lupine plants was correlated with the enhanced level of the post-stress signals in leaves, including distal NO crosstalk with H_2O_2 (Arasimowicz-Jelonek et al. 2012). Using biochemical and genetic approaches in the root system, Wang et al. (2010a, b) proposed a pathway for the regulation of NO biosynthesis that involves the modulation of NIA2 by MPK6. With the increase of intracellular H_2O_2 levels, MPK6 is activated, which in turn leads to the phosphorylation of NIA2 at Ser-627. Phosphorylation of NIA2 by MPK6 dramatically.

Increases the activity of NIA2 and the production of NO and also results in morphological changes. SNP treatment resulted in a rapid $[Ca^{2+}]_{cyt}$ accumulation and the appearance of PCD in tobacco protoplasts. EGTA, $LaCl_3$ or CsA largely prevent NO-induced PCD that is mediated through mitochondrial pathway and regulated by Ca^{2+} (Wang et al. 2010a, b). Moreover, NO is involved in PCD induction via interacting with the pathways of phytohormones (Wang et al. 2010a, b). NO treatments induce ethylene production in tobacco. NO and ethylene act together to regulate O_3-induced AOX expression (Ederli et al. 2006). Transcript profiling indicated a role for NO in attenuation of certain classes of O_3 induced genes, many of which were related to SA biosynthesis or SA signaling (Ahlfors et al. 2009a, b).

17.8 NO Signaling Network in Response to PCD

Based on understanding of related knowledge, we propose NO signaling network in response to PCD in plants (Fig. 17.1). Different signals (developmental, pathogen, invasion, and abiotic stress) trigger NO production. Subsequently, NO promotes the expression of PCD-associated genes (such as VPE, AOX, HSP70, APX) via several pathways. One is cGMP-dependent pathway: NO and cGMP mediate the auxin response during adventitious root formation in cucumber (Pagnussat et al. 2003). Moreover, NO regulates the apoptotic signal cascade through protein *S*-nitrosylation (Wang et al. 2010a, b). Lin et al. (2011) suggested that *S*-nitrosylation was involved in light-dependent leaf cell death in *noe1* rice. NO targets identified only in *noe1* plants included glyceraldehyde 3-phosphate dehydrogenase (GAPDH) and thioredoxin, which have been reported to be involved in *S*-nitrosylation regulated cell death in animals. The last one is MAPK cascade. Asai et al. (2002) identified a complete plant MAP kinase cascade (MEKK1, MKK4/MKK5, and MPK3/MPK6) and WRKY22/WRKY29 transcription factors that function downstream of the flagellin receptor FLS2. Signaling events initiated by diverse pathogens converge into a conserved MAPK cascade. An MAPK signaling cascade is activated during the adventitious rooting process induced by IAA in a NO-mediated but cGMP-independent pathway (Pagnussat et al. 2004). NO mediated caspase-3-like protease activation under Cd^{2+} stress conditions. Pretreatment with cPTIO effectively inhibited Cd^{2+}-induced MAPK activation. Cd^{2+}-induced caspase-3-like activity was significantly suppressed in the *mpk6* mutant, suggesting that MPK6 was required for caspase-3-like protease activation (Ye et al. 2013). NO contributed caspase-3-like protease activation in Cd^{2+} induced *Arabidopsis thaliana* PCD, which was mediated by MPK6 (Ye et al. 2012). NO could also regulate the activity of Ca^{2+}-dependent protein kinase (CDPK) was addressed by Lanteri et al. (2006) who characterized a 50 kDa NO-dependent CDPK in cucumber hypocotyls. These three pathways may work synergistic or solely. In turn, gene expressions provoke some downstream events such as PCD.

17.9 Control of NO Level in Plant Mitochondrion

Previous studies showed mitochondrion has emerged as modulating center of plant PCD and also important sites in controlling NO levels in plants. Nitrite (the source of NO synthesis) inhibited the respiration of isolated *Arabidopsis* mitochondria, in competition with oxygen, an effect that was abolished or potentiated when electron flow occurred via AOX or cytochrome c oxidase (COX), respectively. Electron leakage from external NAD(P)H dehydrogenases contributed the most to NO degradation as higher rates of Amplex Red-detected H_2O_2 production and NO consumption were observed in NAD(P)H-energized mitochondria. Conversely, the NO-insensitive AOX diminished electron leakage from the respiratory chain, allowing the increase of NO half-life without interrupting oxygen consumption. The accumulation of NO derived from nitrite reduction and the superoxide-dependent mechanism of NO degradation in isolated *Arabidopsis* mitochondria are influenced by the external NAD(P)H dehydrogenases and AOX, revealing a role for these alternative proteins of the mitochondrial respiratory chain in the control of NO levels in plant cells (Wulff et al. 2009). Complex III, COX, and AOX are all involved in nitrite to NO reduction. AOX controls NO generation by directly influencing the rate of electron leakage to nitrite (Cverkovska and Vanlerberghe 2012). Robson and Vanlerberghe (2002) found that knocking down of AOX increases the susceptibility of plants to PCD. There exists a negative feedback loop where NO acts to suppress excess mitochondrial reactive nitrogen species (RNS) and presumably ROS via increased AOX expression to modulate the elicitation of PCD. Three mechanisms of AOX-mediated ROS and RNS homeostasis are suggested. First, AOX can modulate the membrane potential and reduce NO levels. Second, aconitase inhibition leads to increase in citrate which induces AOX to maintain electron flow through the electron transport chain and to lower NO concentrations (Gupta et al. 2012). Third, AOX scavenging of NO might help in decreasing ROS production by preventing over-reduction of ubiquinone pool. However, plant lead to PCD or necrotic cell death in response to stress, because NO and ROS generation from nonmitochondrial sources could swamp any AOX-mediated homeostatic mechanisms.

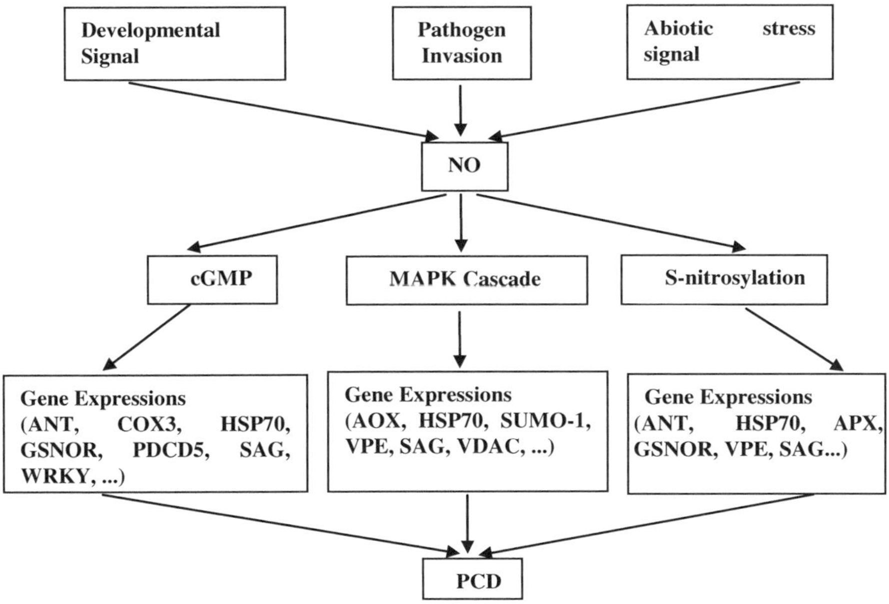

Fig. 17.1 Proposed NO signaling network in response to PCD

17.10 Conclusion and Perspectives

Adverse environmental conditions interferes NO-mediated signal transduction. By direct scavenging of ROS or activating antioxidant enzymes, exogenously applied NO might alleviate metal toxicity in plants. In contrast, NO through *S*-nitrosylation of PCs or promoting metal uptake via iron transporters contributes or even amplifies metal toxicity. The promoting and suppressing effects of NO on cell death is dependent on a variety of factors, such as cell type, cellular redox status, and the flux and dose of local NO (Wang et al. 2010a, b). Cell signaling dysregulation induced by metal not only leads to the death stimulation pathway, but might be able to activate survival signaling towards tolerance response to heavy metal. Active cell death is required for an enhanced effectiveness of protective responses in neighboring cells (Overmyer et al. 2003). In particular, the relationship between NO, ROS signaling and stress-related hormones might play a key role on the dispute on the expression of gene sets responsible for stress tolerance and in the generation of long-distance sensing from roots to shoots. NO is involved in the generation of systemic signal in systemic acquired resistance to pathogens (Vlot et al. 2008). Xiong et al. (2011) showed that tungstate is not completely a specific NR inhibitor in plant NO research. To investigate the roles of NO in plants, it is necessary to search for more NR-deficient mutants and new specific NR inhibitors. The research on transcriptional factors and NO-regulated genes is the key to understand the mechanism of NO in PCD in higher plants. The recognition of the molecular NO targets will be an exciting challenge for future research.

Acknowledgments This work was supported by the National Science Foundation of China (31260296, 30960181, 30560070).

References

Ahlfors R, Brosche M, Kollistl H et al (2009a) Nitric oxide modulates ozone-induced cell death, hormone biosynthesis and gene expression in *Arabidopsis thaliana*. Plant J 58: 1–12

Ahlfors R, Brosche M, Kangasjarvi J (2009b) Ozone and nitric oxide interaction in *Arabidopsis thaliana*. Plant Signaling & Behavior 4:878–879

Arasimowicz-Jelonek M, Floryszak-Wieczorek J, Deckert J et al (2012) Nitric oxide implication in cadmium-induced programmed cell death in roots and signaling response of yellow lupine plants. Plant Physiol Biochem 58:124–134

Asai T, Tena G, Plotnikova J et al (2002) MAP kinase signaling cascade in *Arabidopsis* innate immunity. Nature 415:977–983

Attia K, Li KG, Wei C et al (2005) Overexpression of *OsPDCD5* gene induces programmed cell death in rice. J Integr Plant Biol 47:1115–1122

Beligni MV, Fath A, Bethke PC et al (2002) Nitric oxide acts as an antioxidant and delays programmed cell death in barley aleurone layers. Plant Physiol 129:1642–1650

Brotner CD, Oldenburg NBE, Cidlowski JA (1995) The role of DNA fragmentation in apoptosis. Clinical Cancer Res 11:21–26

Chen X, Wang Y, Li JY et al (2009) Mitochondrial proteome during salt stress-induced programmed cell death in rice. Plant Physiol Biochem 47:407–415

Clarke A, Desikan R, Hurst RD et al (2000) NO way back: nitric oxide and programmed cell death in *Arabidopsis thaliana* suspension cultures. Plant J 24:667–677

Crawford NM (2006) Mechanisms for nitric oxide synthesis in plants. J Exp Bot 57:471–478

Cverkovska M, Vanlerberghe GC (2012) Alternative oxidase modulates leaf mitochondrial concentrations of superoxide and nitric oxide. New Phytol 195:32–39

Delledonne M, Zeier J, Marocco A et al (2001) Signal interactions between nitric oxide and reactive oxygen intermediates in the plant hypersensitive disease resistance response. Proc Natl Acad Sci USA 98:13454–13459

De Michele R, Vurro E, Rigo C et al (2009) Nitric oxide is involved in cadmium-induced programmed cell death in *Arabidopsis* suspension cultures. Plant Physiol 150:217–228

De Pinto MC, Tommasi F, Gara LD (2002) Changes in the antioxidant systems as part of the signaling pathway responsible for the programmed cell death activated by nitric oxide and reactive oxygen species in tobacco Bright-Yellow 2 cells. Plant Physiol 130:698–708

Desikan R, Griffiths R, Hancock J et al (2002) A new role for an old enzyme: Nitrate reductase-mediated nitric oxide generation is required for abscisic acid-induced stomatal closure in *Arabidopsis thaliana*. Proc Natl Acad Sci USA 99:16314–16318

Durner J, Wendehenne D, Klessig DF (1998) Defense gene induction in tobacco by nitric oxide, cyclic GMP, and cyclic ADP-ribose. Proc Natl Acad Sci USA 95:10328–10333

Ederli L, Morettini R, Borgogni A et al (2006) Interaction between nitric oxide and ethylene in the induction of alternative oxidase in ozone-treated tobacco plants. Plant Physiol 142:595–608

Ederli L, Reale L, Madeo L et al (2009) NO release by nitric oxide donors in vitro and in planta. Plant Physiol Biochem 47:42–48

Gabaldon C, Ros Gomez LV, Pedrefio MA et al (2005) Nitric oxide production by the differentiating xylem of *Zinnia elegant*. New Phytol 165:121–130

Gavrieli Y, Sherman Y, Ben-Sasson SA (1992) Identification of programmed cell death *in situ* via specific labeling of nuclear DNA fragmentation. J Cell Biol 119:450–493

Green DR, Reed JC (1998) Mitochondria and apoptosis. Science 281:1309–1312

Guo Y, Cai Z, Gan S (2004) Transcriptome of Arabidopsis leaf senescence. Plant Cell Environ 27:521–549

Gupta KJ, Shah JK, Brotman Y et al (2012) Inhibition of aconitase by nitric oxide leads to induction of alternative oxidase and to a shift of metabolism towards biosynthesis of amino acids. J Exp Bot 63:1773–1784

Hatsugai N, Kuroyanagi M, Yamada K et al (2004) A plant vacuole protease, VPE, mediates virus-induced hypersensitive cell death. Science 305:855–858

Huckelhoven R (2004) BAX Inhibitor-1, an ancient cell death suppressor in animals and plants with prokaryotic relatives. Apoptosis 9:299–307

Jones A (2000) Does the plant mitochondrion integrate cellular stress and regulate programmed cell death? Trends Plant Sci 5:225–230

Kolodziejek I, Koziol-Lipinska J, Waleza M et al (2007) Aspects of programmed cell death during early senescence of barley leaves: possible role of nitric oxide. Protoplasma 232:97–108

Kumar KRR, Kirti PB (2010) A mitogen-activated protein kinase genes, *AhMPK6* from peanut localizes to the nucleus and also induces defense responses upon transient expression in tobacco. Plant Physiol Biochem 48:481–486

Kumar KRR, Srinivasan T, Kirti PB (2009) A mitogen-activated protein kinase genes, *AhMPK3* of peanut: molecular cloning, genomic organization, and heterologous expression conferring resistance against *Spodoptera litura* in tobacco. Mol Genet Genomics 282:65–81

Kuthanova A, Opatrny Z, Fischer L (2008) Is internucleosomal DNA fragmentation an indicator of programmed death in plant cells? J Exp Bot 59:2233–2240

Lamattina L, Garcia-Mata C, Graziano M et al (2003) Nitric oxide: the versatility of an extensive signal molecule. Annu Rev Plant Biol 54:109–139

Lanteri M, Pagnussat GC, Lamattina L (2006) Calcium and calcium-dependent protein kinases are involved in nitric oxide- and auxin-induced adventitious root formation in cucumber. J Exp Bot 57:1341–1351

Li Z, Xing D (2011) Mechanistic study of mitochondria-dependent programmed cell death induced by aluminum phytotoxicity using fluorescence techniques. J Exp Bot 62:331–343

Li Z, Yue HY, Xing D (2012) MAP kinase 6-mediated activation of vacuolar processing enzyme modulates heat shock-induced programmed cell death in *Arabidopsis*. New Phytol 195:85–96

Lin AH, Wang YQ, Tang JY (2011) Nitric oxide and protein *S*-nitrosylation are integral to hydrogen peroxide induced leaf cell death in rice. Plant Physiol 158:451–464

Liu HT, Wang YG, Zhang YM (1999) TFAR19, a novel apoptosis-related gene cloned from human leukemia cell line TF-1 could enhance apoptosis of some tumor cells induced by growth factor withdraw. Biochem Biophys Res Commun 245:203–210

Lombardi L, Ceccarelli N, Picciarelli P et al (2010) Nitric oxide and hydrogen peroxide involvement during programmed cell death of *Sechium edule* nucellus. Physiol Plant 140:89–102

Malerba M, Contran N, Tonelli M et al (2008) Role of nitric oxide in actin depolymerization and programmed cell death induced by fusicoccin in sycamore (*Acer pseudoplatanus*) cultured cells. Physiol Plant 133:449–457

Ma WW, Xu WZ, Xu H et al (2010) Nitric oxide modulates cadmium influx during cadmium-induced programmed cell death in tobacco BY-2 cells. Planta 232:325–335

Murgia I, de Pinto MC, Delledonne M et al (2004) Comparative effects of various nitric oxide donors on ferritin regulation, programmed cell death, and cell redox state in plant cells. J Plant Physiol 161:777–783

Neill S (2005) NO way to die – nitric oxide, programmed cell death and xylogenesis. New Phytol 165:1–2

Overmyer K, Brosche M, Kangasjarvi J (2003) Reactive oxygen species and hormonal control of cell death. Trends Plant Sci 8:335–342

Pagnussat GC, Lanteri ML, Lamattina L (2003) Nitric oxide and cyclic GMP are messengers in the IAA-induced adventitious rooting process. Plant Physiol 132:1241–1248

Pagnussat GC, Lanteri ML, Lombardo MC et al (2004) Nitric oxide mediats the indole acetic acid induction of a mitogen-activated protein kinase cascade involved in adventitious root development. Plant Physiol 135:279–286

Ren DT, Yang HP, Zhang SQ (2002) Cell death mediated by MAPK is associated with hydrogen peroxide production in *Arabidopsis*. J Biol Chem 277:559–565

Robson C, Vanlerberghe G (2002) Transgenic plant cells lacking mitochondrial alternative oxidase have increased susceptibility to mitochondria-dependent and -independent pathways of programmed cell death. Plant Physiol 129:1908–1920

Seo S, Okamoto M, Seto H et al (1995) Tobacco MAP kinase: a possible mediator in wound signal transduction pathway. Science 270:1988–1992

Shimizu S, Narita M, Tsujimoto Y (1999) Bcl-2 family proteins regulate the release of apoptogenic cytochrome c by the mitochondrial channel VDAC. Nature 399:483–487

Van Aken O, Giraud E, Clifton R et al (2009) Alternative oxidase: a target and regulator of stress responses. Physiol Plant 137:354–361

Vlot AC, Klessig DF, Park SW (2008) Systemic acquired resistance: the elusive signal(s). Curr Opin Plant Biol 11:436–442

Wang PC, Du YY, Li Y (2010a) Hydrogen peroxide-mediated activation of MAP kinase 6 modulates nitric oxide biosynthesis and signal transduction in *Arabidopsis*. Plant Cell 22:2981–2998

Wang WZ, Pan JW, Zheng K et al (2009) Ced-9 inhibits Al-induced programmed cell death and promotes Al tolerance in tobacco. Biochem Biophys Res Commun 383:141–145

Wang Y, Lin JS, Wang GX (2010b) Role of calcium in nitric oxide-induced programmed cell death in tobacco protoplasts. Biol Plant 54:471–476

Wang YQ, Chen C, Loake GL et al (2010c) Nitric oxide: promoter or suppressor of programmed cell death? Protein Cell 1:133–142

Wilkins KA, Bancroff J, Bosch M et al (2011) Reactive oxygen species and nitric oxide mediate actin reorganization and programmed cell death in the self-incompatibility response of papaver. Plant Physiol 156:404–416

Wulff A, Oliveira HC, Saviani EE et al (2009) Nitrite reduction and superoxide-dependent nitric oxide degradation by *Arabidopsis* mitochondria: Influence of external NAD(P)H dehydrogenases and alternative oxidase in the control of nitric oxide levels. Nitric Oxide 21:132–139

Xiong J, Fu G, Yang YJ et al (2011) Tungstate: is it really a specific nitrate reductase inhibitor in plant nitric oxide research? J Exp Bot 63:33–41

Ye Y, Li Z, Xing D et al (2012) Sorting out the role of nitric oxide in cadmium-induced *Arabidopsis thaliana* programmed cell death. Plant Signaling & Behavior 7:1493–1494

Ye Y, Li Z, Xing D (2013) Nitric oxide promotes MPK6-mediated caspase-3-like activation in cadmium-induced *Arabidopsis thaliana* programmed cell death. Plant, Cell Environ 36:1–15

Zago E, Morsa S, Dat JF (2006) Nitric oxide- and hydrogen peroxide- responsive gene regulation during cell death induction in tobacco. Plant Physiol 141:404–411

Zhan J, He HY, Wang TJ et al (2013) Aluminum-induced programmed cell death promoted by *AhSAG*, a senescence-associated gene in *Arachis hypoganea* L. Plant Sci 210:108–117

Zhang C, Czymmek KJ, Shapiro AD (2003) Nitric oxide does not trigger early programmed cell death events but may contribute to cell-to-cell signaling governing progression of the *Arabidopsis* hypersensitive response. MPMI 16:962–972

Zhao J, Fujita K, Sakai K (2007) Reactive oxygen species, nitric oxide, and their interactions play different roles in *Cupressus lusitanica* Cell death and phytoalexin biosynthesis. New Phytol 175:215–229

Zheng K, Pan JW, Ye L et al (2007) Programmed cell death-involved aluminum toxicity in yeast alleviated by antiapoptotic members with decreased calcium signals. Plant Physiol 143:38–49

Index

M. N. Khan et al. (eds.), *Nitric Oxide in Plants: Metabolism and Role in Stress Physiology*, DOI: 10.1007/978-3-319-06710-0,